8.25

INTERNATIONAL SERIES OF MONOGRAPHS IN
PURE AND APPLIED MATHEMATICS

Volume 104

LIE ALGEBRAS

LIE ALGEBRAS

by

ZHE-XIAN WAN

Institute of Mathematics, Academia Sinica, Peking

Translated by

CHE-YOUNG LEE

PERGAMON PRESS

OXFORD · NEW YORK · TORONTO · SYDNEY · BRAUNSCHWEIG

Pergamon Press Ltd., Headington Hill Hall, Oxford
Pergamon Press Inc., Maxwell House, Fairview Park, Elmsford, New York 10523
Pergamon of Canada Ltd., 207 Queen's Quay West, Toronto 1
Pergamon Press (Aust.) Pty. Ltd., 19a Boundary Street, Rushcutters Bay, N.S.W. 2011, Australia
Pergamon Press GmbH, Burgplatz 1, Braunschweig 3300, West Germany

---

First edition 1975

Library of Congress Cataloging in Publication Data

Wan, Chê-hsien.
Lie algebras.

(International series of monographs in pure and applied mathematics, v. 104)

Translation of Li tai shu.
Bibliography: p.
1. Lie algebras. I. Lie, Sophus, 1842–1899. II. Title.
QA252.3.W3613 1975 512'.55 74-13832
ISBN 0-08-017952-5

Printed in Hungary

CONTENTS

PREFACE

FROM winter of 1961 to spring of 1963, the author gave a series of lectures in the seminar on Lie groups at the Institute of Mathematics, Academia Sinica. The present book is based on the drafts of these lectures. The contents include the classical theory of complex semisimple Lie algebras, namely, the theory of structure, automorphisms, representations and real forms of such Lie algebras. The purpose of the author's lectures at the Institute of Mathematics was to teach the fundamentals of the theory of Lie algebras to the participants of the seminar in order to study the modern literature on Lie groups and Lie algebras. The main references for these lectures were "The structure of semisimple Lie algebras" by Dynkin and the lecture notes "Théorie des algèbres de Lie et topologie des groupes de Lie" in Séminaire Sophus Lie. The material in Dynkin's paper is accessible to beginners but is not complete enough; the lecture notes in Séminaire Sophus Lie contain more material but presuppose more knowledge. While each of these references has its place, neither of them really meets the needs of beginners. The purpose of the present book, therefore, is to supply an elementary background to the teory of Lie algebras, together with sufficient material to provide a reasonable overview of the subject.

Lie algebras are algebraic structures used for the study of Lie groups; they were introduced by and named after S. Lie. Besides S. Lie, the important contributors to this theory were W. Killing, E. Cartan and H. Weyl. Although discussions of Lie groups in this book have been kept to a minimum in order to facilitate understanding, it should be pointed out that the importance of the classical theory of Lie algebras lies in its applications to the theory of Lie groups. It should also be mentioned that a great part of the material in this book has been generalized to Lie algebras over algebraically closed fields of characteristic zero and some results have been generalized to Lie algebras over arbitrary fields of characteristic zero. In the present book, only Lie algebras over the complex numbers are considered. This is because the theory of such Lie algebras is the most fundamental and requires only knowledge of linear algebra to understand it.

The author wishes to thank the participants of the seminar for their suggestions and discussions which have led to many improvements in the present book. Special thanks are due to Gen-Dao Li, who assisted in the proof reading.

CHAPTER 1

BASIC CONCEPTS

1.1. Lie algebras

Let $\mathfrak{g}$ be a finite dimensional vector space (also called linear space) over the complex field C and suppose that there is a binary operation $[X, Y]$ ($X, Y \in \mathfrak{g}$) defined on $\mathfrak{g}$ which satisfies

I. $[\lambda_1 X_1+\lambda_2 X_2, Y] = \lambda_1[X_1, Y]+\lambda_2[X_2, Y]$, for all X_1, X_2 and Y in $\mathfrak{g}$ and any complex numbers λ_1, λ_2.

II. $[X, Y] = -[Y, X]$, for all X, Y in $\mathfrak{g}$.

III. $[X, [Y, Z]]+[Y, [Z, X]]+[Z, [X, Y]] = 0$, for all X, Y, Z in $\mathfrak{g}$.

Then $\mathfrak{g}$ is called a Lie algebra over the complex numbers; $\mathfrak{g}$ is also called a complex Lie algebra or simply a *Lie algebra*. The operation $[X, Y]$ is called commutation and $[X, Y]$ is called the commutator of X and Y. The dimension $\dim \mathfrak{g}$ of $\mathfrak{g}$ as a vector space is said to be the dimension of the Lie algebra $\mathfrak{g}$.

Condition I states that commutation is linear with respect to the first element. Using II, it can be proved that it is also linear with respect to the second element, i.e.

I′. $[X, \lambda_1 Y_1+\lambda_2 Y_2] = \lambda_1[X, Y_1]+\lambda_2[X, Y_2]$, for all X, Y_1, Y_2 in $\mathfrak{g}$ and complex numbers λ_1, λ_2.

Using II and III, it can be proved that

III′. $[[X, Y], Z]+[[Y, Z], X]+[[Z, X], Y] = 0$.

III can also be written as

III″. $[X, [Y, Z]] = [[X, Y], Z]+[Y, [X, Z]]$.

Condition III is called the *Jacobi identity*. Finally, setting $X = Y$ in II, we have

II′. $[X, X] = 0$, for any $X \in \mathfrak{g}$.

The following are examples of Lie algebras.

EXAMPLE 1. Let $\mathfrak{g}$ be a finite dimensional vector space over C; if for any X and Y in $\mathfrak{g}$, $[X, Y]$ is defined to be the zero vector, then I, II and III certainly hold. $\mathfrak{g}$ thus becomes a Lie algebra; it is also called an *abelian Lie algebra*.

In general, if two elements X, Y of a Lie algebra $\mathfrak{g}$ satisfy $[X, Y] = 0$, then we say that X and Y commute.

EXAMPLE 2. Let V_3 be a three-dimensional vector space over C and e_1, e_2, e_3 form a basis of V_3. For any two elements

$$x = x_1e_1+x_2e_2+x_3e_3$$

and

$$y = y_1e_1+y_2e_2+y_3e_3,$$

define

$$[x, y] = (x_2y_3-x_3y_2)e_1+(x_3y_1-x_1y_3)e_2+(x_1y_2-x_2y_1)e_3,$$

then V_3 becomes a Lie algebra.

EXAMPLE 3. Let $\mathfrak{g}_3$ be the collection of all 3×3 skew symmetric matrices, $\mathfrak{g}_3$ can be considered as a vector space over C. If for any $X, Y \in \mathfrak{g}_3$, $[X, Y]$ is defined to be $XY-YX$, then $\mathfrak{g}_3$ becomes a Lie algebra.

We can choose a basis consisting of the elements

$$M_1 = \begin{bmatrix} 0 & 0 & 0 \\ 0 & 0 & -1 \\ 0 & 1 & 0 \end{bmatrix}, \quad M_2 = \begin{bmatrix} 0 & 0 & 1 \\ 0 & 0 & 0 \\ -1 & 0 & 0 \end{bmatrix}, \quad M_3 = \begin{bmatrix} 0 & -1 & 0 \\ 1 & 0 & 0 \\ 0 & 0 & 0 \end{bmatrix},$$

then $[M_1, M_2] = M_3$, $[M_2, M_3] = M_1$, $[M_3, M_1] = M_2$. An element X in $\mathfrak{g}_3$ can be written as

$$X = \begin{bmatrix} 0 & -x_3 & x_2 \\ x_3 & 0 & -x_1 \\ -x_2 & x_1 & 0 \end{bmatrix} = x_1M_1+x_2M_2+x_3M_3.$$

If

$$Y = \begin{bmatrix} 0 & -y_3 & y_2 \\ y_3 & 0 & -y_1 \\ -y_2 & y_1 & 0 \end{bmatrix} = y_1M_1+y_2M_2+y_3M_3,$$

then

$$[X, Y] = (x_2y_3-x_3y_2)M_1+(x_3y_1-x_1y_3)M_2+(x_1y_2-x_2y_1)M_3.$$

Therefore, the mapping from V_3 to $\mathfrak{g}_3$ defined by

$$x = x_1e_1+x_2e_2+x_3e_3 \to X = x_1M_1+x_2M_2+x_3M_3$$

is one to one and satisfies

(1) if $x \to X$, $y \to Y$, then any $\lambda, \mu \in C$, $\lambda x+\mu y \to \lambda X+\mu Y$;

(2) if $x \to X$, $y \to Y$, then $[x, y] \to [X, Y]$.

That is, V_3 and $\mathfrak{g}_3$ have the same algebraic structure.

In general, a one–one mapping from a Lie algebra $\mathfrak{g}_1$ onto a Lie algebra $\mathfrak{g}_2$ is called an *isomorphism* if it satisfies:

(1) if $X_1 \to Y_1$, $X_2 \to Y_2$, then for any $\lambda, \mu \in C$, $\lambda X_1 + \mu X_2 \to \lambda Y_1 + \mu Y_2$;

(2) if $X_1 \to Y_1$, $X_2 \to Y_2$, then $[X_1, X_2] \to [Y_1, Y_2]$.

We also say that $\mathfrak{g}_1$ and $\mathfrak{g}_2$ are isomorphic and write $\mathfrak{g}_1 \approx \mathfrak{g}_2$. In particular, an isomorphism from $\mathfrak{g}$ onto itself is called an *automorphism.*

One of the fundamental problems of the theory of Lie algebras is to determine all non-isomorphic Lie algebras.

Let $\mathfrak{g}$ be an r-dimensional Lie algebra with basis $X_1, \ldots, X_r$. If

$$[X_i, X_j] = \sum_{k=1}^{r} c_{ij}^k X_k, \qquad 1 \leqslant i, j \leqslant r,$$

then the commutator of any two elements can be obtained by using the r^3 constants c_{ij}^k, i.e. if

$$X = \sum_{i=1}^{r} \lambda_i X_i, \quad Y = \sum_{j=1}^{r} \mu_j X_j$$

then

$$[X, Y] = \sum_{i,j,k=1}^{r} \lambda_i \mu_j c_{ij}^k X_k. \tag{1.1}$$

The r^3 constants c_{ij}^k $(i, j, k = 1, 2, \ldots, r)$ are called structure constants. It is not hard to see that a set of structure constants satisfies:

(1) $$c_{ij}^k = -c_{ji}^k, \qquad 1 \leqslant i, j, k \leqslant r.$$

(2) $$\sum_{s=1}^{r} (c_{ij}^s c_{sk}^l + c_{jk}^s c_{si}^l + c_{ki}^s c_{sj}^l) = 0, \qquad 1 \leqslant i, j, k \leqslant r.$$

Conversely, let $\mathfrak{g}$ be an r-dimensional vector space and c_{ij}^k $(i, j, k = 1, \ldots, r)$ be r^3 constants satisfying (1) and (2). If a basis $X_1, \ldots, X_r$ of $\mathfrak{g}$ is chosen and the commutator of two elements $X = \sum_{i=1}^r \lambda_i X_i$ and $Y = \sum_{j=1}^r \mu_j X_j$ are defined by (1.1), then $\mathfrak{g}$ becomes a Lie algebra.

Obviously, if two Lie algebras are isomorphic, then with respect to a suitable basis, they have the same structure constants. Conversely, Lie algebras with the same set of structure constants are isomorphic. Structure constants are basis dependent, if $Y_1, \ldots, Y_r$ is another basis and

$$[Y_i, Y_j] = \sum_{k=1}^{r} c_{ij}^{\prime k} Y_k, \qquad 1 \leqslant i, j \leqslant r.$$

$$Y_i = \sum_{j=1}^{r} a_i^j X_j, \qquad 1 \leqslant i \leqslant r \quad \text{where } \det(a_i^j) \neq 0,$$

then

$$\sum_{k=1}^{r} c_{ij}^{\prime k} a_k^l = \sum_{s,t=1}^{r} a_i^s a_j^t c_{st}^l, \qquad 1 \leqslant i, j, l \leqslant r. \tag{1.2}$$

Therefore, two Lie algebras are isomorphic if and only if their structure constants c_{ij}^k and $c_{ij}'^k$ satisfy the equations (1.2) where (a_i^j) is a non-singular matrix.

Finally, we give the following example.

EXAMPLE 4. Let $\mathfrak{gl}(n, C)$ be the set of all $n \times n$ matrices. It is known that with respect to matrix addition and scalar multiplication, $\mathfrak{gl}(n, C)$ forms an n^2-dimensional vector space. Now for any $X, Y \in \mathfrak{gl}(n, C)$, define

$$[X, Y] = XY - YX;$$

then $\mathfrak{gl}(n, C)$ forms a Lie algebra.

$\mathfrak{gl}(n, C)$ can also be considered as the set of all linear transformations of an n-dimensional vector space V; then it is usually denoted by $\mathfrak{gl}(V)$. We will sometimes adopt the first viewpoint and sometimes the second viewpoint.

1.2. Subalgebras, ideals and quotient algebras

Let $\mathfrak{g}$ be a Lie algebra and $\mathfrak{m}$, $\mathfrak{n}$ be subsets of $\mathfrak{g}$. Denote by $\mathfrak{m}+\mathfrak{n}$ the linear subspace spanned by elements of the form $M+N$ ($M \in \mathfrak{m}$, $N \in \mathfrak{n}$) and by $[\mathfrak{m}, \mathfrak{n}]$ the subspace spanned by elements of the form $[M, N]$ ($M \in \mathfrak{m}$, $N \in \mathfrak{n}$). If $\mathfrak{m}, \mathfrak{m}_1, \mathfrak{m}_2, \mathfrak{n}, \mathfrak{p}$ are subspaces of $\mathfrak{g}$, then

(1) $[\mathfrak{m}_1+\mathfrak{m}_2, \mathfrak{n}] \subseteq [\mathfrak{m}_1, \mathfrak{n}]+[\mathfrak{m}_2, \mathfrak{n}]$;
(2) $[\mathfrak{m}, \mathfrak{n}] = [\mathfrak{n}, \mathfrak{m}]$;
(3) $[\mathfrak{m}, [\mathfrak{n}, \mathfrak{p}]] \subseteq [\mathfrak{n}, [\mathfrak{p}, \mathfrak{m}]]+[\mathfrak{p}, [\mathfrak{m}, \mathfrak{n}]]$.

A subspace $\mathfrak{h}$ of $\mathfrak{g}$ is said to be a subalgebra if $[\mathfrak{h}, \mathfrak{h}] \subseteq \mathfrak{h}$, i.e. $X, Y \in \mathfrak{h}$ implies that $[X, Y] \in \mathfrak{h}$. A subspace $\mathfrak{h}$ of $\mathfrak{g}$ is said to be an ideal if $[\mathfrak{g}, \mathfrak{h}] \subseteq \mathfrak{g}$, i.e. $X \in \mathfrak{g}$ and $Y \in \mathfrak{h}$ implies that $[X, Y] \in \mathfrak{h}$. An ideal is a subalgebra. If $\mathfrak{h}_1$, and $\mathfrak{h}_2$ are ideals, then $\mathfrak{h}_1+\mathfrak{h}_2$ and $\mathfrak{h}_1 \cap \mathfrak{h}_2$ are also ideals.

Subalgebras of $\mathfrak{gl}(n, C)$ are called *matrix Lie algebras* or *linear Lie algebras*.

If $\mathfrak{h}$ is an ideal of $\mathfrak{g}$, then the quotient space $\mathfrak{g}/\mathfrak{h}$, which consists of all cosets (congruence classes mod $\mathfrak{h}$), is defined. For $X \in \mathfrak{g}$, denote the congruence class containing X by $\bar{X}$. Define

$$[\bar{X}, \bar{Y}] = \overline{[X, Y]}.$$

It can be proved that this definition is independent of the choices of X and Y. The quotient space $\mathfrak{g}/\mathfrak{h}$ thus becomes a Lie algebra; this algebra is called the *quotient algebra* of $\mathfrak{g}$ with respect to $\mathfrak{h}$.

If $\mathfrak{g}$ is a Lie algebra and $\mathfrak{h}$ is an ideal of $\mathfrak{g}$, then it can be proved that the mapping

$$X \to \bar{X}$$

satisfies the conditions:

(1) if $X \to \bar{X}$, $Y \to \bar{Y}$ then for any $\lambda, \mu \in C$, $\lambda X+\mu Y \to \lambda\bar{X}+\mu\bar{Y}$;
(2) if $X \to \bar{X}$, $Y \to \bar{Y}$ then $[X, Y] \to [\bar{X}, \bar{Y}]$.

In general, a mapping

$$X \to X_1$$

from a Lie algebra $\mathfrak{g}$ to a Lie algebra $\mathfrak{g}_1$ is said to be a homomorphism if it satisfies

(1) if $X \to X_1$, $Y \to Y_1$ then for any $\lambda, \mu \in C$, $\lambda X + \mu Y \to \lambda X_1 + \mu Y_1$;
(2) if $X \to X_1$, $Y \to Y_1$, then $[X, Y] \to [X_1, Y_1]$.

If this mapping is onto, then $\mathfrak{g}_1$ is said to be a homomorphic image of $\mathfrak{g}$.

THEOREM 1. *The mapping $X \to \bar{X}$ from $\mathfrak{g}$ to $\mathfrak{g}/h$ is a homomorphism (called canonical homomorphism) and $\mathfrak{g}/\mathfrak{h}$ is a homomorphic image $\mathfrak{g}$. Conversely, if f is a homomorphism from $\mathfrak{g}$ onto $\mathfrak{g}_1$, with kernel $\mathfrak{h}$, then $\mathfrak{h}$ is an ideal of $\mathfrak{g}$ and the mapping $\bar{f}$ defined by*

$$\bar{f}(\bar{X}) = f(X)$$

is an isomorphism from $\mathfrak{g}/\mathfrak{h}$ onto $\mathfrak{g}_1$ ($\bar{f}$ is called the canonical isomorphism induced by f).

Proof. It is only necessary to prove the second part of the theorem. We first prove that $\mathfrak{h}$ is an ideal of $\mathfrak{g}$. Let $X, Y \in \mathfrak{h}$, i.e. $f(X) = f(Y) = 0$, then

$$f(X+Y) = f(X)+f(Y) = 0+0 = 0,$$
$$f(\lambda X) = \lambda f(X) = \lambda \cdot 0 = 0 \quad \text{for any } \lambda \in C.$$

Therefore $X+Y \in \mathfrak{h}$ and $\lambda X \in \mathfrak{h}$ and $\mathfrak{h}$ is a subspace of $\mathfrak{g}$. If $X \in \mathfrak{g}$ and $Y \in \mathfrak{h}$, then

$$f([X, Y]) = [f(X), f(Y)] = [f(X), 0] = 0.$$

Therefore $[X, Y] \in \mathfrak{h}$ and $\mathfrak{h}$ is an ideal of $\mathfrak{g}$.

We now verify that the definition of $\bar{f}$ does not depend on choices of elements in the congruence classes. Suppose X, Y belong to the same class, i.e. $\bar{X} = \bar{Y}$, then $X - Y = H \in \mathfrak{h}$. Thus

$$f(X-Y) = f(H) = 0$$

and $f(X) = f(Y)$, therefore $\bar{f}(\bar{X}) = \bar{f}(\bar{Y})$.

Finally, we show that $\bar{f}$ is an isomorphism. Let $\bar{X}, \bar{Y} \in \mathfrak{g}/\mathfrak{h}$, then

$$\bar{f}(\bar{X}+\bar{Y}) = f(X+Y) = f(X)+f(Y) = \bar{f}(\bar{X})+\bar{f}(\bar{Y}),$$
$$\bar{f}(\lambda\bar{X}) = f(\lambda X) = \lambda f(X) = \lambda\bar{f}(\bar{X}), \quad \text{for} \quad \lambda \in C,$$
$$\bar{f}([\bar{X}, \bar{Y}] = f([X, Y]) = [f(X), f(Y)] = [\bar{f}(\bar{X}), \bar{f}(\bar{Y})].$$

Thus $\bar{f}$ is a homomorphism. If $\bar{f}(\bar{X}) = f(\bar{Y})$, then

$$f(X-Y) = f(X)-f(Y) = \bar{f}(\bar{X})-\bar{f}(\bar{Y}) = 0.$$

Therefore $X - Y \in \mathfrak{h}$ and $\bar{X} = \bar{Y}$. This proves that $\bar{f}$ is a one–one correspondence, hence it is an isomorphism. This completes the proof of Theorem 1.

To explain the concepts introduced above, we give the following examples.

EXAMPLE 5. All trace zero matrices of $\mathfrak{gl}(n, C)$ form a subalgebra; denote it by A_{n-1} In fact, A_{n-1} is an ideal of $\mathfrak{gl}(n, C)$, for if $X, Y \in \mathfrak{gl}(n, C)$, then

$$Tr[X, Y] = Tr(XY - YX) = 0,$$

therefore $[X, Y] \in A_{n-1}$.

All scalar matrices of $\mathfrak{gl}(n, C)$ form a one-dimensional subalgebra, which is also an ideal of $\mathfrak{gl}(n, C)$; for if λI is a scalar matrix, then for any $X \in \mathfrak{gl}(n, C)$, we have

$$[X, \lambda I] = X\cdot\lambda I - \lambda I\cdot X = 0.$$

All diagonal matrices of $\mathfrak{gl}(n, C)$ form an n-dimensional abelian subalgebra; denote it by $\mathfrak{d}(n, C)$. The set of trace zero diagonal matrices form an $(n-1)$-dimensional abelian subalgebra of A_{n-1}.

EXAMPLE 6. Let M be an $n\times n$ matrix. The set of all complex matrices satisfying

$$XM+MX' = 0$$

form a linear Lie algebra. In fact, from $XM+MX' = 0$ and $YM+MY' = 0$, we have

$$[X, Y]M+M[X, Y]' = (XY-YX)M+M(XY-YX)'$$
$$= XYM-YXM+MY'X'-MX'Y' = -XMY'+YMX'-YMX'+XMY' = 0.$$

Denote this Lie algebra by $\mathfrak{g}(n, M, C)$. It is easy to see that if M_1 and M_2 are congruent then $\mathfrak{g}(n, M_1, C)$ and $\mathfrak{g}(n, M_2, C)$ are isomorphic.

The following are important examples of $\mathfrak{g}(n, M, C)$:

When M is non-singular and symmetric, we obtain the orthogonal algebra. Since any complex non-singular symmetric matrix is congruent to the identity matrix, the orthogonal algebra (with respect to some M) can be considered as consisting of all skew-symmetric matrices. Furthermore, a complex non-singular symmetric matrix is either congruent to

$$\begin{bmatrix} 0 & I_m \\ I_m & 0 \end{bmatrix}, \quad \text{if } n = 2m \text{ is even,}$$

or to

$$\begin{bmatrix} 1 & 0 & 0 \\ 0 & 0 & I_m \\ 0 & I_m & 0 \end{bmatrix}, \quad \text{if } n = 2m+1 \text{ is odd.}$$

Therefore, there are two series of orthogonal Lie algebras. Denote them by B_m $(n = 2m+1)$ and D_m $(n = 2m)$, respectively.

If M is non-singular and skew-symmetric, then n must be an even number $n = 2m$; any such matrix is congruent to

$$\begin{bmatrix} 0 & I_m \\ -I_m & 0 \end{bmatrix},$$

the corresponding algebra is called the symplectic algebra and is denoted by C_m.

The Lie algebras A_n, B_n, C_n, D_n are called the *classical Lie algebras.*

1.3. Simple algebras

Let $\mathfrak{g}$ be a Lie algebra; obviously, $\mathfrak{g}$ and $\{0\}$ are ideals of $\mathfrak{g}$. If $\mathfrak{g}$ does not have any other ideals, then it is said to be a simple Lie algebra.

Obviously, one-dimensional Lie algebras are simple and any abelian Lie algebra of dimension greater than one is not simple. Therefore, except for the one-dimensional Lie algebra, simple Lie algebras are not abelian.

THEOREM 2. *The algebras* A_n $(n \geqslant 1)$, B_n $(n \geqslant 1)$, C_n $(n \geqslant 1)$ *and* D_n $(n \geqslant 3)$ *are simple Lie algebras.*

Proof. We will separately consider the structure formulas of A_n, B_n, C_n and D_n.

(A_n) Let $m = n+1$. All $m \times m$ matrices of trace zero form the Lie algebra A_n, the dimension is n^2+2n. Let

$$H_{\lambda_1, \ldots, \lambda_m} = \begin{bmatrix} \lambda_1 & & \\ & \ddots & \\ & & \lambda_m \end{bmatrix};$$

then the set of all $H_{\lambda_1, \ldots \lambda_m}$ $(\sum_1^m \lambda_i = 0)$ forms an n-dimensional abelian subalgebra $\mathfrak{h}$. Let E_{ik} denote the matrix with one at the ith row and kth column and zeros elsewhere and

$$H_{\lambda_i - \lambda_k} = E_{ii} - E_{kk}, \quad (i \neq k),$$

$$E_{\lambda_i - \lambda_k} = E_{ik}, \quad (i \neq k),$$

then $\mathfrak{h}$ and all $E_{\lambda_i - \lambda_k}$ $(i \neq k,\ i, k = 1, 2, \ldots, m)$ span A_n. $\lambda_i - \lambda_k$ $(i \neq k,\ i, k = 1, \ldots, m)$ are called roots of A_n. If $n \geqslant 2$, then any root of A_n can be obtained by adding roots of A_n to a fixed root. The structure formulas of A_n are

$$\left.\begin{aligned} [H_1, H_2] &= 0, && \text{for any} \quad H_1, H_2 \in \mathfrak{h}, \\ [H_{\lambda_1, \ldots \lambda_m}, E_\alpha] &= \alpha E_\alpha, && \text{for any root } \alpha, \\ [E_\alpha, E_{-\alpha}] &= H_\alpha && \text{for any root } \alpha, \\ [E_\alpha, E_\beta] &= && \begin{cases} 0, & \text{if } \alpha+\beta \text{ is not a root,} \\ \pm E_{\alpha+\beta} & \text{if } \alpha+\beta \text{ is a root.} \end{cases} \end{aligned}\right\} \tag{1.3}$$

(B_n) Let $m = 2n+1$. If

$$S = \begin{bmatrix} 1 & 0 & 0 \\ 0 & 0 & I_n \\ 0 & I_n & 0 \end{bmatrix},$$

then B_n consists of all $m \times m$ matrices X satisfying

$$XS + SX' = 0$$

Decomposing an $m\times m$ matrix in the same way as we decompose S, we have

$$X = \begin{bmatrix} a & u & v \\ w & A_{11} & A_{12} \\ z & A_{21} & A_{22} \end{bmatrix},$$

then $X \in B_n$ if and only if

$$a = 0, \quad w = -v', \quad z = -u', \quad A_{11} = -A_{22}', \quad A_{12} = -A_{12}', \quad A_{21} = -A_{21}';$$

in other words, B_n consists of all matrices of the form

$$\begin{bmatrix} 0 & u & v \\ -v' & A_{11} & A_{12} \\ -u' & A_{21} & -A_{11}' \end{bmatrix}, \; A_{12}' = -A_{12}, \; A_{21} = -A_{21}.$$

The dimension of B_n is $2n^2+n$. Let

$$H_{\lambda_1,\dots,\lambda_n} = \begin{bmatrix} 0 & & & & & & & \\ & \lambda_1 & & & & & & \\ & & \lambda_2 & & & & & \\ & & & \ddots & & & & \\ & & & & \lambda_n & & & \\ & & & & & -\lambda_1 & & \\ & & & & & & -\lambda_2 & \\ & & & & & & & \ddots \\ & & & & & & & & -\lambda_n \end{bmatrix},$$

then the set of all $H_{\lambda_1,\dots\lambda_n}$ forms an n-dimensional abelian subalgebra $\mathfrak{h}$. Let

$$E_{\lambda_i-\lambda_k} = \begin{bmatrix} 0 & & \\ & E_{ik} & \\ & & -E_{ki} \end{bmatrix}, \quad E_{-\lambda_i+\lambda_k} = \begin{bmatrix} 0 & & \\ & E_{ki} & \\ & & -E_{ik} \end{bmatrix}, \quad i < k,$$

$$E_{\lambda_i+\lambda_k} = \begin{bmatrix} 0 & & \\ & 0 & E_{ik}-E_{ki} \\ & & 0 \end{bmatrix}, \quad E_{-\lambda_i-\lambda_k} = \begin{bmatrix} 0 & & \\ & 0 & \\ & -E_{ik}+E_{ki} & 0 \end{bmatrix}, \quad i < k,$$

$$E_{\lambda_i} = \begin{bmatrix} 0 & 0 & e_i \\ -e_i' & 0 & 0 \\ 0 & 0 & 0 \end{bmatrix}, \qquad E_{-\lambda_i} = \begin{bmatrix} 0 & -e_i & 0 \\ 0 & 0 & 0 \\ e_i' & 0 & 0 \end{bmatrix},$$

$$H_{\lambda_i-\lambda_k} = \begin{bmatrix} 0 & & \\ & E_{ii}-E_{kk} & \\ & & -E_{ii}+E_{kk} \end{bmatrix}, \quad i < k,$$

$$H_{\lambda_i+\lambda_k} = \begin{bmatrix} 0 & & \\ & E_{ii}+E_{kk} & \\ & & -E_{ii}-E_{kk} \end{bmatrix}, \quad i < k,$$

$$H_{\lambda_i} = \begin{bmatrix} 0 & & \\ & E_{ii} & \\ & & -E_{ii} \end{bmatrix},$$

where e_i is the row vector of length n with one as the ith component and zeros elsewhere. The set of all linear combinations of $\mathfrak{h}$, $E_{\pm\lambda_i\pm\lambda_k}$ $(i < k)$ and $E_{\pm\lambda_i}$ is B_n. $\pm\lambda_i\pm\lambda_k$ $(i < k)$ and $\pm\lambda_i$ are called roots of B_n. If $n \geqslant 2$, then any root of B_n can be obtained by adding roots of B_n to a fixed root; the structure formulas of B_n are also (1.3).

(C_n) Let $m = 2n$. If

$$K = \begin{bmatrix} 0 & I_n \\ -I_n & 0 \end{bmatrix}$$

then C_n consists of all $m\times m$ matrices X satisfying

$$XK+KX' = 0.$$

Obviously, C_n is the set of matrices of the form

$$\begin{bmatrix} A_{11} & A_{12} \\ A_{21} & -A'_{11} \end{bmatrix}, \qquad A'_{12} = A_{12},\ A'_{21} = A_{21},$$

and the dimension of C_n is $2n^2+n$. Let

$$H_{\lambda_1, \ldots, \lambda_n} = \begin{bmatrix} \lambda_1 & & & & & \\ & \ddots & & & & \\ & & \lambda_n & & & \\ & & & -\lambda_1 & & \\ & & & & \ddots & \\ & & & & & -\lambda_n \end{bmatrix}$$

then the set of all $H_{\lambda_1, \ldots, \lambda_n}$ form an n-dimensional abelian subalgebra $\mathfrak{h}$. Let

$$E_{\lambda_i-\lambda_k} = \begin{bmatrix} E_{ik} & 0 \\ 0 & -E_{ki} \end{bmatrix}, \quad E_{-\lambda_i+\lambda_k} = \begin{bmatrix} E_{ki} & 0 \\ 0 & -E_{ik} \end{bmatrix}, \qquad i < k,$$

$$E_{\lambda_i+\lambda_k} = \begin{bmatrix} 0 & E_{ik}+E_{ki} \\ 0 & 0 \end{bmatrix}, \quad E_{-\lambda_i-\lambda_k} = \begin{bmatrix} 0 & 0 \\ E_{ik}+E_{ki} & 0 \end{bmatrix}, \qquad i < k,$$

$$E_{2\lambda_i} = \begin{bmatrix} 0 & E_{ii} \\ 0 & 0 \end{bmatrix}, \qquad E_{-2\lambda_i} = \begin{bmatrix} 0 & 0 \\ E_{ii} & 0 \end{bmatrix},$$

$$H_{\lambda_i-\lambda_k} = \begin{bmatrix} E_{ii}-E_{kk} & 0 \\ 0 & -E_{ii}+E_{kk} \end{bmatrix}, \qquad i < k,$$

$$H_{\lambda_i+\lambda_k} = \begin{bmatrix} E_{ii}+E_{kk} & 0 \\ 0 & -E_{ii}-E_{kk} \end{bmatrix}, \qquad i < k$$

$$H_{2\lambda_i} = \begin{bmatrix} E_{ii} & 0 \\ 0 & -E_{ii} \end{bmatrix};$$

then the set of all linear combinations of $\mathfrak{h}, E_{\pm\lambda_i\pm\lambda_k}(i < k)$ and $E_{\pm 2\lambda_i}$ is C_n. $\pm\lambda_i\pm\lambda_k(i < k)$ and $\pm 2\lambda_i$ are called roots of C_n. If $n \geqslant 2$, then any root of C_n can be obtained by adding roots of C_n to a fixed root. The structure formulas of C_n are also (1.3).

(D_n) Let $m = 2n$. If

$$S = \begin{bmatrix} 0 & I_n \\ I_n & 0 \end{bmatrix},$$

then D_n consists of all $m \times m$ matrices X satisfying

$$XS + SX' = 0.$$

Obviously, D_n is the set of all matrices of the form

$$\begin{bmatrix} A_{11} & A_{12} \\ A_{21} & -A'_{11} \end{bmatrix}, \quad A'_{12} = -A_{12}, \quad A_{21} = -A'_{21},$$

and the dimension of D_n is $2n^2 - n$. Let

$$H_{\lambda_1, \ldots, \lambda_n} = \begin{bmatrix} \lambda_1 & & & & & \\ & \ddots & & & & \\ & & \lambda_n & & & \\ & & & -\lambda_1 & & \\ & & & & \ddots & \\ & & & & & -\lambda_n \end{bmatrix},$$

then the set of all $H_{\lambda_1, \ldots, \lambda_n}$ forms an n-dimensional abelian subalgebra $\mathfrak{h}$. Let

$$E_{\lambda_i - \lambda_k} = \begin{bmatrix} E_{ik} & 0 \\ 0 & -E_{ki} \end{bmatrix}, \qquad E_{-\lambda_i + \lambda_k} = \begin{bmatrix} E_{ki} & 0 \\ 0 & -E_{ik} \end{bmatrix}, \qquad i < k,$$

$$E_{\lambda_i + \lambda_k} = \begin{bmatrix} 0 & E_{ik} - E_{ki} \\ 0 & 0 \end{bmatrix}, \quad E_{-\lambda_i - \lambda_k} = \begin{bmatrix} 0 & 0 \\ -E_{ik} + E_{ki} & 0 \end{bmatrix}, \qquad i < k,$$

$$H_{\lambda_i - \lambda_k} = \begin{bmatrix} E_{ii} - E_{kk} & 0 \\ 0 & -E_{ii} + E_{kk} \end{bmatrix}, \qquad i < k,$$

$$H_{\lambda_i + \lambda_k} = \begin{bmatrix} E_{ii} + E_{kk} & 0 \\ 0 & -E_{ii} - E_{kk} \end{bmatrix}, \qquad i < k,$$

then the set of all linear combinations of $\mathfrak{h}, E_{\pm\lambda_i \pm \lambda_k}$ $(i < k)$ is D_n. $\pm\lambda_i \pm \lambda_k$ $(i < k)$ are called roots of D_n. If $n \geqslant 3$, then any root of D_n can be obtained by adding roots of D_n to a fixed root. The structure formulas of D_n are also (1.3).

We can now prove that the Lie algebras A_n $(n \geqslant 1)$, B_n $(n \geqslant 1)$, C_n $(n \geqslant 1)$ and D_n $(n \geqslant 3)$ are simple. Let $\mathfrak{g}$ denote the Lie algebra under consideration, i.e. $\mathfrak{g}$ is one of A_n $(n \geqslant 1)$, B_n $(n \geqslant 1)$, C_n $(n \geqslant 1)$ or D_n $(n \geqslant 3)$. Suppose that $\mathfrak{n}$ is a non-zero ideal of $\mathfrak{g}$, we want to show that $\mathfrak{n} = \mathfrak{g}$. Take a non-zero element A of $\mathfrak{n}$ and let

$$A = H_0 + \sum_{\alpha \in \Sigma} \lambda_\alpha E_\alpha,$$

where $H_0 \in \mathfrak{h}$ and Σ is the set of all roots of $\mathfrak{g}$. We can assume that $\lambda_\alpha \neq 0$ for some α, for otherwise $A = H_0 \in \mathfrak{h}$, $H_0 \neq 0$ and from the second equation of (1.3) there exists E_α such that $[H_0, E_\alpha] = \alpha_0 E_\alpha \neq 0$, thus $E_\alpha \in \mathfrak{n}$.

From $A \in \mathfrak{n}$, it follows that

$$\underbrace{[H, \ldots [H, [H, A]] \ldots]}_{r} = \sum_{\alpha \in \Sigma} \lambda_\alpha \alpha^r E_\alpha \in \mathfrak{n}, \qquad r = 1, 2, \ldots . \tag{1.4}$$

If l is the number of roots, then the $l \times l$ determinant

$$V(\lambda_1, \ldots, \lambda_n) = \begin{vmatrix} \alpha_1 & \ldots & \alpha_l \\ \alpha_1^2 & \ldots & \alpha_l^2 \\ \cdot & \cdot & \cdot \\ \cdot & \cdot & \cdot \\ \alpha_1^l & \ldots & \alpha_l^l \end{vmatrix} = \prod_j \alpha_j \prod_{i>k} (\alpha_i - \alpha_k) \not\equiv 0.$$

Therefore there exist $\lambda_1^0, \ldots, \lambda_n^0$ such that $V(\lambda_1^0, \ldots, \lambda_n^0) \neq 0$. Multiplying the rth equation of (1.4) by the cofactor of α^r in $V(\lambda_1^0, \ldots, \lambda_n^0)$ and then adding up the equations, we obtain

$$V(\lambda_1^0, \ldots, \lambda_n^0)\lambda_\alpha E_\alpha \in \mathfrak{n}.$$

Therefore $E_\alpha \in \mathfrak{n}$.

For A_n $(n \geq 2)$, B_n $(n \geq 2)$, C_n $(n \geq 2)$ and D_n $(n \geq 3)$, any root can be obtained by adding roots to α, therefore it follows from the fourth equation of (1.3) that for any root α we have $E_\alpha \in \mathfrak{n}$. It then follows from the third equation of (1.3) that for any root α, $H_\alpha \in \mathfrak{n}$, therefore $\mathfrak{n} = \mathfrak{g}$.

For A_1, B_1, C_1, we have $[E_\alpha, E_{-\alpha}] = H_\alpha \in \mathfrak{n}$. Since $\mathfrak{h}$ is one-dimensional, thus $\mathfrak{h} \subseteq \mathfrak{n}$ and $[H, E_\alpha] = -\alpha E_{-\alpha} \in \mathfrak{n}$. It follows that $\mathfrak{g} = \mathfrak{n}$ and this completes the proof of Theorem 2.

1.4. Direct sum

Let $\mathfrak{g}$ be a Lie algebra and $\mathfrak{g}_1, \ldots, \mathfrak{g}_m$ be ideals of $\mathfrak{g}$; if any element X in $\mathfrak{g}$ can be uniquely written as

$$X = X_1 + \ldots + X_m$$

where $X_1 \in \mathfrak{g}_1, \ldots, X_m \in \mathfrak{g}_m$, then $\mathfrak{g}$ is said to be the direct sum of $\mathfrak{g}_1, \ldots, \mathfrak{g}_m$. We also denote direct sum by $\mathfrak{g} = \mathfrak{g}_1 \dotplus \ldots \dotplus \mathfrak{g}_m$.

If $\mathfrak{g}$ is the direct sum of $\mathfrak{g}_1, \ldots, \mathfrak{g}_m$, then for $i \neq j$, $\mathfrak{g}_i \cap \mathfrak{g}_j = \{0\}$. For if $X \in \mathfrak{g}_i \cap \mathfrak{g}_j$, then

$$X = 0 + \ldots + X_i + 0 + \ldots + 0 = 0 + \ldots + X_j + 0 + \ldots + 0$$

are two expressions of X and thus $X = 0$. From $\mathfrak{g}_i \cap \mathfrak{g}_j = \{0\}$ $(i \neq j)$, it follows that $[\mathfrak{g}_i, \mathfrak{g}_j] = \{0\}$ $(i \neq j)$; this is because $\mathfrak{g}_i$ and $\mathfrak{g}_j$ are ideals, thus $[\mathfrak{g}_i, \mathfrak{g}_j] \subseteq \mathfrak{g}_i \cap \mathfrak{g}_j = \{0\}$.

If $\mathfrak{g}$ is the direct sum of the ideals $\mathfrak{g}_1, \ldots, \mathfrak{g}_m$, then any ideal of $\mathfrak{g}_i$ is also an ideal of $\mathfrak{g}$. In fact, if $\mathfrak{h}$ is an ideal of $\mathfrak{g}_i$, then $[\mathfrak{h}, \mathfrak{g}_j] \subseteq [\mathfrak{g}_i, \mathfrak{g}_j] = \{0\}$ $(i \neq j)$, thus

$$[\mathfrak{h}, \mathfrak{g}] \subseteq [\mathfrak{h}, \mathfrak{g}_1] + \ldots + [\mathfrak{h}, \mathfrak{g}_m] = [\mathfrak{h}, \mathfrak{g}_i] \subseteq \mathfrak{h}.$$

From the previous consideration, it follows that if $\mathfrak{g}$ is the direct sum of the ideals $\mathfrak{g}_1, \ldots, \mathfrak{g}_r$, $\mathfrak{h}$ and $\mathfrak{h}$ is the direct sum of the ideals $\mathfrak{g}_{r+1}, \ldots, \mathfrak{g}_m$, then $\mathfrak{g}_1, \ldots, \mathfrak{g}_r, \mathfrak{g}_{r+1}, \ldots, \mathfrak{g}_m$ are ideals of $\mathfrak{g}$ and $\mathfrak{g}$ is the direct sum of them. In fact, it follows from the previous consideration that $\mathfrak{g}_{r+1}, \ldots, \mathfrak{g}_m$ are ideals of $\mathfrak{g}$; if $X \in \mathfrak{g}$, then

$$X = X_1 + \ldots + X_r + H, \quad X_i \in \mathfrak{g}_i \ (1 \leq i \leq r), \quad H \in \mathfrak{h}$$

where $\quad H = X_{r+1} + \ldots + X_m, \quad X_i \in \mathfrak{g}_i \ (r+1 \leq i \leq m).$

Thus $\quad X = X_1 + \ldots + X_r + X_{r+1} + \ldots + X_m.$

If $\quad X = Y_1 + \ldots + Y_r + Y_{r+1} + \ldots + Y_m, \quad Y_i \in \mathfrak{g}_i \ (1 \leq i \leq m),$

then from the assumption that $\mathfrak{g} = \mathfrak{g}_1 \dot{+} \ldots \dot{+} \mathfrak{g}_r \dot{+} \mathfrak{h}$, we have

$$X_1 = Y_1, \ldots, X_r = Y_r, \quad X_{r+1} + \ldots + X_m = Y_{r+1} + \ldots + Y_m.$$

Now from $\mathfrak{h} = \mathfrak{g}_{r+1} \dot{+} \ldots \dot{+} \mathfrak{g}_m$, it follows that

$$X_{r+1} = Y_{r+1}, \ldots, X_m = Y_m.$$

Thus every element X in $\mathfrak{g}$ has a unique expression as the sum of an element from $\mathfrak{g}_i$ $(i = 1, \ldots, m)$, therefore, $\mathfrak{g}$ is the direct sum of $\mathfrak{g}_1, \ldots, \mathfrak{g}_m$.

The following is an example of direct sums.

EXAMPLE 7. $\mathfrak{gl}(n, C)$ is the direct sum of A_{n-1} and the one-dimensional Lie algebra formed by scalar matrices.

We have already shown (Example 5) that A_{n-1} and the set of scalar matrices are ideals of $\mathfrak{gl}(n, C)$. We now show that any element in $\mathfrak{gl}(n, C)$ can be expressed as the sum of an element from A_{n-1} and a scalar matrix and that this expression is unique. Let $\operatorname{Tr} X = \chi$, then

$$X = (X - \chi/nI_n) + \chi/nI_n.$$

Since $\operatorname{Tr}(X - \chi/nI_n) = \chi - n \cdot \chi/n = 0$, thus $X - \chi/nI_n \in A_{n-1}$. Suppose

$$X = X_1 + \lambda_1 I = Y_1 + \lambda_2 I,$$

where $X_1, Y_1 \in A_{n-1}$ and λ_1, λ_2, are complex numbers, then $X_1 - Y_1 = (\lambda_1 - \lambda_2)I$. From $\operatorname{Tr}(X_1 - Y_1) = \operatorname{Tr} X_1 - \operatorname{Tr} Y_1 = 0$, it follows that $\operatorname{Tr}(\lambda_1 - \lambda_2)I = n(\lambda_1 - \lambda_2) = 0$, hence $\lambda_1 = \lambda_2$ and $X_1 = Y_1$. Therefore the expression is unique.

1.5. Derived series and descending central series

Let $\mathfrak{g}$ be a Lie algebra. $[\mathfrak{g}, \mathfrak{g}]$ is called the derived algebra of $\mathfrak{g}$ and is denoted by $\mathcal{D}\mathfrak{g}$. If $\mathfrak{h}$ is an ideal of $\mathfrak{g}$, then so is $\mathcal{D}\mathfrak{h}$. In fact,

$$[\mathfrak{g}, \mathcal{D}\mathfrak{h}] = [\mathfrak{g}, [\mathfrak{h}, \mathfrak{h}]] \subseteq [\mathfrak{h}, [\mathfrak{h}, \mathfrak{g}]] + [\mathfrak{h}, [\mathfrak{g}, \mathfrak{h}]] \subseteq [\mathfrak{h}, \mathfrak{h}] + [\mathfrak{h}, \mathfrak{h}] = \mathcal{D}\mathfrak{h}.$$

The following series of subalgebras is defined by induction:

$$\mathcal{D}^{(0)}\mathfrak{g} = \mathfrak{g}, \ \mathcal{D}^{(1)}\mathfrak{g} = \mathcal{D}\mathfrak{g}, \ \ldots, \ \mathcal{D}^{(n+1)}\mathfrak{g} = \mathcal{D}(\mathcal{D}^{(n)}\mathfrak{g}) \ \ldots.$$

In this way, we obtain a series of ideals satisfying

$$\mathcal{D}^{(0)}\mathfrak{g} \supseteq \mathcal{D}^{(1)}\mathfrak{g} \supseteq \ldots \supseteq \mathcal{D}^{(n)}\mathfrak{g} \supseteq \ldots,$$

this series is called the *derived series* of $\mathfrak{g}$. If there exists a positive integer n such that $\mathcal{D}^{(n)}\mathfrak{g} = \{0\}$, then $\mathfrak{g}$ is said to be a *solvable Lie algebra.*

The following are properties of solvable Lie algebras:

1. Subalgebras of solvable algebras are solvable; homomorphic images of solvable algebras are solvable (in particular, quotient algebras of solvable algebras are solvable).
2. If $\mathfrak{g}$ is a Lie algebra, $\mathfrak{h}$ is an ideal and the Lie algebras $\mathfrak{h}$ and $\mathfrak{g}/\mathfrak{h}$ are solvable, then $\mathfrak{g}$ is solvable.
3. Direct sums of solvable Lie algebras are solvable.

The proofs of these properties are quite simple and are therefore omitted.

If an ideal $\mathfrak{h}$ of a Lie algebra $\mathfrak{g}$ is solvable, then it is called a solvable ideal. If $\mathfrak{g}$ does not contain any solvable ideal except $\{0\}$, then $\mathfrak{g}$ is said to be semisimple. An equivalent definition is that $\mathfrak{g}$ is semisimple if it does not contain any non-zero abelian ideal. In fact, non-zero abelian ideals are certainly solvable. Conversely, if $\mathfrak{n}$ is a non-zero solvable ideal, there then exists a non-negative integer n such that $\mathcal{D}^{(n-1)}\mathfrak{n} \neq \{0\}$ and $\mathcal{D}^{(n)}\mathfrak{n} = \{0\}$. Thus $\mathcal{D}^{(n-1)}\mathfrak{n}$ is a non-zero abelian ideal.

W. Killing and E. Cartan proved that†

"*A semisimple Lie algebra is the direct sum of all the minimal ideals*". (An ideal $\mathfrak{h}$ of $\mathfrak{g}$ is a *minimal ideal* if it is not $\{0\}$ and does not contain any ideals of $\mathfrak{g}$ except itself and $\{0\}$.)

This result explains why the concept of direct sum was introduced above; the proof will be given in Chapter 4.

The one-dimensional Lie algebra is the only Lie algebra which is simple but not semisimple. From now on, the one-dimensional Lie algebra will be excluded from the simple Lie algebras, hence *all simple Lie algebras are semisimple.*

If $\mathfrak{n}_1$ and $\mathfrak{n}_2$ are solvable ideals of a Lie algebra $\mathfrak{g}$, then so is $\mathfrak{n}_1+\mathfrak{n}_2$. In fact, if n_1 and n_2 are non-negative integers such that $\mathcal{D}^{(n_1)}\mathfrak{n}_1 = \{0\}$ and $\mathcal{D}^{(n_2)}\mathfrak{n}_2 = \{0\}$, then

$$\mathcal{D}(\mathfrak{n}_1+\mathfrak{n}_2) = [\mathfrak{n}_1+\mathfrak{n}_2, \mathfrak{n}_1+\mathfrak{n}_2] \subseteq [\mathfrak{n}_1, \mathfrak{n}_1]+\mathfrak{n}_2 = \mathcal{D}\mathfrak{n}_1+\mathfrak{n}_2.$$

Now if $\mathcal{D}^{(m)}(\mathfrak{n}_1+\mathfrak{n}_2) \subseteq \mathcal{D}^{(m)}\mathfrak{n}_1+\mathfrak{n}_2$ for some integer m, then

$$\mathcal{D}^{(m+1)}(\mathfrak{n}_1+\mathfrak{n}_2) \subseteq [\mathcal{D}^{(m)}\mathfrak{n}_1+\mathfrak{n}_2, \mathcal{D}^{(m)}\mathfrak{n}_1+\mathfrak{n}_2] \subseteq [\mathcal{D}^{(m)}\mathfrak{n}_1, \mathcal{D}^{(m)}\mathfrak{n}_1]+\mathfrak{n}_2 = \mathcal{D}^{(m+1)}\mathfrak{n}_1+\mathfrak{n}_2.$$

Hence

$$\mathcal{D}^{(n_1)}(\mathfrak{n}_1+\mathfrak{n}_2) \subseteq \mathfrak{n}_2.$$

Therefore

$$\mathcal{D}^{(n_1+n_2)}(\mathfrak{n}_1+\mathfrak{n}_2) \subseteq \mathcal{D}^{(n_2)}\mathfrak{n}_2 = \{0\}$$

and $\mathfrak{n}_1+\mathfrak{n}_2$ is solvable. It follows that $\mathfrak{g}$ has a unique maximal solvable ideal $\mathfrak{r}$; $\mathfrak{r}$ is called the radical of $\mathfrak{g}$.

† W. Killing, Die Zusammensetzung der stetigen endlichen Transformations-gruppen I, II, III, IV, *Math. Ann.* **31** (1888), 252–90; **33** (1889), 1–48; **34** (1889), 57–122; **36** (1890), 161–89. E. Cartan, Sur la structure des groupes de transformations finis et continus, Thèse, Paris, 1894.

THEOREM 2. *If* $\mathfrak{g}$ *is not solvable, then* $\mathfrak{g}/r$ *is semisimple.*

Proof. Let $\bar{\mathfrak{g}}_1$ be a solvable ideal of $\mathfrak{g}/\mathfrak{r}$. If $\mathfrak{g}_1$ is the set that is mapped to $\bar{\mathfrak{g}}_1$ under the canonical mapping, then $\mathfrak{g}_1$ is solvable. Therefore $\mathfrak{g}_1 \subseteq \mathfrak{r}$ and $\bar{\mathfrak{g}}_1 = \{0\}$, hence $\mathfrak{g}/\mathfrak{r}$ is semisimple.

If $\mathfrak{g}$ is a Lie algebra, let $C^{(0)}\mathfrak{g} = \mathfrak{g}$, $C^{(1)}\mathfrak{g} = [\mathfrak{g}, C^{(0)}\mathfrak{g}]$, ..., $C^{(n+1)}\mathfrak{g} = [\mathfrak{g}, C^{(n)}\mathfrak{g}]$, ... It can be proved that $C^{(n+1)}\mathfrak{g} \subseteq C^{(n)}\mathfrak{g}$ and that all $C^{(n)}\mathfrak{g}$ ($n = 0, 1, 2, \ldots$) are ideals of $\mathfrak{g}$. In fact, $C^{(1)}\mathfrak{g} \subseteq C^{(0)}\mathfrak{g} = \mathfrak{g}$ and from $C^{(n)}\mathfrak{g} \subseteq C^{(n-1)}\mathfrak{g}$ it follows that

$$C^{(n+1)}\mathfrak{g} = [\mathfrak{g}, C^{(n)}\mathfrak{g}] \subseteq [\mathfrak{g}, C^{(n-1)}\mathfrak{g}] = C^{(n)}\mathfrak{g}.$$

If $C^{(n-1)}\mathfrak{g}$ is an ideal of $\mathfrak{g}$, then

$$[\mathfrak{g}, C^{(n)}\mathfrak{g}] \subseteq [\mathfrak{g}, C^{(n-1)}\mathfrak{g}] = C^{(n)}\mathfrak{g};$$

hence $C^{(n)}\mathfrak{g}$ is also an ideal of $\mathfrak{g}$. In this way, we obtain a series of subalgebras satisfying

$$C^{(0)}\mathfrak{g} \supseteq C^{(1)}\mathfrak{g} \supseteq \ldots \supseteq C^{(n)}\mathfrak{g} \supseteq \ldots$$

and each $C^{(n)}\mathfrak{g}$ is an ideal of $\mathfrak{g}$. This series is called the descending central series. If there exists a positive integer n such that $C^{(n)}\mathfrak{g} = \{0\}$, then $\mathfrak{g}$ is said to be nilpotent, i.e. $\mathfrak{g}$ is nilpotent iff for some positive integer n

$$[X_n, [\ldots[X_2, X_1]\ldots]] = 0, \quad \text{for any} \quad X_1, \ldots, X_n \in \mathfrak{g}.$$

It can be proved that $\mathcal{D}^{(n)}\mathfrak{g} \subseteq C^{(n)}\mathfrak{g}$. In fact, from $\mathcal{D}^{(n)}\mathfrak{g} \subsetneq C^{(n)}\mathfrak{g}$ it follows that

$$\mathcal{D}^{(n+1)}\mathfrak{g} = [\mathcal{D}^{(n)}\mathfrak{g}, \mathcal{D}^{(n)}\mathfrak{g}] \subsetneq [\mathfrak{g}, C^{(n)}\mathfrak{g}] = C^{(n+1)}\mathfrak{g}.$$

The following are results on nilpotent Lie algebras; the proofs are quite simple and are therefore omitted.

1. If $\mathfrak{g}$ is nilpotent, then it is solvable.
2. Subalgebras of nilpotent Lie algebras are nilpotent. Homomorphic images of nilpotent Lie algebras are nilpotent, in particular, quotient algebras of nilpotent Lie algebras are nilpotent.
3. Direct sums of nilpotent algebras are nilpotent.

Finally, examples of solvable and nilpotent Lie algebras are given.

EXAMPLE 8. The set of all upper triangular matrices of $\mathfrak{gl}(n, C)$, i.e. matrices of the form

$$\begin{bmatrix} x_{11} & & \cdots & x_{1n} \\ 0 & x_{22} & \cdots & x_{2n} \\ & & & \vdots \\ 0 & & \cdots\ 0 & x_{nn} \end{bmatrix}$$

is a Lie algebra $\mathfrak{t}(n, C)$. Elements in $\mathfrak{t}(n, C)$ that have equal diagonal entries form a Lie algebra $\mathfrak{n}(n, C)$. We will show that $\mathfrak{t}(n, C)$ is solvable and $\mathfrak{n}(n, C)$ is nilpotent.

Denote by $\mathfrak{n}_i$ $(1 \leq i \leq n)$ the set of elements of $\mathfrak{n}(n, C)$ of the form

$$\begin{bmatrix} 0 & 0 & \dots & x_{1i} & \cdot & \dots & x_{1n} \\ 0 & 0 & \dots & 0 & x_{2,i+1} & \dots & x_{2n} \\ & & & & & & \vdots \\ \vdots & & & & & & x_{n-i+1,n} \\ & & & & & & 0 \\ & & & & & & \vdots \\ 0 & 0 & & \dots & \cdot & & 0 \end{bmatrix} \tag{1.5}$$

and let $\mathfrak{n}_{n+1}$ denote the set consisting of the zero matrix. It can be proved by induction that

$$C^{(i-1)}\mathfrak{n}(n, C) \subseteq \mathfrak{n}_i, \quad i = 1, 2, \dots, n+1. \tag{1.6}$$

When $i = 1$, $C^{(0)}\mathfrak{n}(n, C) = \mathfrak{n}(n, C) = \mathfrak{n}_1$, thus (1.6) holds. Suppose (1.6) holds for some positive integer $i \leq n$, and denote the matrix (1.5) by A. Let

$$B = \begin{bmatrix} \lambda & \lambda_{12} & \dots & \lambda_{1n} \\ 0 & & & \vdots \\ & & & \lambda_{n-1,n} \\ 0 & \dots & 0 & \lambda \end{bmatrix}$$

be an element of $\mathfrak{n}(n, C)$, then

$$AB = \begin{bmatrix} 0 & 0 & \dots & \lambda x_{1i} & & * \\ 0 & 0 & \dots & 0 & \lambda x_{2,i+1} & \\ & & & & & \lambda x_{n-i+1,n} \\ & & & & & 0 \\ 0 & 0 & \dots & & & 0 \end{bmatrix}$$

$$BA = \begin{bmatrix} 0 & 0 & \dots & \lambda x_{1i} & & * \\ 0 & 0 & \dots & 0 & \lambda x_{2,i+1} & \\ & & & & & \lambda x_{n-i+1,n} \\ & & & & & 0 \\ 0 & 0 & \dots & & & 0 \end{bmatrix}$$

Therefore $[A, B] \subseteq \mathfrak{n}_{i+1}$ and

$$C^{(i)}\mathfrak{n}(n, C) = [\mathfrak{n}(n, C), C^{(i-1)}\mathfrak{n}(n, C)] \subseteq [\mathfrak{n}(n, C), \mathfrak{n}_i] \subseteq \mathfrak{n}_{i+1}$$

Hence (1.6) holds for $i+1$. Since $\mathfrak{n}_{n+1} = 0$, thus $C^{(n)}\mathfrak{n}(n, C) = \{0\}$ and $\mathfrak{n}(n, C)$ is nilpotent.

From $[\mathfrak{t}(n, C), \mathfrak{t}(n, C)] \subseteq \mathfrak{n}(n, C)$, it follows that $\mathcal{D}\mathfrak{t}(n, C)$ is also nilpotent and hence solvable, thus $\mathfrak{t}(n, C)$ is also solvable.

In Chapter 2, we will prove that a solvable linear Lie algebra is isomorphic to a subalgebra of $\mathfrak{t}(n, C)$ and that a nilpotent linear Lie algebra whose elements are $n\times n$ matrices must be isomorphic to some $\mathfrak{n}(n_1, C)\dot{+} \dots \dot{+}\mathfrak{n}(n_m, C)$ where $n_1+\dots+n_m = n$.

1.6. Killing form

Let $\mathfrak{g}$ be a Lie algebra and for $A \in \mathfrak{g}$, define

$$\operatorname{ad} AX = [A, X], \quad X \in \mathfrak{g};$$

then ad A is a linear transformation on $\mathfrak{g}$ satisfying

$$\operatorname{ad} A[X, Y] = [\operatorname{ad} A\, X, Y]+[X, \operatorname{ad} AY] \quad \text{(this is Section 1.1, Condition III')}.$$

ad A is said to be the *inner derivation* induced by A, it is sometimes denoted by $\operatorname{ad}_{\mathfrak{g}} A$.

The mapping

$$A \to \operatorname{ad} A$$

is a homomorphism for $\mathfrak{g}$ to $\mathfrak{gl}(n, C)$. To show this, we need to show that

(1) $\operatorname{ad}(\lambda A+\mu B) = \lambda \operatorname{ad} A+\mu \operatorname{ad} B$,
(2) $\operatorname{ad}[A, B] = [\operatorname{ad} A, \operatorname{ad} B]$.

For any $X \in \mathfrak{g}$, we have

$$\operatorname{ad}(\lambda A+\mu B) X = [\lambda A+\mu B, X] = \lambda[A, X]+\mu[B, X] = \lambda \operatorname{ad} AX+\mu \operatorname{ad} BX,$$

therefore (1) is satisfied. Now

$$\begin{aligned}\operatorname{ad}[A, B]X &= [[A, B], X] = [[A, X], B]+[A, [B, X]]\\ &= [\operatorname{ad} AX, B]+[A, \operatorname{ad} BX] = -\operatorname{ad} B \operatorname{ad} AX+\operatorname{ad} A \operatorname{ad} BX\\ &= [\operatorname{ad} A, \operatorname{ad} B]\, X,\end{aligned}$$

hence (2) is also satisfied. The image of the mapping $A\to \operatorname{ad}A$ is denoted by ad $\mathfrak{g}$; the kernel of this mapping is the set of all $A\in\mathfrak{g}$ satisfying $[A, X] = 0$ for all $X\in\mathfrak{g}$. These elements are called *central elements*; the set of all central elements form an ideal of $\mathfrak{g}$ called the center.

The characteristic polynomial

$$f(\lambda \mid X) = |\lambda I_r-\operatorname{ad} X| = \lambda^r+a_1(X)\lambda^{r-1}+\dots+a_{r-n_x}(X)\lambda^{n_x}$$

of the transformation ad X is called the Killing polynomial of X; n_x of the eigenvalues of ad X are zero. Since ad $X X = 0$, hence $n_x \geqslant 1$. Let

$$n = \min_{n\in\mathfrak{g}} n_x,$$

then n is called the rank of $\mathfrak{g}$. If $X \in \mathfrak{g}$ and $n_x = n$ then X is said to be a regular element; if $n_x > n$, then X is said to be a singular element.

For $X, Y \in \mathfrak{g}$, define

$$(X, Y) = \operatorname{Tr} \operatorname{ad} X \operatorname{ad} Y.$$

It is easy to see that (X, Y) has the following properties:

(1) $(X, Y) = (Y, X)$,

(2) $(\lambda_1 X_1 + \lambda_2 X_2, Y) = \lambda_1(X_1, Y) + \lambda_2(X_2, Y)$.

Thus (X, Y) is a symmetric bilinear function defined on $\mathfrak{g}$. (X, Y) is called the Killing form of $\mathfrak{g}$ and is also known as the Cartan inner product, it also satisfies

(3) $(\operatorname{ad} AX, Y) + (X, \operatorname{ad} AY) = 0$.

In fact,

$$(\operatorname{ad} AX, Y) + (X, \operatorname{ad} AY) = \operatorname{Tr} \operatorname{ad} [A, X] \operatorname{ad} Y + \operatorname{Tr} \operatorname{ad} X \operatorname{ad} [A, Y]$$

$$= \operatorname{Tr} \operatorname{ad} A \operatorname{ad} X \operatorname{ad} Y - \operatorname{Tr} \operatorname{ad} X \operatorname{ad} A \operatorname{ad} Y + \operatorname{Tr} \operatorname{ad} X \operatorname{ad} A \operatorname{ad} Y - \operatorname{Tr} \operatorname{ad} X \operatorname{ad} Y \operatorname{ad} A = 0.$$

Property (3) is the invariance of the Killing form under inner derivations. From (3), the following result can be obtained.

LEMMA 1. *Let $\mathfrak{g}$ be a Lie algebra and $\mathfrak{h}$ be an ideal of $\mathfrak{g}$. If*

$$\mathfrak{h}' = \{X \mid X \in \mathfrak{g} \quad \textit{and} \quad (X, Y) = 0, \quad \textit{for all} \quad Y \in \mathfrak{h}\},$$

then $\mathfrak{h}'$ is also an ideal of $\mathfrak{g}$.

Proof. $\mathfrak{h}'$ is obviously a subspace of $\mathfrak{g}$. If $X \in \mathfrak{h}'$ and $A \in \mathfrak{g}$, then for any $Y \in \mathfrak{h}$, we have

$$([A, X], Y) = -(X, [A, Y]) = 0.$$

Therefore $[A, Y] \in \mathfrak{h}$ and $[A, X] \in \mathfrak{h}'$. Thus $\mathfrak{h}'$ is an ideal of $\mathfrak{g}$.

The Killing form also has the following two properties:

(4) If $\mathfrak{h}$ is an ideal of $\mathfrak{g}$, X, Y are in $\mathfrak{h}$ and $(X, Y)_{\mathfrak{h}}$ denotes the Killing form of $\mathfrak{h}$, then

$$(X, Y)_{\mathfrak{h}} = (X, Y).$$

Proof. Choose a basis

$$X_1, \ldots, X_s, X_{s+1}, \ldots, X_r$$

of $\mathfrak{g}$ where $X_1, \ldots, X_s$ is a basis of $\mathfrak{h}$. If $X \in \mathfrak{h}$, then $\operatorname{ad}_{\mathfrak{g}} XX_i$ $(1 \leqslant i \leqslant r)$ is a linear combination of $X_1, \ldots, X_s$, i.e.

$$\operatorname{ad}_{\mathfrak{g}} XX_i = \sum_{j=1}^{s} x_{ij} X_j, \qquad i = 1, \ldots, r.$$

Thus the matrix of $\operatorname{ad}_{\mathfrak{g}} X$ has form

$$\operatorname{ad}_{\mathfrak{g}} X = \begin{bmatrix} x_{11} & \ldots & x_{1s} & \ldots & x_{1r} \\ \cdot & \cdot & \cdot & \cdot & \cdot \\ x_{s1} & \ldots & x_{ss} & \ldots & x_{sr} \\ 0 & \ldots & 0 & \ldots & 0 \\ \cdot & \cdot & \cdot & \cdot & \cdot \\ 0 & \ldots & 0 & \ldots & 0 \end{bmatrix}.$$

Similarly,
$$\operatorname{ad}_{\mathfrak{g}} YX_i = \sum_{j=1}^{s} y_{ij}Y_j, \qquad i = 1, \ldots, r$$
and
$$\operatorname{ad}_{\mathfrak{g}} Y = \begin{bmatrix} y_{11} & \ldots & y_{1s} & \ldots & y_{1r} \\ \cdot & \cdot & \cdot & \cdot & \cdot \\ y_{s1} & \ldots & y_{ss} & \ldots & y_{sr} \\ 0 & \ldots & 0 & \ldots & 0 \\ \cdot & \cdot & \cdot & \cdot & \cdot \\ 0 & \ldots & 0 & \ldots & 0 \end{bmatrix}.$$

Notice that
$$\operatorname{ad}_{\mathfrak{h}} X = \begin{bmatrix} x_{11} & \ldots & x_{1s} \\ \cdot & \cdot & \cdot \\ x_{s1} & \ldots & x_{ss} \end{bmatrix}, \quad \operatorname{ad}_{\mathfrak{h}} Y = \begin{bmatrix} y_{11} & \ldots & y_{1s} \\ \cdot & \cdot & \cdot \\ y_{s1} & \ldots & y_{ss} \end{bmatrix},$$
therefore
$$(X, Y)_{\mathfrak{h}} = \operatorname{Tr} \operatorname{ad}_{\mathfrak{h}} X \operatorname{ad}_{\mathfrak{h}} Y = \operatorname{Tr} \begin{bmatrix} x_{11} & \ldots & x_{1s} \\ \cdot & \cdot & \cdot \\ x_{s1} & \ldots & x_{ss} \end{bmatrix} \begin{bmatrix} y_{11} & \ldots & y_{1s} \\ \cdot & \cdot & \cdot \\ y_{s1} & \ldots & y_{ss} \end{bmatrix}$$
$$= \operatorname{Tr} \begin{bmatrix} x_{11} & \ldots & x_{1s} & \ldots & x_{1r} \\ \cdot & \cdot & \cdot & \cdot & \cdot \\ x_{s1} & \ldots & x_{ss} & \ldots & x_{sr} \\ 0 & \ldots & 0 & \ldots & 0 \\ \cdot & \cdot & \cdot & \cdot & \cdot \\ 0 & \ldots & 0 & \ldots & 0 \end{bmatrix} \begin{bmatrix} y_{11} & \ldots & y_{1s} & \ldots & y_{1r} \\ \cdot & \cdot & \cdot & \cdot & \cdot \\ y_{s1} & \ldots & y_{ss} & \ldots & y_{sr} \\ 0 & \ldots & 0 & \ldots & 0 \\ \cdot & \cdot & \cdot & \cdot & \cdot \\ 0 & \ldots & 0 & \ldots & 0 \end{bmatrix} = (X, Y).$$

(5) An automorphism σ of $\mathfrak{g}$ preserves the Killing form, i.e.
$$(\sigma(X), \sigma(Y) = (X, Y), \quad \text{for all} \quad X, Y \in \mathfrak{g}.$$

Proof. If $X_1, \ldots, X_r$ is a basis of $\mathfrak{g}$, then $\sigma(X_1), \ldots, \sigma(X_r)$ is also a basis. Let
$$[X, X_i] = \sum_{j=1}^{r} a_{ij}X_j, \quad a_{ij} \in C$$
then since σ is an automorphism, we have
$$[\sigma(X), \sigma(X_i)] = \sum_{j=1}^{r} a_{ij}\sigma(X_j).$$

Thus the matrix of ad X with respect to $X_1, \ldots, X_r$ is the same as the matrix of ad $\sigma(X)$ with respect to $\sigma(X_1), \ldots, \sigma(X_r)$. Hence
$$(\sigma(X), \sigma(Y)) = \operatorname{Tr} \operatorname{ad} \sigma(X) \operatorname{ad} \sigma(Y) = \operatorname{Tr} \operatorname{ad} X \operatorname{ad} Y = (X, Y).$$

Another important result of E. Cartan is
"$\mathfrak{g}$ is semisimple iff the Killing form of $\mathfrak{g}$ is non-degenerate"; the proof will be given in Chapter 4.

To conclude this chapter, we prove the following theorem.

THEOREM 3. *If the Killing form of* $\mathfrak{g}$ *is non-degenerate, then* $\mathfrak{g}$ *is semisimple; furthermore,* $\mathfrak{g}$ *is the direct sum of all the minimal ideals and the minimal ideals are orthogonal with respect to the Killing form.*

Proof. We first prove that $\mathfrak{g}$ is semisimple. Suppose $\mathfrak{a}$ is an abelian ideal of $\mathfrak{g}$ and $A \in \mathfrak{a}$, then for any $X, Y \in \mathfrak{g}$

$$\big[A, \big[X, [A, [X, Y]]\big]\big] = 0,$$

hence

$$(\text{ad } A \text{ ad } X)^2 Y = 0.$$

Therefore $(\text{ad } A \text{ ad } X)^2 = 0$ and $(A, X) = 0$. Since the Killing form is non-degenerate, A must be zero. Thus $\mathfrak{a} = \{0\}$ and $\mathfrak{g}$ does not contain any non-zero abelian ideal, hence $\mathfrak{g}$ is semisimple.

By induction on the dimension of $\mathfrak{g}$, it can be proved that $\mathfrak{g}$ is the direct sum of all the minimal ideals. If $\mathfrak{g}$ is simple, the result is obvious. Suppose $\mathfrak{g}$ is not simple and $\mathfrak{g}_1$ is a minimal ideal of $\mathfrak{g}$. Let

$$\mathfrak{h}_1 = \{X \mid (X, Y) = 0 \quad \text{for all} \quad Y \in \mathfrak{g}_1\},$$

then from Lemma 1, $\mathfrak{h}_1$ is an ideal of $\mathfrak{g}$. Since the Killing form is non-degenerate, hence

$$\dim \mathfrak{g} = \dim \mathfrak{g}_1 + \dim \mathfrak{h}_1.$$

If it can be proved that $\mathfrak{g}_1 \cap \mathfrak{h}_1 = \{0\}$, then $\mathfrak{g} = \mathfrak{g}_1 \dot{+} \mathfrak{h}_1$. Since $\mathfrak{g}_1$ is minimal, if $\mathfrak{g}_1 \cap \mathfrak{h}_1 \neq \{0\}$, then $\mathfrak{g}_1 \cap \mathfrak{h}_1 = \mathfrak{g}_1$ and $\mathfrak{g}_1 \subseteq \mathfrak{h}_1$. From minimality of $\mathfrak{g}_1$ and semisimplicity of $\mathfrak{g}$, it follows that $[\mathfrak{g}_1, \mathfrak{g}_1] = \mathfrak{g}_1$. Therefore any $X \in \mathfrak{g}_1$ can be expressed as

$$X = \sum_{i=1}^{l} [X_i, Y_i], \quad X_i, Y_i \in \mathfrak{g}_1,$$

Thus for any $Y \in \mathfrak{g}$,

$$(X, Y) = \sum_{i=1}^{l} ([X_i, Y_i], Y) = -\sum_{i=1}^{l} (Y_i, [X_i, Y]) = 0;$$

this is because $Y_i \in \mathfrak{g}_1 \subset \mathfrak{h}_1$ and $[X_i, Y] \in \mathfrak{g}_1$. Since this contradicts the assumption that the Killing form is non-degenerate, thus $\mathfrak{g}_1 \cap \mathfrak{h}_1 = \{0\}$ and

$$\mathfrak{g} = \mathfrak{g}_1 \dot{+} \mathfrak{h}_1.$$

We now prove that the Killing form $(X, Y)_{\mathfrak{h}}$ is non-degenerate. Suppose there exists $A \in \mathfrak{h}_1$ such that $(A, Y)_{\mathfrak{h}_1} = 0$ for all $Y \in \mathfrak{h}_1$. For any $X \in \mathfrak{g}$, write $X = Z + Y (Z \in \mathfrak{g}_1, Y \in \mathfrak{h}_1)$, then

$$(A, X) = (A, Z) + (A, Y) = (A, Y) = (A, Y)_{\mathfrak{h}_1} = 0.$$

This contradicts the non-degeneracy of the Killing form, hence the Killing form of $\mathfrak{h}_1$ is non-degenerate. Since $\dim \mathfrak{h}_1 < \dim \mathfrak{g}$, it follows from the induction assumption that $\mathfrak{h}_1$ is the direct sum of all its minimal ideals (they are also minimal ideals of $\mathfrak{g}$), i.e.

$$\mathfrak{h}_1 = \mathfrak{g}_2 \dot{+} \cdots \dot{+} \mathfrak{g}_m.$$

Therefore,

$$\mathfrak{g} = \mathfrak{g}_1 \dot{+} \mathfrak{g}_2 \dot{+} \cdots \dot{+} \mathfrak{g}_m.$$

Finally, we prove that $\mathfrak{g}_1, \ldots, \mathfrak{g}_m$ are all minimal ideals of $\mathfrak{g}$. Suppose that $\mathfrak{g}^*$ is a minimal ideal and let

$$\mathfrak{g}_i^* = \left\{X_i | X_i \in \mathfrak{g}_i \quad \text{such that for some} \quad X \in \mathfrak{g}^*, X = \sum_{j=1}^{m} X_j, X_i \in \mathfrak{g}_j\right\}.$$

Then for some i, $\mathfrak{g}_i^* \neq \{0\}$. If $\mathfrak{g}_1^* \neq \{0\}$, then from $X = \sum_{i=1}^{m} X_i \ (X_i \in \mathfrak{g}_i)$, it follows that for any $Y \in \mathfrak{g}$,

$$[X, Y] = \sum_{i=1}^{m} [X_i, Y], \quad [X_i, Y] \in \mathfrak{g}_i.$$

Therefore $\mathfrak{g}_i^*$ is an ideal of $\mathfrak{g}_i$ and in particular $\mathfrak{g}_1^* = \mathfrak{g}_1$.
From

$$\mathfrak{g}_1 = [\mathfrak{g}_1, \mathfrak{g}_1] = [\mathfrak{g}_1^*, \mathfrak{g}_1] = [\mathfrak{g}^*, \mathfrak{g}_1] \subset \mathfrak{g}^*$$

and the minimality of $\mathfrak{g}^*$, it follows that $\mathfrak{g}_1 = \mathfrak{g}^*$.

Finally, let $X_i \in \mathfrak{g}_i, X_j \in \mathfrak{g}_j \ (i \neq j)$. For any $X \in \mathfrak{g}$, we have

$$\operatorname{ad} X_i \operatorname{ad} X_j X = [X_i, [X_j, X]] \in \mathfrak{g}_i \cap \mathfrak{g}_j,$$

thus $\operatorname{ad} X_i \operatorname{ad} X_j X = 0$, $\operatorname{ad} X_i \operatorname{ad} X_j = 0$ and $(X_i, X_j) = 0$. Therefore $\mathfrak{g}_1, \ldots \mathfrak{g}_m$ are mutually orthogonal.

CHAPTER 2

NILPOTENT AND SOLVABLE LIE ALGEBRAS

2.1. Preliminaries

Let A be a linear transformation acting on a finite-dimensional linear space. A subspace V_1 of V is said to be invariant under A if for any $x \in V_1$, Ax is also in V_1. If V_1 is invariant under A and for any $x \in V_1$ we define

$$A_1 x = Ax$$

then A_1 is a linear transformation on V_1; this transformation is said to be induced by A. A_1 is sometimes also denoted by A.

Suppose V_1 is invariant under A and consider the quotient space V/V_1. For any $\bar{x} \in V/V_1$, define

$$\bar{A}\bar{x} = \overline{Ax};$$

then it can be proved that this definition is independent of the choice of x in $\bar{x}$. In fact, if $\bar{x} = \bar{y}$, then $x-y \in V_1$ and $Ax-Ay \in V_1$, hence $\overline{Ax} = \overline{Ay}$. It is easy to show that $\bar{A}$ is a linear transformation on V/V_1; this transformation is said to be induced by A and is sometimes also denoted by A.

Suppose A is a nilpotent linear transformation, i.e. $A^m = 0$ for some positive integer m. Let m be such that $A^{m-1} \neq 0$ and $A^m = 0$, then there exists $x \in V$ such that $x \neq 0$ $A^{m-1}x = 0$. Since $A(A^{m-1}x) = 0$, zero is an eigenvalue of A.

If A is nilpotent and V_1 is invariant under A, then $\bar{A}$ (defined on V/V_1) is also nilpotent. From nilpotency of A and by induction, it can be proved that all eigenvalues of a nilpotent linear transformation are zero. Conversely, if all eigenvalues of a linear transformation are zero, it must be nilpotent.

Let $\mathfrak{h}$ be a set of linear transformations on V. A subspace V_1 of V is said to be invariant under $\mathfrak{h}$ or simply an invariant subspace if for any $x \in V_1$ and $H \in \mathfrak{h}$ it is always true that $Hx \in V_1$. Obviously, V itself and $\{0\}$ are invariant subspaces under $\mathfrak{h}$; if these are the only invariant subspaces, then V is said to be irreducible under $\mathfrak{h}$ and $\mathfrak{h}$ is said to be a set of irreducible transformations.

2.2. Engel's theorem

LEMMA 1. *Let X be a nilpotent linear transformation defined on V and let $X^k = 0$. Define a mapping* ad X *from* $\mathfrak{gl}(V)$ *to* $\mathfrak{gl}(V)$ *by*

$$Y \to \operatorname{ad} XY = [X, Y], \quad Y \in \mathfrak{gl}(V);$$

then ad X *is nilpotent.*

Proof. We have

$$(\operatorname{ad} X)^m Y = \sum_{i+j=m} \pm X^i Y X^j.$$

Now $X^k = 0$ implies that $(\operatorname{ad} X)^{2k-1} Y = 0$, for any $Y \in \mathfrak{gl}(V)$. Thus $(\operatorname{ad} X)^{2k-1} = 0$.

THEOREM 1 *(Engel). Let V be a finite-dimensional linear space and $\mathfrak{g}$ be a Lie subalgebra of $\mathfrak{gl}(V)$. If every element of $\mathfrak{g}$ is nilpotent and $V \neq \{0\}$, then there exists a non-zero $x \in V$, such that $Xx = 0$ for all $X \in \mathfrak{g}$.*

Proof. Use induction on the dimension of $\mathfrak{g}$. If $\dim \mathfrak{g} = 0$, then the theorem is certainly true. Suppose now that the theorem is true for Lie algebras of dimension $< r$, we proceed to prove that it is also true for Lie algebras of dimension r. We first show that $\mathfrak{g}$ contains an ideal $\mathfrak{h}$ of dimension $r-1$.

Let $\mathfrak{h}$ be a subalgebra of $\mathfrak{g}$ of dimension m and $m < r$. For any $X \in \mathfrak{h}$, consider $\operatorname{ad}_{\mathfrak{gl}(V)} X$. $\operatorname{ad}_{\mathfrak{gl}(V)} X$ maps $\mathfrak{gl}(V)$ into itself and thus induces a mapping $\operatorname{ad}_{\mathfrak{g}} X$ from $\mathfrak{g}$ into itself. Since $X \in \mathfrak{h}$, hence $\mathfrak{h}$ is an invariant subspace under $\operatorname{ad}_{\mathfrak{g}} X$ and $\operatorname{ad}_{\mathfrak{g}} X$ induces a mapping $\sigma(X)$ on the quotient space $\mathfrak{g}/\mathfrak{h}$. From Lemma 1, $\operatorname{ad}_{\mathfrak{gl}(V)} X$ is nilpotent $(\forall X \in \mathfrak{h})$, hence $\sigma(X)$ is also nilpotent $(\forall X \in \mathfrak{h})$. Since the set $\sigma(\mathfrak{h})$ of all $\sigma(X)$ $(\forall X \in \mathfrak{h})$ is a homomorphic image of $\mathfrak{h}$, $\sigma(\mathfrak{h})$ is also a Lie algebra. Now $\dim \mathfrak{h} < r$, thus $\dim \sigma(\mathfrak{h}) < r$. From the induction assumption, it follows that there exists some non-zero element $Y + \mathfrak{h}$ in $\mathfrak{g}/\mathfrak{h}$ such that $\sigma(X)(Y+\mathfrak{h}) = 0$ $(\forall X \in \mathfrak{h})$. Thus $[X, Y] \in \mathfrak{h}$ $(\forall X \in \mathfrak{h})$ and $\mathfrak{h}$ and Y generate a subalgebra of $\mathfrak{g}$ of dimension $m+1$ which contains $\mathfrak{h}$ as an ideal.

From the previous argument, starting with $\mathfrak{h} = \{0\}$, we can obtain an ideal $\mathfrak{h}$ of $\mathfrak{g}$ dimension $r-1$. According to the induction assumption, there exists $x \in V$, $x \neq 0$, such that $Xx = 0$ $(\forall X \in \mathfrak{h})$. Let

$$U = \{x \mid x \in V \quad \text{and} \quad Xx = 0 \quad (\forall X \in \mathfrak{h})\}$$

then $U \neq \{0\}$. If $A \in \mathfrak{g}$ and $A \notin \mathfrak{h}$, then U is invariant under A. In fact, if $x \in U$ and $X \in \mathfrak{h}$,

$$XAx = AXx + [X, A]\,x = 0,$$

since $[X, A] \in \mathfrak{h}$. Now A is nilpotent, hence the transformation induced by A on U is also nilpotent and there exists a non-zero $x \in U$ such that $Ax = 0$. Therefore $Xx = 0$ $(\forall X \in \mathfrak{g})$ and this completes the proof of the theorem.

COROLLARY 1. *Let $\mathfrak{g}$ be a subalgebra of $\mathfrak{gl}(Y)$. If every element of $\mathfrak{g}$ is nilpotent, then $\mathfrak{g}$ is a nilpotent Lie algebra.*

Proof. According to Theorem 1, there exists a non-zero $x_1 \in V$ such that $Xx_1 = \sigma(\forall X \in \mathfrak{g})$. The transformations on $V/\{x_1\}$ induced by elements of $\mathfrak{g}$ also form a nilpotent Lie algebra, hence there exists $x_2 \in V(x_2 \notin \{x_1\})$ such that

$$Xx_2 \equiv 0 \quad (\text{mod } \{x_1\}.$$

By repeating the same argument, a basis $x_1, \ldots, x_n$ can be obtained which satisfies

$$Xx_i \equiv 0 \quad (\text{mod } x_1, \ldots, x_{i-1}),$$

for all $X \in \mathfrak{g}$. with respect to this basis, matrices of elements of $\mathfrak{g}$ have the form

$$\begin{bmatrix} 0 & x_{12} & \ldots & x_{1n} \\ 0 & 0 & & \\ & & & x_{n-1,n} \\ 0 & \ldots & 0 & 0 \end{bmatrix}$$

thus $\mathfrak{g} \subseteq \mathfrak{n}(n, C)$ and $\mathfrak{g}$ is nilpotent.

COROLLARY 2 *(Engel). A Lie algebra $\mathfrak{g}$ is nilpotent iff for any $X \in \mathfrak{g}$*, ad X *is nilpotent.*

Proof. If $\mathfrak{g}$ is nilpotent, then there exists some positive integer m such that

$$[X_m, [\ldots[X_2, X_1]\ldots]] = 0$$

holds for any $X_1, \ldots, X_m \in \mathfrak{g}$. In particular,

$$\underbrace{[X, \ldots[X,}_{m-1} Y]\ldots] = 0$$

holds for any $X, Y \in \mathfrak{g}$, hence $(\text{ad } X)^{m-1}Y = 0$. Thus $(\text{ad}X)^{m-1} = 0\ (\forall X \in \mathfrak{g})$.

Conversely, suppose that all ad $X(X \in \mathfrak{g})$ are nilpotent. Since ad X is a linear transformation on $\mathfrak{g}$, according to Engel's theorem, there exists non-zero $X_1 \in \mathfrak{g}$ such that $(\text{ad } X)X_1 = 0$ $(\forall X \in \mathfrak{g})$. Hence the center $\mathfrak{c}$ of $\mathfrak{g}$ is not $\{0\}$. Let $\hat{\mathfrak{g}} = \mathfrak{g}/\mathfrak{c}$; if $\bar{X} \in \bar{\mathfrak{g}}$, then $\text{ad}_{\mathfrak{g}}\bar{X}$ is also nilpotent. In fact, if $(\text{ad } X)^m = 0$, then

$$\underbrace{[X, [\ldots[X,}_{m} Y]\ldots]] = 0 \quad (\forall Y \in \mathfrak{g}),$$

and thus

$$[\bar{X}, [\ldots[\bar{X}, \bar{Y}]\ldots]] = 0 \quad (\forall \bar{Y} \in \bar{\mathfrak{g}}).$$

Using induction on the dimension of $\mathfrak{g}$, it follows that $\bar{\mathfrak{g}}$ is nilpotent, i.e. there exists a positive integer m such that

$$[\bar{X}_m, \ldots[\bar{X}_2, \bar{X}_1]\ldots] = 0$$

holds for any $\bar{X}_1, \ldots, \bar{X}_m$. Thus

$$[X_m, [\ldots[X_2, X_1]\ldots] \in \mathfrak{c}$$

for any $X_1, \ldots, X_m \in \mathfrak{g}$ and

$$[X_{m+1}, [X_m, \ldots[X_2, X_1]\ldots]] = 0$$

for any $X_1, \ldots, X_{m+1} \in \mathfrak{g}$. Hence $\mathfrak{g}$ is nilpotent.

2.3. Lie's theorem

LEMMA 2. *Let V be a finite-dimensional linear space, $\mathfrak{g}$ be a subalgebra of $\mathfrak{gl}(V)$ and $\mathfrak{n}$ be an ideal of $\mathfrak{g}$. If $\varphi(A)$ is a complex-valued linear function defined on $\mathfrak{n}$ and*

$$K = \{x \mid x \in V, \quad Ax = \varphi(A)\, x(\forall\, A \in \mathfrak{n})\},$$

then K is an invariant subspace of $\mathfrak{g}$.

Proof. Let $A \in \mathfrak{n}$, $X \in \mathfrak{g}$, $x \in K$; then

$$AXx = [A, X]x + XAx = \varphi([A, X]x + \varphi(A)Xx. \tag{2.1}$$

If $K = \{0\}$, then the lemma is naturally true. Suppose that $K \neq \{0\}$, we will show that $\varphi([A, X]) = 0$, and the invariance of $\mathfrak{g}$ under K will then follow from (2.1).

Let $x \in K$ and $x \neq 0$. Define $x_k = X^k x$. Since V is finite-dimensional, there exists a non-negative integer p such that $x_0, x_1, \ldots, x_p$ are linearly independent and $x_1, \ldots x_{p+1}$ are linearly dependent. Thus the subspace W spanned by $x_0, \ldots, x_p$ is invariant under X. By induction, it can be proved that: for all $A \in \mathfrak{n}$,

$$Ax_q = \varphi(A)x_q \quad (\text{mod } x_0, \ldots, x_{g-1}). \tag{2.2}$$

In fact, when $q = 0$, (2.2) naturally holds. Suppose now that (2.2) holds for q, then

$$\begin{aligned} Ax_{q+1} &= AXx_q = [A, X]x_q + XAx_q \\ &\equiv \varphi([A, X])x_q + X\varphi(A)x_q \\ &\quad (\text{mod } x_0, x_1, \ldots, x_{q-1}, Xx_0, \ldots, Xx_{q-1}) \\ &\equiv \varphi(A)x_{q+1}\ (\text{mod } x_0, x_1, \ldots, x_q). \end{aligned}$$

Thus (2.2) also holds for $q+1$. Hence W is an invariant subspace of $\mathfrak{n}$.

Since W is invariant under X and $\mathfrak{n}$ the trace $\mathrm{Tr}_W(A)$ ($A \in \mathfrak{n}$) can be found to be

$$\mathrm{Tr}_W(A) = (p+1)\,\varphi(A).$$

Since $\mathfrak{n}$ is an ideal of $\mathfrak{g}$, hence $[A, X] \in \mathfrak{n}$ ($\forall\, A \in \mathfrak{n}$) and thus

$$\mathrm{Tr}_W([A, X]) = (p+1)\,\varphi([A, X]). \tag{2.3}$$

On the other hand,

$$\mathrm{Tr}_W([A, X]) = \mathrm{Tr}_W AX - \mathrm{Tr}_W XA = 0, \tag{2.4}$$

thus it follows from (2.8) and (2.4) that $\varphi([A, X]) = 0$.

THEOREM 2 *(Lie). Let $\mathfrak{g}$ be a solvable linear Lie algebra on a finite-dimensional linear space V, then there is a non-zero vector x in V such that $Xx = \varphi(X)x$ for all $X \in \mathfrak{g}$, i.e. x is a common eigenvector for all transformations in $\mathfrak{g}$.*

Proof. Use induction on dim $\mathfrak{g}$. When dim $\mathfrak{g} = 0$, the theorem is naturally true. Now suppose that the theorem is true for solvable linear Lie algebras of dimensions smaller than r.

Since $\mathfrak{g}$ is solvable, $\mathcal{D}\mathfrak{g} \neq \mathfrak{g}$ and any subspace that lies between $\mathfrak{g}$ and $\mathcal{D}\mathfrak{g}$ is an ideal. Hence $\mathfrak{g}$ has an ideal $\mathfrak{n}$ of dimension $r-1$ and $\mathfrak{g}$ is generated by $\mathfrak{n}$ and an element $X \notin \mathfrak{n}$.

Naturally, $\mathfrak{n}$ is solvable. By the induction assumption, there exists a non-zero $x \in V$ such that

$$Ax = \varphi(A)x, \quad \forall A \in \mathfrak{n}.$$

Obviously, $\varphi(A)$ is a linear function defined on $\mathfrak{n}$. Let

$$K = \{x \mid x \in V \quad \text{and} \quad Ax = \varphi(A)x \ (\forall A \in \mathfrak{n})\},$$

then $K \neq \{0\}$. According to Lemma 2, K is an invariant subspace of $\mathfrak{g}$; in particular, K is an invariant subspace of X. Thus a non-zero eigenvector of X can be found in K; it is a common eigenvector of all linear transformations in $\mathfrak{g}$.

COROLLARY 1. *Let $\mathfrak{g}$ be a solvable subalgebra of the Lie algebra $\mathfrak{gl}(V)$, then there exists a basis of V with respect to which matrices of elements in $\mathfrak{g}$ have the form*

$$\begin{bmatrix} x_{11} & x_{12} & \cdots & x_{1n} \\ 0 & x_{22} & \cdots & x_{2n} \\ \vdots & & \ddots & \vdots \\ 0 & \cdots & 0 & x_{nn} \end{bmatrix}.$$

Thus solvable linear algebras are subalgebras of $\mathfrak{t}(n, C)$.

Proof. Similar to the proof of Corollary 1 to Theorem 1.

COROLLARY 2. *If $\mathfrak{g}$ is a Lie algebra, then $\mathfrak{g}$ is solvable iff $\mathcal{D}\mathfrak{g}$ is nilpotent.*

Proof. If $\mathcal{D}\mathfrak{g}$ is nilpotent, then it is solvable and hence $\mathfrak{g}$ is solvable.

Conversely, let $\mathfrak{g}$ be solvable and consider $\mathrm{ad}_X \mathfrak{g}$ $(X \in \mathfrak{g})$. Set

$$\mathrm{ad}_\mathfrak{g} \mathfrak{g} = \{\mathrm{ad}_\mathfrak{g} X \mid X \in \mathfrak{g}\},$$

then as a homomorphic image of $\mathfrak{g}$, $\mathrm{ad}_\mathfrak{g} \mathfrak{g}$ is a solvable linear Lie algebra. According to Corollary 1, a basis of $\mathfrak{g}$ exists with respect to which the matrix of $\mathrm{ad}_\mathfrak{g} X$ $(\forall X \in \mathfrak{g})$ has the form

$$\begin{bmatrix} x_{11} & x_{12} & \cdots & x_{1r} \\ 0 & x_{22} & \cdots & x_{2r} \\ \vdots & & \ddots & \vdots \\ 0 & \cdots & 0 & x_{rr} \end{bmatrix}.$$

Thus all $\mathrm{ad}_\mathfrak{g} X (X \in \mathcal{D}\mathfrak{g})$ are nilpotent matrices and hence all $\mathrm{ad}_{\mathcal{D}\mathfrak{g}} X\,(X \in \mathcal{D}\mathfrak{g})$ are nilpotent. By Corollary 2 to Theorem 2, $\mathcal{D}\mathfrak{g}$ is nilpotent.

Corollary 3. *Let $\mathfrak{g}$ be an irreducible linear Lie algebra acting on the linear space V. If $\mathfrak{n}$ is a solvable ideal of $\mathfrak{g}$, then any linear transformation A of $\mathfrak{n}$ has form $\varphi(A)I$, where I is the identity matrix.*

Proof. By Lie's theorem, there exists a non-zero $x \in V$ such that

$$Ax = \varphi(A)x \quad \text{for all } A \in \mathfrak{n}.$$

Let

$$K = \{x \mid x \in V \text{ and } Ax = \varphi(A)\, x\ (\forall A \in \mathfrak{n})\},$$

then $K \neq \{0\}$. According to Corollary 2, K is an invariant subspace of $\mathfrak{g}$. Since $\mathfrak{g}$ is irreducible, hence $K = V$. Thus $A = \varphi(A)\, I\ (\forall A \in \mathfrak{n})$.

Since $\mathfrak{gl}(n, C)$ and A_{n-1} are irreducible, it follows from this Corollary that the only non-zero solvable ideal of $\mathfrak{gl}(n, C)$ is the set of scalar matrices and that A_{n-1} is semisimple, since it does not contain any non-zero solvable ideal.

Similarly, by using Corollary 3, it can be proved that the Lie algebras B_n ($n \geqslant 1$), C_n ($n \geqslant 1$) and D_n ($n \geqslant 2$) are semisimple.

2.4. Nilpotent linear Lie algebras

Lemma 3. *Let A be a linear transformation on a finite-dimensional linear space V and λ be a complex number. Let $V_A^\lambda = \{v \mid v \in V$ and $(A-\lambda I)^n v = 0$ for some positive integer $n\}$, then*

1° *V_A^λ is an invariant subspace of A.*
2° *$V_A^\lambda \neq \{0\}$ iff λ is an eigenvalue of A; moreover, if $V_A^\lambda \neq \{0\}$, then the only eigenvalue of A in V_A^λ is λ.*
3° *If $V_A^\lambda \neq \{0\}$, then $v \in V_A^\lambda$ implies that $(A-\lambda I)^{\dim V_A^\lambda} v = 0$.*
4° *The multiplicity of the eigenvalue λ of A is $\dim V_A^\lambda$.*
5° *$V = \sum_{\lambda \in \Delta}^{\cdot} V_A^\lambda$ (direct sum), where Δ is the collection of all distinct eigenvalues of A.*

Proof. 1° is obvious. To prove 2°, let λ be an eigenvalue of A, then there exists non-zero $v \in V$ such that $(A-\lambda I)v = 0$. Thus $v \in V_A^\lambda$ and $V_A^\lambda \neq \{0\}$. Conversely, let $V_A^\lambda \neq \{0\}$ and μ be an eigenvalue of A in V_A^λ, i.e. for some non-zero $v \in V$, $Av = \mu v$. From the definition of V_A^λ, it follows that $(A-\lambda I)^n v = 0$ for some positive integer n; substitute $Av = \mu v$ in this equation, we get $(\mu-\lambda)^n\ v = 0$, thus $\mu = \lambda$.

From the fact that any matrix is similar to an upper triangular matrix and 2°, it follows that there exists a basis of V_A^λ with respect to which the mapping induced by A on V_A^λ has the matrix

$$\begin{bmatrix} \lambda & & * \\ & \ddots & \\ & & \lambda \end{bmatrix}$$

Hence 3° holds.

To prove 4°, first notice that if W is a subspace of V which is invariant under A, then the multiplicity of the eigenvalue λ of A is the sum of the multiplicities of λ of the induced mappings A_W (on W) and $A_{V/W}$ (on V/W). Thus it suffices to show that λ is not an eigenvalue of A_{V/V_A^λ}. Now suppose that there is a $v \in V$ and $v \notin V_A^\lambda$ such that $Av \equiv \lambda v \pmod{V_A^\lambda}$, then $(A-\lambda)v \equiv 0 \pmod{V_A^\lambda}$ and from 3°,

$$(A-\lambda I)^{\dim V_A^\lambda+1} v = 0.$$

This contradicts the definition of V_A^λ, hence no such v exists.

To prove 5°, let $\lambda_1, \ldots, \lambda_s$ be the different eigenvalues of A. We first show that $\sum_1^s V_A^{\lambda_i}$ is a direct sum. Suppose

$$v_1 + \ldots + v_s = 0, \qquad v_i \in V_A^{\lambda_i}. \tag{2.5}$$

Let
$$f_k(A) = \prod_{\substack{i=1\\ i\neq k}}^{s} (A-\lambda_i I)^{\dim V_A^{\lambda_i}}, \qquad k = 1, 2, \ldots, s$$

and apply $f_k(A)$ to the two sides of (2.5), then according to 3°,

$$f_k(A)v_k = 0, \qquad k = 1, \ldots, s. \tag{2.6}$$

On the other hand, we have

$$(A-\lambda_k I)^{\dim V_A^{\lambda_k}} v_k = 0, \quad k = 1, \ldots, s. \tag{2.7}$$

Since $f_k(\lambda)$ and $(\lambda-\lambda_k)^{\dim V_A^{\lambda_k}}$ are relatively prime, there exist polynomials $p_k(\lambda)$ and $q_k(\lambda)$ such that

$$p_k(\lambda) f_k(\lambda) + q_k(\lambda)(\lambda-\lambda_k)^{\dim V_A} = 1.$$

Substituting A into this equation, we have

$$p_k(A) f_k(A) + q_k(A)(A-\lambda_k I)^{\dim V_A^{\lambda_k}} = I.$$

Applying both sides of this equation to v_k and by (2.6) and (2.7), we get

$$0 = v_k, \quad k = 1, \ldots, s.$$

Thus $\sum_i^s V_A^{\lambda_i}$ is a direct sum.

To show $V = \sum_1^s V_A^{\lambda_i}$, let n_i be the multiplicity of λ_i, then $\dim V = n_1 + \ldots + n_s$. From 4° $\dim V_A^{\lambda_i} = n_i$ $(1 \leq i \leq s)$, thus $V = \sum_i^s V_A^{\lambda_i}$. This completes the proof of the theorem.

LEMMA 4. *If A and B are linear transformations on a linear space V and*

$$[A, \ldots [A, [A, B] \ldots]] = 0,$$

then V_A^λ is invariant under B.

Proof. Let

$$\underbrace{[A, \ldots [A, B] \ldots]}_{k} = 0$$

and use induction on k to prove the lemma. When $k = 0$, the lemma is naturally true (since $B = 0$). Now suppose the lemma is true for $k-1$ and let $C = [A, B]$, then

$$\underbrace{[A, [A, \ldots [A}_{k-1}, C]]\ldots] = 0$$

and according to the induction assumption, V_A^λ is invariant under C.

Now $(A-\lambda I)B = B(A-\lambda I)+[A, B]$ and by induction, it can be proved that

$$(A-\lambda I)^n B = B(A-\lambda I)^n + \sum_{s=0}^{n-1} (A-\lambda I)^{n-s-1}[A, B](A-\lambda I)^s. \tag{2.8}$$

In fact, suppose (2.8) is true for n, then

$$(A-\lambda I)^{n+1}B = B(A-\lambda I)^{n+1}+[A, B](A-\lambda I)^n + \sum_{s=0}^{n-1} (A-\lambda I)^{n-s}[A,B](A-\lambda I)^s$$

$$= B(A-\lambda I)^{n+1} + \sum_{s=0}^{n} (A-\lambda I)^{n-s}[A,B](A-\lambda I)^s.$$

Therefore (2.8) is also true for $n+1$. Let $n = 2 \dim V_A^\lambda$ in (2.8) and then apply both sides of (2.8) to $v \in V_A^\lambda$, we have

$$(A-\lambda I)^n Bv = B(A-\lambda I)^n v + \sum_{s=0}^{n-1} (A-\lambda I)^{n-s-1}[A, B](A-\lambda I)^s v.$$

Now

$$(A-\lambda I)^n v = 0.$$

If $s \geqslant \dim V_A^\lambda$, then

$$(A-\lambda I)^s v = 0.$$

If $s < \dim V_A^\lambda$, then $n-s-1 \geqslant \dim V_A^\lambda$. By the induction assumption, $[A, B](A-\lambda I)^s v = C(A-\lambda I)^s v \in V_A^\lambda$, thus

$$(A-\lambda I)^{n-s-1}[A, B](A-\lambda I)^s v = 0.$$

This proves that

$$(A-\lambda I)^n Bv = 0,$$

i.e. $Bv \in V_A^\lambda$.

Now let $\mathfrak{h}$ be a subalgebra of $\mathfrak{gl}(V)$ for some linear space V. A complex-valued function $\varphi(H)$ defined on $\mathfrak{h}$ is said to be a *weight* of $\mathfrak{h}$ if there exists a non-zero $v \in V$ such that

$$Hv = \varphi(H)v \quad (\forall H \in \mathfrak{h}).$$

v is said to be a weight vector corresponding to φ.

According to Lie's theorem, any solvable linear Lie algebra (in particular, any nilpotent linear Lie algebra) has at least one weight.

THEOREM 3. *Let h be a nilpotent linear Lie algebra acting on the linear space V and $\varphi(H)$ be a complex-valued linear function, defined on $\mathfrak{h}$. Let*
$V_{\mathfrak{h}}^{\varphi} = \{v| \in V$ *and for some integer* n, $(H-\varphi(H)I)^n v = 0\ (\forall H \in \mathfrak{h})\}$
($V_{\mathfrak{h}}^{\varphi}$ is said to be weight subspace of φ), then

(1) $V_{\mathfrak{h}}^{\varphi} = \bigcap_{A\in\mathfrak{h}} V_A^{\varphi(A)}$ *and* $V_{\mathfrak{h}}^{\varphi}$ *is an invariant subspace of* $\mathfrak{h}$.

(2) $V_{\mathfrak{h}}^{\varphi} \neq 0$ *iff* φ *is a weight of* $\mathfrak{h}$. φ *is the only weight of* $\mathfrak{h}$ *in* $V_{\mathfrak{h}}^{\varphi}$.

(3) *If* $V_{\mathfrak{h}}^{\varphi} \neq 0$, *then* $(H-\varphi(H)I)^{\dim V_{\mathfrak{h}}^{\varphi}}\, v = 0\ (\forall H \in \mathfrak{h}, v \in V_{\mathfrak{h}}^{\varphi})$.

(4) $V = \sum_{\varphi\in\Delta}^{\cdot} V_{\mathfrak{h}}^{\varphi}$, *where* Δ *is the set of all weights of* $\mathfrak{h}$.

Proof. Obviously, $V_{\mathfrak{h}}^{\varphi} \subseteq \bigcap_{A\in\mathfrak{h}} V_A^{\varphi(A)}$. Let $V' = \bigcap_{A\in\mathfrak{h}} V_A^{\varphi(A)}$. Since $\mathfrak{h}$ is nilpotent, for any $B \in \mathfrak{h}$ there exists a positive integer m such that $(\operatorname{ad} A)^m B = 0$. From Lemma 4, $V_A^{\varphi(A)}$ is invariant under all $B \in \mathfrak{h}$, hence V' is an invariant subspace of $\mathfrak{h}$. The transformations on V' induced by elements of $\mathfrak{h}$ form a nilpotent Lie algebra. Since this Lie algebra is also solvable, by Lie's theorem, there exists a basis of V' with respect to which the matrix of the transformation induced by A ($A \in \mathfrak{h}$) is upper triangular with $\varphi_1(A), \ldots, \varphi_s(A)$ ($s = \dim V'$) on the main diagonal. Since $V' = \bigcap_{A\in\mathfrak{h}} V_A^{\varphi(A)}$, hence $\varphi(A) = \varphi_1(A) = \ldots = \varphi_s(A)$. Therefore $V' \subseteq V_{\mathfrak{h}}^{\varphi}$ and $V_{\mathfrak{h}}^{\varphi} = \bigcap_{A\in\mathfrak{h}} V_{\mathfrak{h}}^{\varphi(A)}$.

Invariance of $V_{\mathfrak{h}}^{\varphi}$ under $\mathfrak{h}$ can be proved from the fact that $V_{\mathfrak{h}}^{\varphi} = \bigcap_{A\in\mathfrak{h}} V_{\mathfrak{h}}^{\varphi(A)}$.

The proof of (2) is similar to the proof of 2° of Lemma 3.

(3) follows from Corollary 1 to Lie's theorem and the fact that φ is the only weight of $\mathfrak{h}$ in $V_{\mathfrak{h}}^{\varphi}$.

To prove (4), use induction on the dimension of V. If $\varphi(H)$ is the only eigenvalue of $H \in \mathfrak{h}$ ($\forall H \in \mathfrak{h}$), then by Lie's theorem, $\varphi(H)$ is a weight on $\mathfrak{h}$ and $V = V_{\mathfrak{h}}^{\varphi}$. If some $H \in \mathfrak{h}$ has eigenvalues $\lambda_1, \ldots, \lambda_s$ and at least two of them are distinct ($s \geqslant 2$), then by 5° of Lemma 3, $V = \sum_{i=1}^{\cdot s} V_H^{\lambda_i}$. Since $V_H^{\lambda_i}$ $(1 \leqslant i \leqslant s)$ are invariant subspaces of $\mathfrak{h}$ and $\dim V_A^{\lambda_i} < \dim V$, by induction assumption,

$$V_H^{\lambda_i} = \sum_{\varphi\in\Delta_i}^{\cdot} V_{\mathfrak{h}}^{\varphi},$$

where Δ_i is the set of all weights of $\mathfrak{h}$ in $V_H^{\lambda_i}$. Since all $\lambda_1, \ldots, \lambda_s$ are distinct, the sets $\Delta_1, \ldots \Delta_s$ are mutually disjoint. If $\Delta = \Delta_1 \cup \ldots \cup \Delta_s$, then

$$V = \sum_{\varphi\in\Delta}^{\cdot} V_{\mathfrak{h}}^{\varphi}$$

COROLLARY 1. *If $\mathfrak{h}$ is a nilpotent linear Lie algebra acting on an n-dimensional linear space V, then $\mathfrak{h}$ is isomorphic to a subalgebra of $\mathfrak{n}(n_1, C) \dotplus \ldots \dotplus \mathfrak{n}(n_m, C)$, where $n_1, \ldots, n_m$ are some positive integers satisfying $n_1 + \ldots + n_m = n$ and m is the number of distinct weights on $\mathfrak{h}$.*

Proof. Let $\varphi_1, \ldots, \varphi_m$ be all distinct weights of $\mathfrak{h}$ and $\dim V_{\mathfrak{h}}^{\varphi_i} = n_i$. The Lie algebra $\mathfrak{h}^i$ induced by $\mathfrak{h}$ on $V_{\mathfrak{h}}^{\varphi_i}$ is a homomorphic image of $\mathfrak{h}$ and is thus nilpotent. By Corollary 1 to Lie's theorem, there exists a basis of $V_{\mathfrak{h}}^{\varphi_i}$ with respect to which the matrix of the transformation induced by H ($\forall H \in \mathfrak{h}$) has the form

$$\begin{bmatrix} a_{11}^{(i)} & \cdots & a_{1n_i}^{(i)} \\ & \ddots & \vdots \\ 0 & & a_{n_in_i}^{(i)} \end{bmatrix}$$

and $a_{11}^{(i)} = \ldots = a_{n_in_i}^{(i)} = \varphi(H)$. Thus Lemma 1 follows.

COROLLARY 2. *Let $\mathfrak{h}$ be a nilpotent linear Lie algebra acting on the linear space V and $\varphi_1, \ldots, \varphi_n$ be all weights defined on $\mathfrak{h}$. If $H \in \mathcal{D}\mathfrak{h}$, then*

$$\varphi_1(H) = \ldots = \varphi_m(H) = 0.$$

CHAPTER 3

CARTAN SUBALGEBRAS

3.1. Cartan subalgebras

Let $\mathfrak{g}$ be a Lie algebra and $\mathfrak{h}$ be a nilpotent subalgebra. The linear transformations $\operatorname{ad}_{\mathfrak{g}} H$ $(H \in \mathfrak{h})$ form a nilpotent linear Lie algebra acting on $\mathfrak{g}$. Denote this Lie algebra by $\operatorname{ad}_{\mathfrak{g}} \mathfrak{h}$ or simply by $\operatorname{ad} \mathfrak{h}$. According to Theorem 2 of Chapter 3, $\mathfrak{g}$ can be decomposed into

$$\mathfrak{g} = \sum_{\varphi \in \Delta} \mathfrak{g}^{\varphi}_{\operatorname{ad} \mathfrak{h}} \tag{3.1}$$

where Δ is the set of weights of $\operatorname{ad} \mathfrak{h}$ which is sometimes called the set of weights of $\mathfrak{h}$. The following are properties of this decomposition:

(1) $[\mathfrak{g}^{\alpha}_{\operatorname{ad} \mathfrak{h}}, \mathfrak{g}^{\beta}_{\operatorname{ad} \mathfrak{h}}] \subseteq \mathfrak{g}^{\alpha+\beta}_{\operatorname{ad} \mathfrak{h}}$, for $\alpha, \beta \in \Delta$. In particular, $\mathfrak{g}^{0}_{\operatorname{ad} \mathfrak{h}}$ is a subalgebra of $\mathfrak{g}$. If $\alpha+\beta \notin \Delta$, then

$$[\mathfrak{g}^{\alpha}_{\operatorname{ad} \mathfrak{h}}, \mathfrak{g}^{\beta}_{\operatorname{ad} \mathfrak{h}}] = 0$$

Proof. We have the formula

$$(\operatorname{ad} H-\alpha(H)I-\beta(H)I)\,[X, Y] = [(\operatorname{ad} H-\alpha(H)I)X, Y]+[X, (\operatorname{ad} H-\beta(H)I)Y],$$

by induction, it can be proved that

$$(\operatorname{ad} H-\alpha(H)I-\beta(H)I)^k\,[X, Y]$$
$$= \sum_{s=0}^{k} c^k_s[(\operatorname{ad} H-\alpha(H)I)^s X, (\operatorname{ad} H-\beta(H)I)^{k-s} Y],$$

where c^k_s $(s = 0, \ldots, k)$ are the binomial coefficients. If $X \in \mathfrak{g}^{\alpha}_{\operatorname{ad} \mathfrak{h}}$, $Y \in \mathfrak{g}^{\beta}_{\operatorname{ad} \mathfrak{h}}$, then for a large k, the right-hand side of the formula above is zero, therefore $[X, Y] \in \mathfrak{g}^{\alpha+\beta}_{\operatorname{ad} \mathfrak{h}}$.

(2) $(\mathfrak{g}^{\alpha}_{\operatorname{ad} \mathfrak{h}}, \mathfrak{g}^{\beta}_{\operatorname{ad} \mathfrak{h}}) = 0, \quad \text{if } \alpha+\beta \neq 0.$

Proof. Take a basis from each $\mathfrak{g}^{\alpha}_{\operatorname{ad} \mathfrak{h}}$; the union of these bases is a basis of $\mathfrak{g}$. If $X \in \mathfrak{g}^{\alpha}_{\operatorname{ad} \mathfrak{h}}$, $Y \in \mathfrak{g}^{\beta}_{\operatorname{ad} \mathfrak{h}}$, then according to (1), we have

$$\operatorname{ad} X \operatorname{ad} Y\, \mathfrak{g}^{\varphi}_{\operatorname{ad} \mathfrak{h}} \subseteq \mathfrak{g}^{\varphi+\alpha+\beta}_{\operatorname{ad} \mathfrak{h}}.$$

Thus with respect to the basis chosen above, all diagonal elements of ad X ad Y are zero, hence

$$(X, Y) = \text{Tr ad } X \text{ ad } Y = 0.$$

(3) $\mathfrak{h} \subseteq \mathfrak{g}^0_{\text{ad } \mathfrak{h}}$.

Proof. Let $H \in \mathfrak{h}$. Since $\mathfrak{h}$ is nilpotent, hence for some positive integer m we have $(\text{ad } A)^m H = 0$ $(\forall A \in \mathfrak{h})$. Thus $H \in \mathfrak{g}^0_{\text{ad } \mathfrak{h}}$.

Definition. If $\mathfrak{h} = \mathfrak{g}^0_{\text{ad } \mathfrak{h}}$, then $\mathfrak{h}$ is said to be a Cartan subalgebra of $\mathfrak{g}$ and (3.1) is said to be the Cartan decomposition of $\mathfrak{g}$ with respect to the Cartan subalgebra $\mathfrak{h}$. The set of non-zero weights of $\mathfrak{h}$ is denoted by Σ and is called the set of roots of $\mathfrak{g}$ with respect to $\mathfrak{h}$. If φ is a root, then $\mathfrak{g}^{\varphi}_{\text{ad } \mathfrak{h}}$ is said to be the *root subspace* corresponding to φ.

Obviously, any Cartan subalgebra $\mathfrak{h}$ of $\mathfrak{g}$ contains the center $\mathfrak{c}$ of $\mathfrak{g}$. In fact, if $X \in \mathfrak{c}$, then $[H, X] = 0$ $(\forall H \in \mathfrak{h})$, i.e. $\text{ad } HX = 0$ $(\forall H \in \mathfrak{h})$. Therefore $X \in \mathfrak{h}^0_{\text{ad } \mathfrak{h}} = \mathfrak{h}$.

THEOREM 1. *A Cartan subalgebra $\mathfrak{h}$ of $\mathfrak{g}$ is a maximal nilpotent subalgebra.*

Proof. Suppose $\mathfrak{h}'$ is a nilpotent subalgebra and $\mathfrak{h}' \supseteq \mathfrak{h}$. Since $\mathfrak{h}'$ is nilpotent, there exists some positive integer m such that

$$\left[H'_m, \ldots [H'_3, [H'_2, H'_1] \ldots]\right] = 0,$$

for any m elements $H'_1, \ldots, H'_m \in \mathfrak{h}$. In particular,

$$[\underbrace{H, \ldots [H, [H}_{m-1}, H'] \ldots]] = 0,$$

for any $H \in \mathfrak{h}$ and $H' \in \mathfrak{h}'$, i.e. $(\text{ad } H)^{m-1} H' = 0$. Thus $H' \in \mathfrak{g}^0_{\text{ad } \mathfrak{h}} = \mathfrak{h}$ $(\forall H' \in \mathfrak{h}')$ and $\mathfrak{h}' \subseteq \mathfrak{h}$. Now it is clear that $\mathfrak{h}' = \mathfrak{h}$.

THEOREM 2. Suppose $\mathfrak{g} = \mathfrak{g}_1 \dotplus \mathfrak{g}_2$, then the direct sum of a Cartan subalgebra of $\mathfrak{g}_1$ and a Cartan subalgebra of $\mathfrak{g}_2$ is a Cartan subalgebra of $\mathfrak{g}$. Conversely, every Cartan subalgebra of $\mathfrak{g}$ is obtained in this way.

Proof. Let $\mathfrak{h}_i$ be a Cartan subalgebra of $\mathfrak{g}_i$ $(i = 1, 2)$, we have

$$\mathfrak{g}^0_{i \text{ ad } \mathfrak{h}_i} = \mathfrak{h}_i, \quad i = 1, 2.$$

Let $\mathfrak{h} = \mathfrak{h}_1 \dotplus \mathfrak{h}_2$, then $\mathfrak{h}$ is nilpotent. We now prove that

$$\mathfrak{g}^0_{\text{ad } \mathfrak{h}} = \mathfrak{h}.$$

Suppose that $X \in \mathfrak{g}^0_{\text{ad } \mathfrak{h}}$, then $(\text{ad } H)^m X = 0$ $(\forall H \in \mathfrak{h})$ for some positive integer m. Write $X = X_1 + X_2$ $(X_i \in \mathfrak{g}_i, i = 1, 2)$ and $H = H_1 + H_2$ $(H_i \in \mathfrak{h}_i, i = 1, 2)$, then from

$$(\text{ad } H)^m X = (\text{ad } H)^m X_1 + (\text{ad } H)^m X_2 = 0,$$

it follows that

$$(\text{ad } H_1)^m X_1 = (\text{ad } H_2)^m X_2 = 0, \quad \text{for all} \quad H_1 \in \mathfrak{h}_1, H_2 \in \mathfrak{h}_2.$$

Hence $X_i \in \mathfrak{g}^0_{i\,\mathrm{ad}\,\mathfrak{h}_i} = \mathfrak{h}_i$ $(i = 1, 2)$ and

$$X = X_1 + X_2 \in \mathfrak{h}_1 \dotplus \mathfrak{h}_2 = \mathfrak{h}.$$

Conversely, let $\mathfrak{h}$ be a Cartan subalgebra of $\mathfrak{g}$, i.e.

$$\mathfrak{g}^0_{\mathrm{ad}\,\mathfrak{h}} = \mathfrak{h}.$$

If

$$\mathfrak{h}_1 = \{H_1 \mid H_1 \in \mathfrak{g}_1 \text{ and for some } H \in \mathfrak{h},\ H = H_1 + H_2, \text{ where } H_2 \in \mathfrak{h}_2\}$$

and

$$\mathfrak{h}_2 = \{H_2 \mid H_2 \in \mathfrak{g}_2 \text{ and for some } H \in \mathfrak{h},\ H = H_2 + H_1, \text{ where } H_1 \in \mathfrak{h}_1\}$$

then $\mathfrak{h}_i$ is a subalgebra of $\mathfrak{g}_i$ $(i = 1, 2)$. We will show that $\mathfrak{h} = \mathfrak{h}_1 \dotplus \mathfrak{h}_2$. Suppose $H_1 \in \mathfrak{h}_1$ and for some $H \in \mathfrak{h}$, $H = H_1 + H_2$ where $H_2 \in \mathfrak{g}_2$. For any $H' \in \mathfrak{h}$,

$$(\mathrm{ad}\, H')^m H = 0, \quad \text{for some positive integer } m.$$

From
$$(\mathrm{ad}\, H')^m H = (\mathrm{ad}\, H')^m H_1 + (\mathrm{ad}\, H')^m H_2 = 0$$

and
$$(\mathrm{ad}\, H')^m H_i \in \mathfrak{g}_i \qquad (i = 1, 2),$$

it follows that $(\mathrm{ad}\, H')^m H_1 = 0$. Therefore $H_1 \in \mathfrak{h}$ and $\mathfrak{h}_1 \subseteq \mathfrak{h}$. Similarly, we can prove $\mathfrak{h}_2 \subseteq \mathfrak{h}$. On the other hand, let $H \in \mathfrak{h}$ and write $H = H_1 + H_2$ $(H_i \in \mathfrak{g}_i,\ i = 1, 2)$; then $H_i \in \mathfrak{h}_i$ $(i = 1, 2)$ and thus $\mathfrak{h} = \mathfrak{h}_1 \dotplus \mathfrak{h}_2$. Since $\mathfrak{h}$ is nilpotent, $\mathfrak{h}_1$ and $\mathfrak{h}_2$ are also nilpotent. Let $X_1 \in \mathfrak{g}^0_{1\,\mathrm{ad}\,\mathfrak{h}_1}$, then for some positive integer m, $(\mathrm{ad}\, H_1)^m X_1 = 0$ $(\forall H_1 \in \mathfrak{h}_1)$. Write any $H \in \mathfrak{h}$ as $H = H_1 + H_2$ $(H_i \in \mathfrak{h}_i,\ i = 1, 2)$, then $(\mathrm{ad}\, H)^m X_1 = (\mathrm{ad}\, H_1)^m X_1 = 0$. Hence $X_1 \in \mathfrak{g}^0_{\mathrm{ad}\,\mathfrak{h}} = \mathfrak{h}$ and $X_1 \in \mathfrak{h}_1$. Therefore $\mathfrak{h}_1$ is a Cartan subalgebra of $\mathfrak{g}_1$ and similarly, $\mathfrak{h}_2$ is a Cartan subalgebra of $\mathfrak{g}_2$.

Cartan subalgebras and Cartan decompositions of the classical Lie algebras are now given as examples.

Let $\mathfrak{g}$ be one of the Lie algebras A_n $(n \geqslant 1)$, B_n $(n \geqslant 1)$, C_n $(n \geqslant 1)$ and D_n $(n \geqslant 1)$. In Section 1.3 an n-dimensional abelian subalgebra was defined for every classical Lie algebra. From the structure formulas (1.3) (see Section 1.3), it can be seen that

$$\mathfrak{g} = \mathfrak{h} \dotplus \sum_{\alpha \in \Sigma} \mathfrak{g}^\alpha,$$

where Σ is the set of roots of $\mathfrak{g}$ and $\mathfrak{g}^\alpha$ is the one-dimensional subspace of E_α. We have

$$\mathfrak{g}^0_{\mathrm{ad}\,\mathfrak{h}} = \mathfrak{h}, \quad \mathfrak{g}^\alpha_{\mathrm{ad}\,\mathfrak{h}} = \mathfrak{g}^\alpha,$$

therefore $\mathfrak{h}$ is a Cartan subalgebra of $\mathfrak{g}$. Notice that

(1) $\mathfrak{h}$ is abelian,
(2) $\mathfrak{g}^\alpha$ $(\alpha \in \Sigma)$ are all one-dimensional.

It will be shown that these two properties are common to Cartan subalgebras and Cartan decompositions of all semisimple Lie algebras.

Finally, another characterization of Cartan subalgebras is given.

THEOREM 3. *Let* $\mathfrak{g}$ *be a Lie algebra and* $\mathfrak{h}$ *be a nilpotent subalgebra of* $\mathfrak{g}$. *If*

$$\mathfrak{n}(\mathfrak{h}) = \{X \in \mathfrak{g} \mid [X, \mathfrak{h}] \subset \mathfrak{h}\}$$

($\mathfrak{n}(\mathfrak{h})$ *is called the* normalizer *of* $\mathfrak{h}$ *in* $\mathfrak{g}$), *then* $\mathfrak{h}$ *is a Cartan subalgebra of* $\mathfrak{h}$ *iff* $\mathfrak{h} = \mathfrak{n}(\mathfrak{h})$.

Proof. First prove that $\mathfrak{n}(\mathfrak{h}) \subseteq \mathfrak{g}^0_{\mathrm{ad}\,\mathfrak{h}}$. Suppose $X \in \mathfrak{n}(\mathfrak{h})$, then $[X, \mathfrak{h}] \subseteq \mathfrak{h}$ or equivalently, $(\mathrm{ad}\ H)X \in \mathfrak{h}$ $(\forall H \in \mathfrak{h})$. Since $\mathfrak{h}$ is nilpotent, hence for some positive integer m, we have $(\mathrm{ad}\ H^m)X = 0$. Thus $X \in \mathfrak{g}^0_{\mathrm{ad}\,\mathfrak{h}}$ and $\mathfrak{n}(\mathfrak{h}) \subseteq \mathfrak{g}^0_{\mathrm{ad}\,\mathfrak{h}}$.

Now let $\mathfrak{h}$ be a Cartan subalgebra of $\mathfrak{g}$, i.e. $\mathfrak{h} = \mathfrak{g}^0_{\mathrm{ad}\,\mathfrak{h}}$. Hence $\mathfrak{n}(\mathfrak{h}) = \mathfrak{g}^0_{\mathrm{ad}\,\mathfrak{h}} = \mathfrak{h}$. On the other hand, it is obvious that $\mathfrak{h} \subseteq \mathfrak{n}(\mathfrak{h})$, thus $\mathfrak{h} = \mathfrak{n}(\mathfrak{h})$.

Conversely, let $\mathfrak{h} = \mathfrak{n}(\mathfrak{h})$. According to Lie's theorem, there is a basis of $\mathfrak{g}^0_{\mathrm{ad}\,\mathfrak{h}}$ with respect to which all the matrices of the transformations induced by $\mathrm{ad}\,H$ $(H \in \mathfrak{h})$ on $\mathfrak{g}^0_{\mathrm{ad}\,\mathfrak{h}}$ have form

$$\begin{bmatrix} 0 & & & \\ & 0 & & * \\ & & \ddots & \\ & & & 0 \end{bmatrix}.$$

Hence if $X \in \mathfrak{g}^0_{\mathrm{ad}\,\mathfrak{h}}$, then $(\mathrm{ad}\,H_1)(\mathrm{ad}\,H_2) \ldots (\mathrm{ad}\,H_s)X = 0$ $(\forall H_1, \ldots, H_s \in \mathfrak{h})$ where $s = \dim\ \mathfrak{g}^0_{\mathrm{ad}\,\mathfrak{h}}$. Thus $[H_1, (\mathrm{ad}\ H_2) \ldots (\mathrm{ad}\ H_s X)] = 0$ $(\forall H_1, \ldots, H_s \in \mathfrak{h})$ and it follows that $(\mathrm{ad}\,H_2) \ldots (\mathrm{ad}\,H_s)X \in \mathfrak{n}(\mathfrak{h}) = \mathfrak{h}$ $(\forall H_2, \ldots, H_s \in \mathfrak{h})$; now from $[H_2, (\mathrm{ad}\,H_3) \ldots (\mathrm{ad}\,H_s)X] \in \mathfrak{h}$ it follows that $(\mathrm{ad}\,H_3) \ldots (\mathrm{ad}\,H_s)X \in \mathfrak{n}(\mathfrak{h}) = \mathfrak{h}$ $(\forall H_3, \ldots, H_s \in \mathfrak{h})$. Repeating the argument, we finally get $X \in \mathfrak{n}(\mathfrak{h}) = \mathfrak{h}$. Thus $\mathfrak{g}^0_{\mathrm{ad}\,\mathfrak{h}} \subseteq \mathfrak{h}$ and $\mathfrak{h} = \mathfrak{g}^0_{\mathrm{ad}\,\mathfrak{h}}$, i.e. $\mathfrak{h}$ is a Cartan subalgebra of $\mathfrak{g}$.

3.2. Existence of Cartan subalgebras

Let $\mathfrak{g}$ be a Lie algebra and $X \in \mathfrak{g}$. The dimension of $\mathfrak{g}^0_{\mathrm{ad}\,X}$ is n_X (see Section 1.6). Recall that the rank n of $\mathfrak{g}$ is defined to be $n = \min_{X \in \mathfrak{g}} n_X$, thus the rank of a nilpotent Lie algebra is equal to the dimension; X is said to be a regular element if $n = n_X$.

LEMMA 1. *Let* X_0 *be a fixed element of* $\mathfrak{g}$. *If for any* $Y \in \mathfrak{g}$, *there exist infinitely many* μ *such that* $\mathrm{ad}\,(X_0 + \mu Y)$ *are all nilpotent, then* $\mathfrak{g}$ *is a nilpotent Lie algebra.*

Proof. Write the Killing polynomial of X as

$$f(\lambda \mid X) = \lambda^r + a_1(X)\lambda^{r-1} + \ldots + a_r(X).$$

$\mathrm{ad}\,X$ is nilpotent iff all roots of this polynomial are zero, i.e.

$$a_1(X) = \ldots = a_r(X) = 0.$$

For any $Y \in \mathfrak{g}$, the Killing polynomial $X_0 + \mu Y$ is

$$f(\lambda \mid X_0 + \mu Y) = \lambda^r + a_1(X_0 + \mu Y)\lambda^{r-1} + \ldots + a_r(X_0 + \mu Y).$$

If $\mathrm{ad}\,(X_0 + \mu Y)$ is nilpotent for infinitely many μ, then

$$a_1(X_0 + \mu Y) = \ldots = a_r(X_0 + \mu Y) = 0$$

for infinitely many μ, thus

$$a_1(X_0+\mu Y) \equiv a_2(X_0+\mu Y) \equiv \dots \equiv a_r(X_0+\mu Y) \equiv 0.$$

Let $Y = X - X_0$ and $\mu = 1$, then

$$a_1(X) \equiv \dots \equiv a_r(X) \equiv 0.$$

Thus ad X ($X \in \mathfrak{g}$) are all nilpotent. By Engel's theorem, $\mathfrak{g}$ is a nilpotent Lie algebra.

THEOREM 4. *If X_0 is a regular element, then $\mathfrak{g}^0_{\mathrm{ad}\,X_0}$ is a Cartan subalgebra.*

Proof. We have

$$\mathfrak{g} = \mathfrak{g}^0_{\mathrm{ad}\,X_0} + \dot{\sum_{\lambda \in \Sigma}} \mathfrak{g}^\lambda_{\mathrm{ad}\,X_0},$$

where Σ is the set of non-zero eigenvalues of ad X_0. Let $\tilde{\mathfrak{g}} = \dot{\sum}_{\lambda\in\Sigma} \mathfrak{g}^\lambda_{\mathrm{ad}\,X_0}$. We will prove that $\mathfrak{g}^0_{\mathrm{ad}\,X_0}$ is nilpotent. Let $X \in \mathfrak{g}^0_{\mathrm{ad}\,X_0}$, then $[X, \tilde{\mathfrak{g}}] = \tilde{\mathfrak{g}}$. Let $D(X)$ denote the determinant of the transformation induced by ad X on $\tilde{\mathfrak{g}}$. We know that $D(X_0) \neq 0$. For any $Y \in \mathfrak{g}^0_{\mathrm{ad}\,X_0}$, $D(X_0+\mu Y)$ is a polynomial in μ; since $D(X_0) \neq 0$, there exist infinitely many μ such that $D(X_0+\mu Y) \neq 0$. Now the rank of $\mathfrak{g}$ is n, so for these infinitely many μ, the eigenvalues of ad $(X_0+\mu Y)$ in $\mathfrak{g}^0_{\mathrm{ad}\,X_0}$ are zero, i.e. the eigenvalues of $\mathrm{ad}_{\mathfrak{g}^0_{\mathrm{ad}\,X_0}}(X_0+\mu Y)$ are zero. Thus $\mathrm{ad}_{\mathfrak{g}^0_{\mathrm{ad}\,X_0}}(X_0+\mu Y)$ is nilpotent and from Lemma 1, $\mathfrak{g}^0_{\mathrm{ad}\,X_0}$ is a nilpotent subalgebra.

Now let $\mathfrak{h} = \mathfrak{g}^0_{\mathrm{ad}\,X_0}$ and $X \in \mathfrak{g}^0_{\mathrm{ad}\,X_0}$, then for some positive integer m,

$$(\mathrm{ad}\,H)^m X = 0, \qquad \forall H \in \mathfrak{h}.$$

In particular, $X_0 \in \mathfrak{h}$ implies that

$$(\mathrm{ad}\,X_0)^m X = 0.$$

thus $X \in \mathfrak{g}^0_{\mathrm{ad}\,\mathfrak{h}} = \mathfrak{h}$. Therefore $\mathfrak{g}^0_{\mathrm{ad}\,\mathfrak{h}} \subseteq \mathfrak{h}$ and $\mathfrak{g}^0_{\mathrm{ad}\,\mathfrak{h}} = \mathfrak{h}$, hence $\mathfrak{h}$ is a Cartan subalgebra of $\mathfrak{g}$.

COROLLARY 1. *Every Lie algebra $\mathfrak{g}$ contains a Cartan subalgebra.*

Proof. Since $\mathfrak{g}$ must contain a regular element, the result follows from Theorem 4.

COROLLARY 2. *Let $\mathfrak{h}$ be a nilpotent subalgebra of $\mathfrak{g}$ that contains a regular element of $\mathfrak{g}$, then $\mathfrak{g}^0_{\mathrm{ad}\,\mathfrak{h}}$ is a Cartan subalgebra of $\mathfrak{g}$.*

Proof. Let X_0 be a regular element of $\mathfrak{g}$ contained in $\mathfrak{h}$, then $\mathfrak{g}^0_{\mathrm{ad}\,X_0}$ is a Cartan subalgebra of $\mathfrak{g}$. Let $\mathfrak{h}_1 = \mathfrak{g}^0_{\mathrm{ad}\,X_0}$. Since $\mathfrak{h}$ is a nilpotent subalgebra, $\mathfrak{h} \subseteq \mathfrak{g}^0_{\mathrm{ad}\,X_0} = \mathfrak{h}_1$. From $X_0 \in \mathfrak{h} \subseteq \mathfrak{h}_1$, it follows that

$$\mathfrak{h}_1 = \mathfrak{g}^0_{\mathrm{ad}\,X_0} \supseteq \mathfrak{g}^0_{\mathrm{ad}\,\mathfrak{h}} \supseteq \mathfrak{g}^0_{\mathrm{ad}\,\mathfrak{h}_1} = \mathfrak{g}_1.$$

Thus $\mathfrak{g}^0_{\mathrm{ad}\,\mathfrak{h}} = \mathfrak{g}^0_{\mathrm{ad}\,X_0}$ is a Cartan subalgebra of $\mathfrak{g}$.

THEOREM 5. Let $\mathfrak{h}$ be a maximal nilpotent subalgebra. If $\mathfrak{h}$ contains a regular element X_0 of $\mathfrak{g}$, then $\mathfrak{h}$ is a Cartan subalgebra of $\mathfrak{g}$ and $\mathfrak{h} = \mathfrak{g}^0_{\mathrm{ad}\,X_0}$.

Proof. According to Theorem 4, $\mathfrak{g}^0_{\mathrm{ad}\,X_0}$ is a Cartan subalgebra of $\mathfrak{g}$. Since $\mathfrak{h}$ is a nilpotent subalgebra, thus $\mathfrak{h} \subseteq \mathfrak{g}^0_{\mathrm{ad}\,X_0}$. From maximality of $\mathfrak{h}$, it follows that $\mathfrak{h} = \mathfrak{g}^0_{\mathrm{ad}\,X_0}$.

3.3. Preliminaries†

Let V be an n-dimensional vector space over C and $e_1, \ldots, e_n$ be a basis of V. If $x_1, \ldots, x_n$ are n independent variables over C, then an arbitrary element of V can be written as

$$x = x_1e_1 + \ldots + x_ne_n.$$

Now let W be an m-dimensional vector space over C and $\tilde{e}_1, \ldots, \tilde{e}_m$ be a basis of W. If $y_1, \ldots, y_m$ are m independent variables over C, then an arbitrary element w can be written as

$$w = y_1\tilde{e}_1 + \ldots + y_m\tilde{e}_m.$$

A mapping f from V to W defined by

$$x = x_1e_1 + \ldots + x_ne_n \to f(x) = x' = y_1'\tilde{e}_1 + \ldots + y_m'\tilde{e}_m$$

is said to be a polynomial mapping if

$$\begin{aligned} y_1 &= \varphi_1(x_1, \ldots, x_n), \\ &\vdots \\ y_m' &= \varphi_m(x_1, \ldots, x_n) \end{aligned}$$

are polynomials of $x_1, \ldots, x_n$. The mapping φ defined by

$$\Sigma C_{i_1 \ldots i_m} y_1^{i_1} \ldots y_m^{i_m} \xrightarrow{\varphi} \Sigma C_{i_1 \ldots i_m} \varphi_1^{i_1} \ldots \varphi_m^{i_m}$$

is then a homomorphism from the ring of polynomials $C[y_1, \ldots, y_m]$ to $C[x_1, \ldots, x_n]$ satisfying

$$\varphi(1) = 1,$$
$$\varphi(\lambda\psi(y_1, \ldots, y_m)) = \lambda\varphi(\psi(y_1, \ldots, y_m)), \quad \forall \psi \in C[y_1, \ldots, y_m], \lambda \in C.$$

φ is said to be a C-homomorphism or the C-homomorphism determined by the polynomial mapping f.

Conversely, suppose that φ is a C-homomorphism from $C[y_1, \ldots, y_m]$ to $C[x_1, \ldots, x_n]$ and let

$$\varphi(y_i) = \varphi_i(x_1, \ldots, x_n), \qquad 1 \leqslant i \leqslant m,$$

then the φ_i's are polynomials and the mapping

$$x = x_1e_1 + \ldots + x_ne_n \to x' = \varphi_1\tilde{e}_1 + \ldots + \varphi_m\tilde{e}_m$$

is a polynomial mapping from V to W. Obviously, φ is the C-homomorphism determined by this polynomial mapping. Thus the C-homomorphisms from $C[y_1, \ldots, y_m]$ to $C[x_1, \ldots, x_n]$ are in one–one correspondence with polynomial mappings from V to W.

The following are examples of polynomial mappings.

† This and the following section presuppose more knowledge of algebra. For example, the reader should understand the material in Vol. 1 of *Algebra* by Van der Waerden. By taking for granted that Theorem 6 of the next section is true, the reader can omit these two sections.

EXAMPLE 1. Linear mappings from V to W are polynomial mappings.

EXAMPLE 2. Let $V = (0)$ and

$$f:0 \to w = b_1\tilde{e}_1 + \ldots + b_m\tilde{e}_m$$

be a mapping from (0) to W. f is obviously a polynomial mapping and the C-homomorphism determined by f is from $C[y_1, \ldots, y_m]$ to C and defined by

$$\varphi : Q(y_1, \ldots, y_m) \to Q(b_1, \ldots, b_m).$$

Define $Q(w) = Q(y_1, \ldots, y_m)$, then the vectors in W are in one–one correspondence with C-homomorphisms from $C[y_1, \ldots, y_m]$ to C.

EXAMPLE 3. Let $W = C$. Let f be a mapping from V to $W = C$ and

$$f : v \to P(v) \in C.$$

If f is a polynomial mapping, then $P(v)$ is a polynomial in $x_1, \ldots, x_n$:

$$P(v) = P(x_1, \ldots, x_n).$$

If $v = a_1e_1 + \ldots + a_ne_n$ then

$$P(v) = P(a_1, \ldots, a_n).$$

Thus f is determined by the polynomial $P(x_1, \ldots, x_n)$. We then denote f by $\tilde{P}$.

We now consider the decompositions of polynomial mappings. Let

$$x = x_1e_1 + \ldots + x_ne_n \xrightarrow{f} x' = \varphi_1\tilde{e}_1 + \ldots + \varphi_m\tilde{e}_m$$

be a polynomial mapping from V to W, then $\varphi_1, \ldots, \varphi_m$ are polynomials in $x_1, \ldots, x_n$. Write

$$\varphi_i = \sum_k \varphi_{ik},$$

where φ_{ik} is the homogeneous component of φ_i of degree k. Define

$$x = x_1e_1 + \ldots + x_ne_n \xrightarrow{f_k} \varphi_{1k}\tilde{e}_1 + \ldots + \varphi_{mk}\tilde{e}_m,$$

then f_k $(k = 0, 1, 2, \ldots)$ are polynomial mappings from V to W. Obviously,

$$f = \sum_k f_k.$$

f_k is called the homogeneous component of f of degree k.

LEMMA 1. *If f is polynomial mapping from V to W and g is a polynomial mapping from W to U, then $h = g \circ f$ is a polynomial mapping from V to U. Furthermore, if $f = \sum_{k \geq k_0 \geq 1} f_k$ and $g = \sum_{l \geq l_0} g_l$ are decompositions of f and g into homogeneous components, then the homogeneous components h_j of h satisfy*

$$h_j = \begin{cases} 0, & j < k_0 l_0 \\ g_{l_0} \circ f_{k_0}, & j = l_0 k_0. \end{cases}$$

Proof. Let $\tilde{\tilde{e}}_1, \ldots, \tilde{\tilde{e}}_p$ be a basis of U and

$$f : x = x_1 e_1 + \ldots + x_n e_n \to f(x) = \varphi_1 \tilde{e}_1 + \ldots + \varphi_m \tilde{e}_m,$$
$$g : y = y_1 \tilde{e}_1 + \ldots + y_m \tilde{e}_m \to g(y) = \psi_1 \tilde{\tilde{e}}_1 + \ldots + \psi_p \tilde{\tilde{e}}_p,$$

then

$$g \circ f(x) = \xi_1 \tilde{\tilde{e}}_1 + \ldots + \xi_p \tilde{\tilde{e}}_p$$

and

$$\xi_i = \psi_i(\varphi_1, \ldots, \varphi_m), \qquad i = 1, 2, \ldots, p.$$

ψ_i $(i = 1, \ldots, p)$ are obviously polynomials of $x_1, \ldots, x_n$, thus the first part of the theorem is proved. For the second part, observe that since $k_0 \geqslant 1$, from

$$\xi_i = \sum_{l \geqslant l_0} \psi_{il}\left(\sum_{k \geqslant k_0} \varphi_{1k}, \ldots, \sum_{k \geqslant k_0} \varphi_{mk}\right)$$

it follows that

$$\xi_{ij} = \begin{cases} 0, & j < k_0 l_0 \\ \psi_{il_0}(\varphi_{1k_0}, \ldots, \varphi_{mk_0}), & j = k_0 l_0. \end{cases}$$

Hence

$$h_j = \begin{cases} 0, & j < k_0 l_0 \\ g_{l_0} \circ f_{k_0} & j = k_0 l_0. \end{cases}$$

This completes the proof of the lemma.

We now define differentials of polynomial mappings. Let f be a polynomial mapping from V to W. For any $x, v \in V$, let

$$\Delta f(x, v) = f(x+v) - f(v).$$

When v is fixed, the mapping

$$x \to \Delta f(x, v)$$

is also a polynomial mapping from V to W. This polynomial mapping does not have components of degree zero and the homogeneous component of degree one of it, denoted by $(df)_v$, is said to be the differential of f at v. $(df)_v$ is a linear mapping from V to W.

LEMMA 2. *Let f be a polynomial mapping from V to W, g be a polynomial mapping from W to U and $h = g \circ f$. If $v \in V$ and $w = f(v)$, then*

$$(dh)_v = (dg)_w \circ (df)_v .$$

Proof. Let $x \in V$, then

$$\begin{aligned} \Delta h(x, v) &= h(x+v) - h(v) = g(f(x+v)) - g(f(v)) \\ &= g(f(v) + \Delta f(x, v)) - g(f(v)) \\ &= g(w + \Delta f(x, v)) - g(w) \\ &= \Delta g(\Delta f(x, v), w). \end{aligned}$$

Comparing the first degree components of the two sides of this equation, noticing that Δf and Δg do not have components of degree zero, and using Lemma 1, we have

$$(dh)_v = (dg)_w \circ (df)_v.$$

LEMMA 3. *Let f be a polynomial mapping from V to W and suppose that for some v, $(df)_v$ is onto, then the C-homomorphism φ from $C[y_1, \ldots, y_m]$ to $C[x_1, \ldots, x_n]$ determined by f is a C-isomorphism.*

Proof. Observe that φ does not change constants of $C[y_1, \ldots, y_n]$.

Let P be a polynomial of degree $\geqslant 1$ in $C[y_1, \ldots, y_m]$. We will prove that $\varphi(P) \neq 0$. P determines a polynomial mapping $\tilde{P}$ from W to C defined by

$$\tilde{P}: b_1\tilde{e}_1 + \ldots + b_m\tilde{e}_m \to P(b_1, \ldots, b_m)$$

and $h = \tilde{P} \circ f$ is a polynomial mapping from V to C. If x is any vector of V and $f(x) = \varphi_1\tilde{e}_1 + \ldots + \varphi_m\tilde{e}_m$, then

$$h(x) = \tilde{P} \circ f(x) = P(f(x)) = P(\varphi_1, \ldots, \varphi_m) = \varphi(P).$$

Thus to show that $\varphi(P) \neq 0$, it suffices to show that $h = \tilde{P} \circ f \neq 0$. According to the proof of Lemma 2, we have

$$\Delta h(x, v) = \Delta\tilde{P}(\Delta f_1 x, v), w),$$

where $w = f(v)$. Denote the component of $\Delta\tilde{P}$ of lowest degree by $(\Delta\tilde{P})_{k_0}$. Since the degree of P is $\geqslant 1$, thus $\Delta\tilde{P} \neq 0$ and thus $(\Delta\tilde{P})_{k_0} \neq 0$. By Lemma 1, the term of lowest degree of $\Delta h(x, v)$ is $(\Delta\tilde{P})_{k_0} \circ (df)_v$. Since $(df)_v$ is onto, thus $\Delta\tilde{P}_{k_0} \circ (df)_v \neq 0$. Hence $\Delta h(x, v) \neq 0$ and $h = \tilde{P} \circ f \neq 0$. This completes the proof of the lemma.

LEMMA 4. *Let f be a polynomial mapping from an n-dimensional vector space V to an m-dimensional vector space W and for some v_0,* let $(df)_{v_0}$ be *onto. Let $e_1, \ldots, e_n$ be a basis of V, $\tilde{e}_1, \ldots, \tilde{e}_m$ be a basis of W and $P(x_1, \ldots, x_n) \in C[x_1, \ldots, x_n]$. For any $v = a_1e_1 + \ldots + a_ne_n$, let $P(v) = P(a_1, \ldots, a_n)$.* If *$P(x_1, \ldots, x_n) \not\equiv 0$, then there exists $Q(y_1, \ldots, y_m) \not\equiv 0$ in $C[y_1, \ldots, y_m]$ such that if $Q(w) \neq 0$ for some $w \in W$, then there exists a $v \in V$ satisfying $w = f(v)$ and $P(v) \neq 0$.*

Proof. Under the assumptions, the C-homomorphism φ from $C[y_1, \ldots, y_m]$ to $C[x_1, \ldots, x_n]$ determined by f is an isomorphism. Embed $C[y_1, \ldots, y_m]$ into $C[x_1, \ldots, x_n]$ by φ. For any $v \in V$, the mapping φ_v defined by

$$P(x_1, \ldots, x_n) \to P(v)$$

is a C-homomorphism from $C[x_1, \ldots, x_n]$ to C. The restriction of φ_v to $C[y_1, \ldots, y_m]$ is a C-homomorphism $\varphi_v \circ \varphi$ from $C[y_1, \ldots, y_m]$ to C, thus this mapping is determined by some $w \in W$, i.e.

$$\varphi_v \circ \varphi(Q) = Q(w), \quad \text{for a } w \in W \text{ and all } Q \in C[y_1, \ldots, y_m].$$

Let $x = x_1e_1 + \ldots + \ldots + x_ne_n \in V$ and

$$f(x) = \varphi_1\tilde{e}_1 + \ldots + \varphi_m\tilde{e}_m,$$

where $\varphi_1, \ldots, \varphi_m$ are polynomials of $x_1, \ldots, x_n$, then

$$\begin{aligned}\varphi_v \circ \varphi(Q) &= \varphi_v(Q(\varphi_1, \ldots, \varphi_m)) \\ &= Q(\varphi_1(v), \ldots, \varphi_m(v)) = Q(f(v)).\end{aligned}$$

Since the C-homomorphism $\varphi_v \circ \varphi$ is uniquely determined by a vector of W, thus

$$w = f(v).$$

Thus it suffices to prove that there exists a $Q(y_1, \ldots, y_m)$ such that any C-homomorphism from $C[y_1, \ldots, y_m]$ to C that does not map Q to 0 can be extended to a C-homomorphism from $C[x_1, \ldots, x_n]$ to C that does not map P to 0.

Let $A = C[x_1, \ldots, x_n]$ and $B = \varphi(C[y_1, \ldots, y_m])$. Suppose that A is generated by B and the elements $z_1, \ldots, z_h$. Let A_k be the ring generated by B and $z_{k+1}, \ldots, z_h$, then $A_0 = A$, $A_h = B$ and A_{k-1} is generated by A_k and z_k. Suppose some $P \not\equiv 0$ in A_0 is given. We want to prove that for any $P_{k-1} \not\equiv 0$ in A_{k-1}, there exists $P_k \not\equiv 0$ in A_k such that any C-homomorphism from A_k to C that does not map P_k to zero can be extended to a C-homomorphism from A_{k-1} to C that does not map P_{k-1} to zero ($k = 1, \ldots, h$).

For a fixed k, we consider two cases:

(1) z_k does not satisfy a polynomial whose coefficients are elements of A_k. In this case, A_{k-1} is the ring of polynomials in z_k with coefficients in A_k. Obviously, if ψ is a C-homomorphism from A_k to C, then ψ can be extended to a C-homomorphism $\bar{\psi}$ from A_{k-1} to C by setting $\bar{\psi}(z_k) = b$ for some $b \in C$. Now $P_{k-1} \not\equiv 0$ and thus there is a coefficient P_k of P_{k-1} which is not identically zero. If ψ is a C-homomorphism from A_k to C such that $\psi(P_k) \neq 0$, apply ψ to the coefficients of P_{k-1}, then we obtain a polynomial $\bar{P}_{k-1}$ of z_k. Now $\psi(P_k) \neq 0$ implies that $\bar{P}_{k-1} \neq 0$, thus there exists some $b \in C$ such that $\bar{P}_{k-1}(b) \neq 0$. By setting $\bar{\psi}(z_k) = b$, we obtain an extension $\bar{\psi}$ of ψ which satisfies $\bar{\psi}(P_{k-1}) = \bar{P}_{k-1}(b) \neq 0$.

(2) z_k satisfies a polynomial whose coefficients are elements of A_k. Suppose ψ is a C-homomorphism from A_k to C. If ψ can be extended to a C-homomorphism $\bar{\psi}$ from A_{k-1} to C and $\bar{\psi}(z_k) = b$, then b must satisfy any polynomials $\bar{P}$ obtained by applying ψ to the coefficients of a polynomial P satisfied by z_k. Let $P(X)$ be a polynomial with coefficients in A_{k-1} of lowest degree which is satisfied by z_k and write $P(X) = a_0 + a_1X + \ldots + a_nX^n$, where $a_i \in A_k$ and $a_n \neq 0$. If $Q(X)$ is another polynomial, then by dividing $Q(X)$ by $P(X)$, we get

$$a_n^l Q(X) = U(X)P(X) + V(X),$$

where l is sufficiently large, the coefficients of U and V are in A_k and $\deg V < \deg P$. If z_k satisfies $Q(X)$, then $V(z_k) = 0$ and

$$a_n^l Q(X) = U(X)P(X).$$

Hence if $\psi(a_n) \neq 0$, then from $\bar{P}(b) = 0$ it follows that $\psi(a_n)^l \bar{Q}(b) = 0$ and $\bar{Q}(b) = 0$. Thus if $\psi(a_n) \neq 0$, then ψ can be extended to a homomorphism $\bar{\psi}$ from A_{k-1} to C iff $\bar{\psi}(z_k) = b$ is a root of $\bar{P}$.

Now the field of quotients of A_{k-1} is algebraic over the field of quotients of A_k, thus every element in A_{k-1} satisfies a polynomial with coefficients in A_k. Let $H(X)$ be a polynomial of lowest degree satisfied by P_{k-1} and write

$$H(X) = b_0 + b_1 X + \ldots + b_m X^m,$$

where $b_i \in A_k$ and $b_m \neq 0$, hence $b_0 \neq 0$. Thus

$$\bar{H}(X) = \psi(b_0) + \psi(b_1)X + \ldots + \psi(b_m)X^m.$$

From $H(P_{k-1}) = 0$, it follows that $\bar{\psi}(H(P_{k-1})) = 0$, i.e.

$$\bar{H}(\bar{\psi}(P_{k-1})) = \psi(b_0) + \psi(b_1)\,\bar{\psi}(P_{k-1}) + \ldots + \psi(b_m)\,\bar{\psi}(P_{k-1})^m = 0.$$

Thus if $\psi(b_0) \neq 0$, then $\bar{\psi}(P_{k-1}) \neq 0$. By setting $P_k = a_n b_0$, it is clear that if ψ does not map P_k to zero, then $\bar{\psi}$ does not map P_{k-1} to zero. This completes the proof of the lemma.

3.4. Conjugacy of Cartan subalgebras

LEMMA 5. *Let $\mathfrak{g}$ be a Lie algebra and $X \in \mathfrak{g}$. If ad X is a nilpotent linear transformation, then*

$$\sigma(X) = \exp \operatorname{ad} X = \sum_{m=0}^{\infty} \frac{1}{m!} (\operatorname{ad} X)^m \quad \textit{(there are only finitely many terms)}$$

is an automorphism of $\mathfrak{g}$.

Proof. Let $X, Y \in \mathfrak{g}$, we have

$$\operatorname{ad} X[Y, Z] = [\operatorname{ad} XY, Z] + [Y, \operatorname{ad} XZ].$$

By induction, it can be proved that

$$(\operatorname{ad} X)^m [Y, Z] = \sum_{k=0}^{m} \frac{m!}{k!\,(m-k)!} [(\operatorname{ad} X)^k Y, (\operatorname{ad} X)^{m-k} Z].$$

Thus

$$\begin{aligned}\sigma(X)[Y, Z] &= \sum_{m=0}^{\infty} \frac{1}{m!} \sum_{k=0}^{m} \frac{m!}{k!\,(m-k)!} [(\operatorname{ad} X)^k Y, (\operatorname{ad} X)^{m-k} Z] \\ &= \left[\sum_{k=0}^{\infty} \frac{1}{k!} (\operatorname{ad} X)^k Y, \sum_{l=0}^{\infty} \frac{1}{l!} (\operatorname{ad} X)^l Z\right] \\ &= [\sigma(X)Y, \sigma(X)Z].\end{aligned}$$

Obviously, $\sigma(X)$ is an invertible linear transformation, thus $\sigma(X)$ is an automorphism of $\mathfrak{g}$.

THEOREM 6 *(Cartan–Chevelley).*† (1) *Let $\mathfrak{h}$ be a Cartan subalgebra of $\mathfrak{g}$. If $H \in \mathfrak{h}$ and $\alpha(H) \neq 0$ for any non-zero root α, then H is a regular element of $\mathfrak{g}$.*

† C. Chevelley, An algebraic proof of a property of Lie groups, *Amer. J. Math.* **63** (1941), 785–93; C. Chevelley, *Théorie des Groupes de Lie*, Tome III, Paris, 1955.

(2) *If A denotes the group of automorphisms of* $\mathfrak{g}$ *generated by* $\sigma(X)$ *(*ad X *is nilpotent), then any two Cartan subalgebras of* $\mathfrak{g}$ *are conjugate under A, i.e. one of the algebras is transformed to the other by an element of A.*

Proof. Let $\mathfrak{h}$ be a Cartan subalgebra of $\mathfrak{g}$ and Σ be the set of all non-zero roots, then $\mathfrak{g}$ has the Cartan decomposition

$$\mathfrak{g} = \mathfrak{h} \dotplus \sum_{\alpha\in\Sigma}{}^{\cdot}\, \mathfrak{g}^{\alpha}, \quad \mathfrak{h} = \mathfrak{g}^{0}_{\mathrm{ad}\,\mathfrak{h}}.$$

Now $[\mathfrak{g}^{\alpha}, \mathfrak{g}^{\beta}] \subseteq \mathfrak{g}^{\alpha+\beta}$, thus if $X \in \mathfrak{g}^{\alpha}$, then ad X maps $\mathfrak{g}^{\beta}$ to $\mathfrak{g}^{\alpha+\beta}$ and $(\mathrm{ad}\,X)^{k}\mathfrak{g}^{\beta} \subseteq \mathfrak{g}^{k\alpha+\beta}$. Since Σ is a finite set, ad X must be nilpotent, hence $\sigma(X) = \exp \mathrm{ad}\, X$ is an automorphism of $\mathfrak{g}$.

Let $\Sigma = \{\alpha_1, \ldots, \alpha_m\}$ and write $\mathfrak{g} = \mathfrak{h} \dotplus \mathfrak{g}^{\alpha_1} \dotplus \ldots \dotplus \mathfrak{g}^{\alpha_m}$. Consider the mapping f from $\mathfrak{g}$ to $\mathfrak{g}$ defined by

$$f(H, X_1, \ldots, X_m) = \sigma(X_1) \ldots \sigma(X_m)H, \ (H \in \mathfrak{h},\ X_1 \in \mathfrak{g}^{\alpha_1}, \ldots X_m \in \mathfrak{g}^{\alpha_m}).$$

Since ad $X_1, \ldots,$ ad X_m are nilpotent, f is a polynomial mapping. Now compute the differential of f at the point $(H_0, 0, \ldots, 0)$:

$$\begin{aligned}
&\Delta f(H, X_1, \ldots, X_m; H_0, 0, \ldots, 0)\\
&= f(H+H_0, X_1, \ldots, X_m) - f(H_0, 0, \ldots, 0)\\
&= \sigma(X_1) \ldots \sigma(X_m)(H+H_0) - H_0\\
&= \sum_{k_1, \ldots, k_m} \frac{1}{\Pi k_i!}\left[X_1^{k_1}, \left[X_2^{k_2}, \ldots [X_m^{k_m}, H] \ldots\right]\right]\\
&\quad + \sum_{k_1, \ldots, k_m} \frac{1}{\Pi k_i!}\left[X_1^{k_1}, \left[X_2^{k_2}, \ldots [X_m^{k_m}, H_0] \ldots\right]\right] - H_0,
\end{aligned}$$

where $[X^k, Y]$ is understood to be $(\mathrm{ad}\,X)^k Y$. When $X_1, \ldots, X_m, H$ and $\left[X_1^{k_1}, [\ldots[X_m^{k_m}, H]\ldots\right]]$ are written as linear combinations of a basis of $\mathfrak{g}$ formed by the union of a basis of $\mathfrak{h}$, a basis of $\mathfrak{g}^{\alpha_1}, \ldots$ and a basis of $\mathfrak{g}^{\alpha_m}$, the coefficients of $\left[X_1^{k_1} \ldots [X_m^{k_m}, H] \ldots\right]$ are polynomials of coefficients of $X_1, \ldots, X_m$ and H. Now notice that $\left[X_1^{k_1}, [X_2^{k_2} \ldots [X_m^{k_m}, H] \ldots\right]$ is homogeneous of degree $k_1+k_2+ \ldots +k_m+1$ and $\left[X_1^{k_1}, [X_2^{k_2}, \ldots [X_m^{k_m}, H_0] \ldots\right]$ is homogeneous of degree $k_1+ \ldots +k_m$. Thus

$$df(H_1, X_1, \ldots, X_m; H_0, 0, \ldots, 0) = H + \sum_{i=1}^{m} [X_i, H_0].$$

Now choose $H_0 \in \mathfrak{h}$ such that $\prod_{i=1}^{m} \alpha_i(H_0) \neq 0$. If

$$df(H, X_1, \ldots, X_m; H_0, 0, \ldots, 0) = 0,$$

i.e.

$$H + \sum_{i=1}^{m} [X_i, H_0] = 0,$$

then from $[X_i, H_0] \in \mathfrak{g}^{\alpha_i}$, it follows that $H = 0$ and $[X_i, H_0] = 0$ $(i = 1, \ldots, m)$. Since the determinant of ad H_0 in $\mathfrak{g}^{\alpha_i}$ is $(\alpha_i(H_0))^{\dim \mathfrak{g}^{\alpha_i}} \neq 0$, ad H_0 is non-singular in $\mathfrak{g}^{\alpha_i}$. Thus $[X_i, H_0] =$

0 implies that $X_i = 0$. We have proved that if $H_0 \in \mathfrak{h}$ and

$$\prod_{i=1}^{m} \alpha_i(H_0) \neq 0,$$

then $(df)_{(H_0, 0, \ldots, 0)}$ maps $\mathfrak{g}$ onto itself.

We can now apply Lemma 4 of Section 3.3. Let G be the group of automorphisms generated by $\sigma(X)$ $(X \in \mathfrak{g}^{\alpha})$. Notice that

$$\prod_{i=1}^{m} \alpha_i(H)$$

can be considered as a non-zero polynomial defined on $\mathfrak{g}$; in fact, if $H_1, \ldots, H_n$ is a basis of $\mathfrak{h}$, $\alpha_i(H_k) = \alpha_{ik}$ and $H = x_1H_1 + \ldots + x_nH_n$, then

$$\prod_{i=1}^{m} \alpha_i(H_1) = \prod_{i=1}^{m} (\alpha_{i1}x_1 + \ldots + \alpha_{in}x_n).$$

Thus there exists a polynomial $Q \not\equiv 0$ defined on $\mathfrak{g}$ such that if $X \in \mathfrak{g}$ and $Q(X) \neq 0$, then X is transformed by G to an $H \in \mathfrak{h}$ satisfying

$$\prod_{i=1}^{m} \alpha_i(H) \neq 0.$$

Suppose the Killing polynomial of $\mathfrak{g}$ is

$$|\lambda I - \operatorname{ad} X| = \lambda^r + \varphi_1(X)\lambda^{r-1} \ldots + \varphi_{r-n}(X)\lambda^n,$$

where $\varphi_{r-n}(X) \not\equiv 0$. If $X_0 \in \mathfrak{g}$ and $\varphi_{r-n}(X_0) \neq 0$, then X_0 is a regular element of $\mathfrak{g}$. Obviously, automorphisms of $\mathfrak{g}$ transform regular elements to regular elements, in particular, elements of G transform regular elements to regular elements. Since $Q(X)\varphi_{r-n}(X) \neq 0$, hence $Q(X_0)\varphi_{r-n}(X_0) \neq 0$ for some $X_0 \in \mathfrak{g}$; X_0 is then a regular element of $\mathfrak{g}$ and $Q(X_0) \neq 0$, thus X_0 is the image of some $H_0 \in \mathfrak{h}$ under G and $\prod_{i=1}^{m} \alpha_i(H_0) \neq 0$. Thus $\mathfrak{h}$ contains a regular element H_0. Now it is clear that if $H \in \mathfrak{h}$ and $\prod_{i=1}^{m} \alpha_i(H) \neq 0$, then H is a regular element of $\mathfrak{g}$; according to Theorem 5, we have

$$\mathfrak{g}^0_{\operatorname{ad} H_0} = \mathfrak{h} = \mathfrak{g}^0_{\operatorname{ad} H}.$$

We have proved that if a regular element X_0 satisfies $Q(X_0) \neq 0$, then X_0 is the image of some regular $H_0 \in \mathfrak{h}$ under G. Now suppose that $\mathfrak{h}'$ is another Cartan subalgebra of $\mathfrak{g}$, then we get a polynomial $Q'(x) \not\equiv 0$ and a group G' of automorphisms. Similarly, it can be proved that if a regular element X_0 satisfies $Q'(X_0) \neq 0$, then it is the image under G' of some regular element in $\mathfrak{h}'$. Obviously, G and G' are contained in A. Now if X_0 is a regular element and $Q'(X_0)\,Q(X_0) \neq 0$, then X_0 is conjugate under A to a regular element $H_0 \in \mathfrak{h}$ and a regular element $H_0' \in \mathfrak{h}'$. Hence H_0 and H_0' are conjugate under A. From $\mathfrak{h} = \mathfrak{g}^0_{\operatorname{ad} H_0}$ and $\mathfrak{h}' = \mathfrak{g}^0_{\operatorname{ad} H_0'}$, it follows that $\mathfrak{h}$ and $\mathfrak{h}'$ are conjugate under A. This completes the proof of Theorem 6.

CHAPTER 4

CARTAN'S CRITERION

4.1. Preliminaries

Let $\mathfrak{g}$ be a Lie algebra. In this section, $\mathfrak{h}$ is a fixed Cartan subalgebra. Let dim $\mathfrak{g} = r$ and dim $\mathfrak{h} = n$. Under $\text{ad}_{\mathfrak{g}}\, \mathfrak{h}$, $\mathfrak{g}$ has the following decomposition:

$$\mathfrak{g} = \mathfrak{h} \dotplus \sum_{\varphi \in \Sigma} \mathfrak{g}^{\varphi} = \sum_{\varphi \in \Delta} \mathfrak{g}^{\varphi}$$

where Δ is the set of all weights of $\text{ad}_{\mathfrak{g}}\, \mathfrak{h}$ and Σ is the set of non-zero weights which is also called the set of roots of $\mathfrak{g}$ with respect to $\mathfrak{h}$ or simply the root system of $\mathfrak{g}$; $\mathfrak{h} = \mathfrak{g}^{0}_{\text{ad}\, \mathfrak{h}}$ and $\mathfrak{g}^{\varphi} = \mathfrak{g}^{\varphi}_{\text{ad}\, \mathfrak{h}}$. If $\nu_{\varphi} = \dim \mathfrak{g}^{\varphi}$, then we have the following:

LEMMA 1. *If* $X, Y, \in \mathfrak{h}$, *then*

$$(X, Y) = \sum_{\varphi \in \Sigma} \nu_{\varphi} \varphi(X)\, \varphi(Y). \tag{4.1}$$

Proof. Let $\varphi \in \Delta$ and $X \in H$, then the eigenvalues of ad X in the subspace $\mathfrak{g}^{\varphi}$ are all equal to $\varphi(X)$, thus all eigenvalues of $(\text{ad}\, X)^2$ in $\mathfrak{g}^{\varphi}$ are equal to $\varphi(X)^2$. Therefore

$$\text{Tr}_{\mathfrak{g}^{\varphi}} (\text{ad}\, X)^2 = \nu_{\varphi} \varphi(X)^2$$

and

$$(X, X) = \text{Tr}_{\mathfrak{g}} (\text{ad}\, X)^2 = \sum_{\varphi \in \Delta} \nu_{\varphi} \varphi(X)^2 = \sum_{\varphi \in \Sigma} \nu_{\varphi} \varphi(X)^2.$$

If $X, Y \in \mathfrak{h}$, then

$$\begin{aligned}(X+Y, X+Y) &= \sum_{\varphi \in \Sigma} \nu_{\varphi} \varphi(X+Y)^2 \\ &= \sum_{\varphi \in \Sigma} \nu_{\varphi} \varphi(X)^2 + 2 \sum_{\varphi \in \Sigma} \nu_{\varphi} \varphi(X)\, \varphi(Y) + \sum_{\varphi \in \Sigma} \nu_{\varphi} \varphi(Y)^2 \\ &= (X, X) + 2 \sum_{\varphi \in \Sigma} \nu_{\varphi} \varphi(X)\, \varphi(Y) + (Y, Y).\end{aligned} \tag{4.2}$$

On the other hand,

$$(X+Y, X+Y) = (X, X) + 2(X, Y) + (Y, X). \tag{4.3}$$

Thus (4.1) follows from comparing (4.2) and (4.3).

LEMMA 2. *Let $\varphi \in \Delta$, $\alpha \in \Delta$ and $-\alpha \in \Delta$. If p and q are non-negative integers such that $[\mathfrak{g}^{-\alpha}, \mathfrak{g}^{\varphi-p\alpha}] = [\mathfrak{g}^{\alpha}, \mathfrak{g}^{\varphi+q\alpha}] = 0$ and $r_{\varphi,\alpha}$ is defined by*

$$r_{\varphi,\alpha} = \frac{-\sum_{k=-p}^{q} k\nu_{\varphi+k\alpha}}{\sum_{k=-p}^{q} \nu_{\varphi+k\alpha}}$$

then for any $Z \in [\mathfrak{g}^{\alpha}, \mathfrak{g}^{-\alpha}]$,

$$\varphi(Z) = r_{\varphi,\alpha}\alpha(Z). \tag{4.4}$$

In particular, $\varphi(Z) = 0$ for any $Z \in [\mathfrak{h}, \mathfrak{h}] = \mathcal{D}\mathfrak{h}$.

Proof. Since both sides of (4.4) are linear functions of Z, hence it suffices to prove (4.4) for $Z = [X_\alpha, X_{-\alpha}]$, where $X_\alpha \in \mathfrak{g}^\alpha$ and $X_{-\alpha} \in \mathfrak{g}^{-\alpha}$. Since $[\mathfrak{g}^{-\alpha}, \mathfrak{g}^{\varphi-p\alpha}] = [\mathfrak{g}^{\alpha}, \mathfrak{g}^{\varphi+q\alpha}] = 0$, the subspace

$$\tilde{\mathfrak{g}} = \sum_{k=-p}^{q} \mathfrak{g}^{\varphi+k\alpha}$$

is invariant under ad X_α and ad $X_{-\alpha}$. Now the trace of ad Z = ad $[X_\alpha, X_{-\alpha}]$ = ad X_α ad $X_{-\alpha}$ − ad $X_{-\alpha}$ ad X_α in $\tilde{\mathfrak{g}}$ is zero, i.e.

$$\mathrm{Tr}_{\tilde{\mathfrak{g}}}\, \mathrm{ad}\, Z = 0. \tag{4.5}$$

On the other hand, since $Z \in \mathfrak{h}$, the eigenvalues of ad Z in $\mathfrak{g}^{\varphi+k\alpha}$ are all equal to $(\varphi+k\alpha)(Z) = \varphi(Z)+k\alpha(Z)$, thus

$$\mathrm{Tr}_{\tilde{\mathfrak{g}}}\, \mathrm{ad}\, Z = \sum_{k=-p}^{q} \nu_{\varphi+k\alpha}(\varphi(Z)+k\alpha(Z)) \tag{4.6}$$

From (4.5) and (4.6) it follows that

$$\sum_{k=-p}^{q} \nu_{\varphi+k\alpha}(\varphi(Z)+k\alpha(Z)) = 0,$$

hence

$$\varphi(Z) = \frac{-\sum_{k=-p}^{q} k\nu_{\varphi+k\alpha}}{\sum_{k=-p}^{q} \nu_{\varphi+k\alpha}}\, \alpha(Z).$$

4.2. Cartan's criterion for solvable Lie algebras

Let $\mathfrak{g}$ be a Lie algebra. According to Lie's theorem, $\mathfrak{g}$ is solvable iff $\mathcal{D}\mathfrak{g}$ is nilpotent. According to Engel's theorem, $\mathcal{D}\mathfrak{g}$ is nilpotent iff every ad X ($X \in \mathcal{D}\mathfrak{g}$) is nilpotent or equivalently, Killing polynomials of all elements in $\mathcal{D}\mathfrak{g}$ are equal to λ^r. Thus we have the following:

THEOREM 1. *$\mathfrak{g}$ is solvable iff the Killing polynomial of every element in $\mathcal{D}\mathfrak{g}$ is λ^r.*

Now it is clear that the Killing polynomial of any $X \in \mathcal{D}\mathfrak{g}$ is λ^r iff

$$a_1(X) \equiv a_2(X) \equiv \dots \equiv a_r(X) \equiv 0.$$

Cartan weakened the condition of this theorem and obtained the following:

THEOREM 2 *(Cartan's criterion for solvable Lie algebras). $\mathfrak{g}$ is solvable iff $(X, X) = 0$ for all $X \in \mathcal{D}\mathfrak{g}$.*

Proof. Suppose $\mathfrak{g}$ is solvable, then

$$a_1(X) = a_2(X) = \dots = a_r(X) = 0, \quad \text{for all } X \in \mathcal{D}\mathfrak{g}.$$

Now for all $X \in \mathfrak{g}$, we have

$$(X, X) = a_1(X)^2 - 2a_2(X),$$

hus

$$(X, X) = 0, \quad \text{for all } X \in \mathcal{D}\mathfrak{g}.$$

Conversely, suppose $(X, X) = 0$ for all $X \in \mathcal{D}\mathfrak{g}$. We will prove that $\mathfrak{g}$ is solvable by using induction on dim $\mathfrak{g}$. Let $\alpha_1, -\alpha \in \Delta$ and $Z \in [\mathfrak{g}^\alpha, \mathfrak{g}^{-\alpha}]$; we will first show that $\varphi(Z) = 0$ for all $\varphi \in \Delta$. Since $Z \in \mathfrak{h}$, we have $(Z, Z) = 0$. According to Lemma 1,

$$(Z, Z) = \sum_{\varphi \in \Delta} \nu_\varphi \varphi^2(Z).$$

Thus

$$\sum_{\varphi \in \Delta} \nu_\varphi \varphi(Z)^2 = 0$$

According to Lemma 2, $\varphi(Z)$ is a rational multiple of $\alpha(Z)$, hence it follows from (7) tha

$$\alpha(Z) = 0.$$

From (4.4), we have

$$\varphi(Z) = 0, \quad \text{for all } \varphi \in \Delta.$$

We now show that $\varphi(Z) = 0$ for all $Z \in [\mathfrak{g}, \mathfrak{g}] \cap \mathfrak{h}$. From $\mathfrak{g} = \sum_{\varphi \in \Delta} \mathfrak{g}$, we have

$$[\mathfrak{g}, \mathfrak{g}] = \sum_{\varphi, \psi \in \Delta} [\mathfrak{g}^\varphi, \mathfrak{g}^\psi].$$

Let

$$\tilde{\mathfrak{g}}^\varphi = \sum_{\psi \in \Delta} [\mathfrak{g}^\psi, \mathfrak{g}^{\varphi - \psi}],$$

hen $\tilde{\mathfrak{g}}^\varphi \subseteq \mathfrak{g}^\varphi$ and

$$[\mathfrak{g}, \mathfrak{g}] = \sum_{\varphi \in \Delta}{}^{\cdot} \tilde{\mathfrak{g}}^\varphi$$

is a direct sum. Thus

$$[\mathfrak{g}, \mathfrak{g}] \cap \mathfrak{h} = \mathfrak{g}^0 = \sum_{\psi \in \Delta} [\mathfrak{g}^\psi, \mathfrak{g}^{-\psi}].$$

From this equation and the fact that $\varphi(Z) = 0$ for all $Z \in [\mathfrak{g}^\psi, \mathfrak{g}^{-\psi}]$, it follows that $\varphi(Z) = 0$ for all $Z \in \mathcal{D}\mathfrak{g} \cap \mathfrak{h}$.

Now we claim that $\mathfrak{g} \neq \mathcal{D}\mathfrak{g}$. In fact, if $\mathfrak{g} = \mathcal{D}\mathfrak{g}$, then $\varphi(Z) = 0$ ($\forall Z \in \mathfrak{h}$), i.e. ad $\mathfrak{h}$ only has the weight zero. Thus $\mathfrak{g} = \mathfrak{h}$. Since $\mathfrak{h}$ is nilpotent, we have $\mathcal{D}\mathfrak{h} \neq \mathfrak{h}$, hence $\mathcal{D}\mathfrak{g} \neq \mathfrak{g}$. This contradicts the assumption that $\mathcal{D}\mathfrak{g} = \mathfrak{g}$.

Let (X, Y) denote the Killing form of $\mathcal{D}\mathfrak{g}$. Since $\mathcal{D}\mathfrak{g}$ is an ideal of $\mathfrak{g}$, we have

$$(X, X)_1 = (X, X), \quad \text{for all } X \in \mathcal{D}\mathfrak{g}.$$

Thus

$$(X, X)_1 = 0 \quad \text{for all } X \in \mathcal{D}\mathfrak{g}.$$

In particular,

$$(X, X)_1 = 0 \quad \text{for all } X \in \mathcal{D}(\mathcal{D}\mathfrak{g}).$$

Therefore $\mathcal{D}\mathfrak{g}$ also satisfies the assumption of Theorem 2. Now $\mathcal{D}\mathfrak{g} \neq \mathfrak{g}$, and from the induction assumption it follows that $\mathcal{D}\mathfrak{g}$ is solvable, hence $\mathfrak{g}$ is also solvable.

COROLLARY. *If* $(X, X) = 0$ *for all* $X \in \mathfrak{g}$, *then* $\mathfrak{g}$ *is solvable.*

4.3. Cartan's criterion for semisimple Lie algebras

THEOREM 3. $\mathfrak{g}$ *is semisimple iff the Killing form of* $\mathfrak{g}$ *is non-degenerate.*

Proof. In Chapter 1 (Theorem 3) it was proved that non-degeneracy of the Killing form implies semisimplicity of $\mathfrak{g}$.

Now suppose that (X, Y) is non-degenerate and let

$$\mathfrak{n} = \{X | X \in \mathfrak{g} \quad \text{such that} \quad (X, Y) = 0\ (\forall Y \in \mathfrak{g})\},$$

then $\mathfrak{n} \neq \{0\}$ and by Lemma 1 of Chapter 1, $\mathfrak{n}$ is an ideal of $\mathfrak{g}$. If $(X, Y)_\mathfrak{n}$ is the Killing form of $\mathfrak{n}$, then

$$(X, Y)_\mathfrak{n} = (X, Y) \quad \text{for } X, Y \in \mathfrak{n}.$$

Thus

$$(X, X) = 0 \quad \text{for all } X \in \mathfrak{n}.$$

By Theorem 2, $\mathfrak{n}$ is solvable, hence $\mathfrak{g}$ is not semisimple.

THEOREM 4. *A semisimple Lie algebra is the direct sum of all the minimal ideals. These ideals are simple Lie algebras and are mutually orthogonal with respect to the Killing form.*

COROLLARY 1. *Ideals of semisimple Lie algebras are semisimple. If* $\mathfrak{g}_1$ *is an ideal of* $\mathfrak{g}$, *then there exists a uniquely determined ideal* $\mathfrak{g}_2$ *such that* $\mathfrak{g} = \mathfrak{g}_1 \dotplus \mathfrak{g}_2$.

COROLLARY 2. *If* $\mathfrak{g}$ *is semisimple, then the center of* $\mathfrak{g}$ *is* $\{0\}$ *and* $\mathcal{D}\mathfrak{g} = \mathfrak{g}$.

CHAPTER 5

CARTAN DECOMPOSITIONS AND ROOT SYSTEMS OF SEMISIMPLE LIE ALGEBRAS

5.1. Cartan decompositions of semisimple Lie algebras

THEOREM 1. *Let* $\mathfrak{g}$ *be a semisimple Lie algebra,* $\mathfrak{h}$ *be a Cartan subalgebra,* $\dim \mathfrak{g} = r$ and $\dim \mathfrak{h} = n$. *Suppose that*

$$\mathfrak{g} = \mathfrak{h} \dotplus \sum_{\alpha \in \Sigma}{}^{\cdot} \mathfrak{g}^{\alpha}$$

is a Cartan decomposition of $\mathfrak{g}$, *then*:

(I) $\mathfrak{h}$ *is an abelian subalgebra and the restriction of* (X, Y) *to* $\mathfrak{h}$ *is non-degenerate.*

(II) $\mathfrak{g}$ *has n linearly independent roots and any root space* $\mathfrak{g}^{\alpha}$ *is one-dimensional. Moreover, if* $\alpha \in \Sigma$, *then* $-\alpha \in \Sigma$; *if* $k \neq 1$, *then* $k\alpha \notin \Sigma$.

(III) *If* $\alpha+\beta$ *is a root, then* $[\mathfrak{g}^{\alpha}, \mathfrak{g}^{\beta}] = \mathfrak{g}^{\alpha+\beta}$.

Proof. We will prove the following results:

(1) The restriction of (X, Y) to $\mathfrak{h}$ is non-degenerate.

Let $H \in \mathfrak{h}$. From Section 3.1, it is known that $(H, \mathfrak{g}^{\alpha}) = 0$ for all $\alpha \in \Sigma$. If $(H, \mathfrak{h}) = 0$, then $(H, \mathfrak{g}) = 0$ and hence $H = 0$. Thus the restriction of (X, Y) to $\mathfrak{h}$ is non-degenerate.

(2) For any $\alpha \in \Sigma$, there exists a unique $H_{\alpha} \in \mathfrak{h}$ such that

$$(H, H_{\alpha}) = \alpha(H) \quad \text{for all } H \in \mathfrak{h}.$$

In fact, every $H \in \mathfrak{h}$ defines a linear function on $\mathfrak{h}$ by

$$\varphi_H(X) = (H, X) \quad \text{for all } X \in \mathfrak{h}.$$

Since the restriction of (X, Y) to $\mathfrak{h}$ is non-degenerate, if $H_1 \neq H_2$ then $\varphi_{H_1} \neq \varphi_{H_2}$. Thus

$$H \to \varphi_H$$

maps $\mathfrak{h}$ onto the dual space $\mathfrak{h}^*$. Hence a linear function (in particular, any root $\alpha(H)$) on $\mathfrak{h}$ can be determined by an element in $\mathfrak{h}$.

(3) $\mathfrak{g}$ has n independent roots

If $\mathfrak{g}$ has less than n independent roots, then there is a non-zero $H \in \mathfrak{h}$ such that $\varphi(H) = 0$ for all $\varphi \in \Sigma$. Thus

$$(H, H') = \sum_{\varphi \in \Sigma} \nu_\varphi \varphi(H)\varphi(H') = 0, \quad \text{for all } H' \in \mathfrak{h}.$$

This contradicts with (2).

(4) $\mathfrak{h}$ is an abelian subalgebra.

According to Corollary 2 to Theorem 3 of Chapter 2, $\varphi(H) = 0$ for all $\varphi \in \Sigma$ and $H \in \mathcal{D}\mathfrak{h}$. Thus $(H, \mathfrak{h}) = 0$ and $H = 0$. This proves that $\mathcal{D}\mathfrak{h} = \{0\}$, i.e. $\mathfrak{h}$ is abelian.

(5) If $\alpha \in \Sigma$, then $-\alpha \in \Sigma$.

In fact, if $\alpha \in \Sigma$ and $-\alpha \notin \Sigma$, then from (2) of Section 3.1, it follows that $(\mathfrak{g}^\alpha, \mathfrak{g}^\varphi) = 0$ for all $\varphi \in \Delta$. Thus $(\mathfrak{g}^\alpha, \mathfrak{g}) = 0$. This contradicts the assumption that (X, Y) is non-degenerate.

(6) If $E_\alpha \in \mathfrak{g}^\alpha$, i.e. ad $HE_\alpha = \alpha(H)E_\alpha$ for all $H \in \mathfrak{h}$ and $X_{-\alpha} \in \mathfrak{g}^{-\alpha}$, then

$$[E_\alpha, X_{-\alpha}] = (E_\alpha, X_{-\alpha})H_\alpha. \tag{5.1}$$

In fact $[E_\alpha, X_{-\alpha}] \in \mathfrak{h}$ and $H_\alpha \in \mathfrak{h}$, hence for any $H \in \mathfrak{h}$ we have

$$\begin{aligned}([E_\alpha, X_{-\alpha}], H) &= -(X_{-\alpha}, [E_\alpha, H]) = (X_{-\alpha}, \alpha(H)E_\alpha)\\ &= \alpha(H)(E_\alpha, E_{-\alpha}) = (E_\alpha, X_{-\alpha})(H_\alpha, H)\\ &= ([E_\alpha, X_{-\alpha}]H_\alpha, H).\end{aligned}$$

Now (5.1) follows from (1).

(7) $\alpha(H_\alpha) \neq 0$ for all $\alpha \in \Sigma$.

We claim that there exists $X_{-\alpha} \in \mathfrak{g}^{-\alpha}$ such that $(E_\alpha, X_{-\alpha}) \neq 0$. For otherwise, $(E_\alpha, \mathfrak{g}^{-\alpha}) = 0$ and $(E_\alpha, \mathfrak{g}) = 0$ which contradicts the semisimplicity of $\mathfrak{g}$. Thus there exists an element $X_{-\alpha} \in \mathfrak{g}^{-\alpha}$ such that $(E_\alpha, X_{-\alpha}) = 1$, hence

$$H_\alpha = [E_\alpha, X_{-\alpha}] \in [\mathfrak{g}^\alpha, \mathfrak{g}^{-\alpha}].$$

According to Lemma 2 of Chapter 4, if $\alpha(H_\alpha) = 0$, then $\varphi(H_\alpha) = 0$ for all $\varphi \in \Delta$, hence

$$(H, H_\alpha) = \sum_{\varphi \in \Delta} \nu_\varphi \varphi(H)\varphi(H_\alpha) = 0 \quad \text{for all } H \in \mathfrak{h}.$$

This contradicts (1), thus $\alpha(H_\alpha) \neq 0$.

(8) If $\alpha \in \Sigma$, then dim $\mathfrak{g}^\alpha = 1$.

Let p be the largest positive integer such that $-p\alpha$ is a root and E_α be a root vector corresponding to the root α. Consider

$$\tilde{\mathfrak{g}} = \{E_\alpha\} \dotplus \mathfrak{h} \dotplus \mathfrak{g}^{-\alpha} \dotplus \ \dots \ \dotplus \mathfrak{g}^{-p\alpha},$$

where $\{E_\alpha\}$ denotes the one-dimensional subspace generated by E_α. By (7), we can choose $X_{-\alpha} \in \mathfrak{g}^{-\alpha}$ such that $(E_\alpha, X_{-\alpha}) = 1$, hence by (6), we have $[E_\alpha, X_{-\alpha}] = H_\alpha$. It is easy to see that $\tilde{\mathfrak{g}}$ is invariant under ad E_α and ad $X_{-\alpha}$, hence it is also invariant under ad H_α. Now

$$\mathrm{Tr}_{\tilde{\mathfrak{g}}}\, \mathrm{ad}\, H_\alpha = \mathrm{Tr}_{\tilde{\mathfrak{g}}}\, (\mathrm{ad}\, E_\alpha\, \mathrm{ad}\, X_{-\alpha} - \mathrm{ad}\, X_{-\alpha}\, \mathrm{ad}\, E_\alpha) = 0.$$

On the other hand, if $m_j = \dim \mathfrak{g}^{-j\alpha}$, then

$$\begin{aligned}\mathrm{Tr}_{\tilde{\mathfrak{g}}}\,\mathrm{ad}\, H_\alpha &= \alpha(H_\alpha)-\alpha(H_\alpha)m_1-2\alpha(H_\alpha)m_2- \ \dots \ -p\alpha(H_\alpha)m_p \\ &= \alpha(H_\alpha)(1-m_1- \ \dots \ -pm_p).\end{aligned}$$

By (7), $\alpha(H_\alpha) \neq 0$, hence $(1-m_1-2m_2- \ \dots \ -pm_p) = 0$. Thus $m_1 = 1$ and $m_2 = \dots = m_p = 0$. This proves that $-\alpha$ is a simple root, i.e. $\dim \mathfrak{g}^{-\alpha} = 1$ and $-2\alpha_1, -3\alpha_1, \dots$ are not roots. Now replace $-\alpha$ by α in the argument above, it follows that α is also a simple root and $2\alpha, 3\alpha, \dots,$ are not roots.

We have also proved that

(9) If $\alpha \in \Sigma$ and $k \neq \pm 1$ is an integer, then $k\alpha \notin \Sigma$. Obviously, we also have

(10) Let $\alpha \in \Sigma$, then for any $E_\alpha \in \mathfrak{g}^\alpha$, there exists a unique $E_{-\alpha} \in \mathfrak{g}^{-\alpha}$ such that

$$(E_\alpha, E_{-\alpha}) = 1, \quad [E_\alpha, E_{-\alpha}] = H_\alpha. \tag{5.2}$$

(11) Before we prove (III), we first prove the following lemma.

LEMMA 1. *Let* $\alpha \in \Sigma$, $\varphi \in \Delta$ *and* p, q *be non-negative integers such that* $\varphi+k\alpha \in \Delta\,(-p \leq k \leq q)$ *and* $\varphi-(p+1)\alpha \notin \Delta$, $\varphi+(q+1)\alpha \notin \Delta$, *then*

$$\frac{2(H_\varphi, H_\alpha)}{(H_\alpha, H_\alpha)} = -(q-p)$$

and $\varphi+k\alpha$ ($k > q$ or $k < -p$) *are not roots. In particular,*

$$\varphi-\frac{2(H_\varphi, H_\alpha)}{(H_\alpha, H_\alpha)}\alpha \in \Delta.$$

Moreover, if $\varphi \in \Sigma$ *and* $E_\varphi, E_\alpha, E_{-\alpha}$ *are root vectors corresponding to the roots* φ, α *and* $-\alpha$, $[E_\alpha, E_{-\alpha}] = H_\alpha$ *and*

$$\begin{gathered}(\mathrm{ad}\, E_\alpha)E_\varphi \neq 0, \ \dots, \ (\mathrm{ad}\, E_\alpha)^{q'}E_\varphi \neq 0, \\ (\mathrm{ad}\, E_\alpha)^{q'+1}E_\varphi = 0, \\ (\mathrm{ad}\, E_{-\alpha})E_\varphi \neq 0, \ \dots, \ (\mathrm{ad}\, E_{-\alpha})^{p'}E_\varphi \neq 0, \\ (\mathrm{ad}\, E_{-\alpha})^{p'+1}E_\varphi = 0,\end{gathered}$$

then $p = p'$ and $q = q'$.

Proof. If $\varphi = \alpha$, then by (9), $p = 2$ and $q = 0$. If $\varphi = 0$, then $p = 1$ and $q = 1$. If $\varphi = -\alpha$, then $p = 0$ and $q = 2$. In all these cases, Lemma 1 is true. We can now exclude these cases. By (9), we can assume $\varphi+k\alpha \neq 0$, then $\mathfrak{g}^{\varphi+k\alpha}$ $(-p \leq k \leq q)$ are one-dimensional.

Let $\tilde{\mathfrak{g}} = \sum_{k=-p}^{q} \mathfrak{g}^{\varphi+k\alpha}$, then $\tilde{\mathfrak{g}}$ is invariant under $\mathrm{ad}\, E_{-\alpha}$ and $\mathrm{ad}\, E_\alpha$, because $\varphi-(p+1)\alpha \notin \Delta$ and $\varphi+(q+1)\alpha \notin \Delta$. Thus

$$\begin{aligned}0 = \mathrm{Tr}_{\tilde{\mathfrak{g}}}[\mathrm{ad}\, E_\alpha, \mathrm{ad}\, E_{-\alpha}] = \mathrm{Tr}_{\tilde{\mathfrak{g}}}\, \mathrm{ad}\, H_\alpha &= \sum_{k=-p}^{q} (\varphi+k\alpha)(H_\alpha) \\ &= (p+q+1)\,\varphi(H_\alpha)+\frac{(p+q+1)(q-p)}{2}\,\alpha(H_\alpha).\end{aligned}$$

Thus

$$\frac{2(H_\varphi, H_\alpha)}{(H_\alpha, H_\alpha)} = \frac{2\varphi(H_\alpha)}{\alpha(H_\alpha)} = -(q-p). \tag{5.3}$$

Now if q' is the largest non-negative integer such that $(\operatorname{ad} E_\alpha)^k E_\varphi \neq 0$ $(\forall\, 0 \leqslant k \leqslant q')$, then $\tilde{\mathfrak{g}}' = \sum_{k=-p}^{q'} \mathfrak{g}^{\varphi+k\alpha}$ is also invariant under $\operatorname{ad} E_\alpha$ and $\operatorname{ad} E_{-\alpha}$. By repeating the previous argument, we have

$$\frac{2(H_\varphi, H_\alpha)}{(H_\alpha, H_\alpha)} = -(q'-p). \tag{5.4}$$

By comparing (5.3) and (5.4), we get $q = q'$. Similarly, it can be proved that p is the largest non-negative integer such that $(\operatorname{ad} E_{-\alpha})^k E_\varphi \neq 0$ $(\forall -p \leqslant k \leqslant 0)$.

Finally, if q'' is the largest non-negative integer such that $\varphi+q''\alpha \in \Sigma$, then $q'' \geqslant q$. Suppose that $q'' > q$. Let $\varphi_0 = \varphi+q''\alpha$ and p_0, q_0 be largest non-negative integers such that $\varphi_0+k\alpha \in \Sigma(-p_0 \leqslant k \leqslant q_0)$ and $\varphi_0-(p_0+1)\alpha \notin \Delta$, $\varphi_0+(q_0+1)\alpha \notin \Delta$. Then $q_0 = 0$ and $p_0 < q''-p-1$. By our previous argument, we have

$$\frac{2(H_{\varphi_0}, H_\alpha)}{(H_\alpha, H_\alpha)} = -(q_0-p_0) = p_0 < q''-q-1.$$

Now $(H_{\varphi_0}, H_\alpha) = (H_\varphi, H_\alpha)+q''(H_\alpha, H_\alpha)$, hence

$$\frac{2(H_{\varphi_0}, H_\alpha)}{(H_\alpha, H_\alpha)} = -(q-p)+2q''.$$

Thus $-(q-p)+2q'' < q''-q-1$ and $q''+p < -1$ —— contradiction. Hence $q'' = q$. Similarly, it can be proved that $\varphi+k\alpha$ $(k < -p)$ are not roots.

The roots

$$\beta-p\alpha,\ \ldots,\ \beta-\alpha,\ \beta,\ \beta+\alpha,\ \ldots,\ \beta+q\alpha$$

are called the α-string of roots containing β.

(12) Let α, β and $\alpha+\beta$ be in Σ, then $[\mathfrak{g}^\alpha, \mathfrak{g}^\beta] = \mathfrak{g}^{\alpha+\beta}$. This follows from setting $\beta = \varphi$ in Lemma 1.

The proof of Theorem 1 is now completed.

Corollary. *Let $\mathfrak{g}$ be a semisimple Lie algebra and $\mathfrak{h}$ be a Cartan subalgebra. For every root α, there exists a unique $H_\alpha \in \mathfrak{h}$ such that*

$$(H, H_\alpha) = \alpha(H), \quad \text{for all } H \in \mathfrak{h}.$$

If for every pair of roots $\pm\alpha$, elements $E_\alpha \in \mathfrak{g}^\alpha$ and $E_{-\alpha} \in \mathfrak{g}^{-\alpha}$ that satisfy

$$(E_\alpha, E_{-\alpha}) = 1 \tag{5.5}$$

are chosen, then the structure formulas of $\mathfrak{g}$ *can be written as*

$$\begin{aligned} [H, H'] &= 0, && \text{for all } H, H' \in \mathfrak{h}. \\ [H, E_\alpha] &= \alpha(H)E_\alpha, && \text{for all } H \in \mathfrak{h}. \\ [E_\alpha, E_{-\alpha}] &= 0, && \text{for all } \alpha \in \Sigma. \\ [E_\alpha, E_\beta] &= 0, && \text{if } \alpha+\beta \notin \Sigma. \\ [E_\alpha, E_\beta] &= N_{\alpha\beta}E_{\alpha+\beta}, && N_{\alpha\beta} \neq 0 \quad \text{if } \alpha+\beta \in \Sigma. \end{aligned}$$

Notice that the choice of E_α is not unique, but if E_α has been chosen, then there is a unique $E_{-\alpha}$ that satisfies (5.5).

THEOREM 2. *A subalgebra* $\mathfrak{h}$ *of a semisimple Lie algebra* $\mathfrak{g}$ *is a Cartan subalgebra iff it has the following properties:*

1° $\mathfrak{h}$ *is a maximal abelian subalgebra.*

2° *There exists a basis of* $\mathfrak{g}$ *with respect to which all matrices of* ad H ($H \in \mathfrak{h}$) *are diagonal, i.e. all elementary divisors of* ad H ($H \in \mathfrak{h}$) *are simple.*

Proof. A Cartan subalgebra of $\mathfrak{g}$ certainly has these two properties. Conversely, suppose that $\mathfrak{h}$ is a subalgebra of $\mathfrak{g}$ that has these properties, then $\mathfrak{h}$ is a nilpotent subalgebra. Consider the decomposition

$$\mathfrak{g} = \mathfrak{g}^{0}_{\mathrm{ad}\,\mathfrak{g}} \dotplus \sum_{\alpha \in \Sigma} \mathfrak{g}^{\alpha}_{\mathrm{ad}\,\mathfrak{g}}$$

of $\mathfrak{g}$ under $\mathfrak{h}$. Suppose that there is an $A \in \mathfrak{g}^{0}_{\mathrm{ad}\,\mathfrak{g}}$ which is not in $\mathfrak{h}$. From 2°, it follows that $[H, A] = 0$ $(\forall H \in \mathfrak{h})$. Therefore the linear combination of $\mathfrak{h}$ and A is an abelian subalgebra and this contradicts the maximality of $\mathfrak{h}$. Thus $\mathfrak{h} = \mathfrak{g}^{0}_{\mathrm{ad}\,\mathfrak{g}}$, i.e. $\mathfrak{h}$ is a Cartan subalgebra of $\mathfrak{g}$.

It can be seen from the proof that 2° can be weakened to.

2°′ All elementary divisors corresponding to the eigenvalue zero of ad H $(\forall H \in \mathfrak{h})$ are simple.

Now by Theorem 3, we can rewrite Theorem 2 as

THEOREM 2′. *A Cartan subalgebra of a semisimple Lie algebra is a subalgebra which is maximal with respect to the following two properties*:

1° $\mathfrak{h}$ *is an abelian subalgebra.*

2° *For all* $H \in \mathfrak{h}$, *the elementary divisors of* ad H *are simple.*

Proof. Any Cartan subalgebra is certainly maximal with respect to 1° and 2°.

Conversely, suppose that $\mathfrak{h}_1$ is a subalgebra of $\mathfrak{g}$ with properties 1° and 2°, we will show that $\mathfrak{h}_1$ is contained in some Cartan subalgebra of $\mathfrak{g}$; it will then follow that any subalgebra of $\mathfrak{g}$ which is maximal with respect to 1° and 2° is a Cartan subalgebra. Let

$$\mathfrak{z}(\mathfrak{h}_1) = \{X \in \mathfrak{g} \,|\, [X, \mathfrak{h}_1] = 0\}$$

be the centralizer of $\mathfrak{g}$. $\mathfrak{z}(\mathfrak{h}_1)$ is a subalgebra of $\mathfrak{g}$. If $\mathfrak{h}$ is a Cartan subalgebra of $\mathfrak{z}(\mathfrak{h}_1)$ and $H_1 \in \mathfrak{h}_1$, then for any $X \in \mathfrak{z}(\mathfrak{h}_1)$, we have $[X, H] = 0$. In particular, $[\mathfrak{h}, H_1] = 0$. Since $\mathfrak{h}$ is a maximal nilpotent subalgebra of $\mathfrak{z}(\mathfrak{h}_1)$, hence $H_1 \in \mathfrak{h}$ and $\mathfrak{h}_1 \subseteq \mathfrak{h}$. Now it is only necessary to show that $\mathfrak{h}$ is a Cartan subalgebra of $\mathfrak{h}$. We will first prove that $\mathfrak{n}(\mathfrak{h}) = \mathfrak{h}$. Let

$$\mathfrak{g} = \mathfrak{z}(\mathfrak{h}_1) + \sum_{\alpha_i \neq 0} \mathfrak{g}^{\alpha_i}_{\text{ad}\,\mathfrak{h}_1}$$

be a decomposition of $\mathfrak{g}$ with respect to ad $\mathfrak{h}_1$, then any $X \in \mathfrak{g}$ can be written as

$$X = Z + \sum_i X_i, \quad Z \in \mathfrak{z}(\mathfrak{h}_1),\ X_i \in \mathfrak{g}^{\alpha_i}_{\text{ad}\,\mathfrak{h}_1}$$

If $X \in \mathfrak{n}(\mathfrak{h})$, then $[X, \mathfrak{h}] \subseteq \mathfrak{h}$ and for any $H_1 \in \mathfrak{h}_1$, we have

$$[H_1, X] = \sum_i \alpha_i(H_1) X_i \in \mathfrak{h}.$$

Now choose H_1 such that $\alpha_i(H_1) \neq 0$, then from the above equation we have $X_i = 0$. Hence $X \in \mathfrak{z}(\mathfrak{h}_1)$. Now $\mathfrak{h}$ is a Cartan subalgebra of $\mathfrak{z}(\mathfrak{h}_1)$; by Theorem 3 of Chapter 3, the normalizer of $\mathfrak{h}$ in $\mathfrak{z}(\mathfrak{h}_1)$ is itself. Thus $[X, \mathfrak{h}] \subseteq \mathfrak{h}$ and $X \in \mathfrak{h}$. This proves that $\mathfrak{n}(\mathfrak{h}) = \mathfrak{h}$. By Theorem 3 of Chapter 3, $\mathfrak{h}$ must be a Cartan subalgebra of $\mathfrak{g}$.

5.2. Root systems of semisimple Lie algebras

Let $\mathfrak{g}$ be a semisimple Lie algebra and $\mathfrak{h}$ be a Cartan subalgebra of $\mathfrak{g}$. Let Σ be the set of all roots of $\mathfrak{g}$, Σ is said to be a root system. Any $\alpha \in \Sigma$ uniquely determines an element $H_\alpha \in \mathfrak{h}$ such that

$$(H, H_\alpha) = \alpha(H) \quad \text{for all } H \in \mathfrak{h}.$$

We call the determination of H_α by α the embedding of α into $\mathfrak{h}$. More generally, if $\mathfrak{h}^*$ is the dual space of $\mathfrak{h}$, then any $\mu \in \mathfrak{h}$ determines a unique $H_\mu \in \mathfrak{h}$ such that

$$(H, H_\mu) = \mu(H) \quad \text{for all } H \in \mathfrak{h}.$$

We also call the determination of H_μ by μ the embedding of the linear function μ into $\mathfrak{h}$.

Now for $\lambda, \mu \in \mathfrak{h}^*$, embed them into $\mathfrak{h}$ and obtain H_λ and H_μ. Define

$$(\lambda, \mu) = (H_\lambda, H_\mu),$$

then an inner product is defined on $\mathfrak{h}^*$. Naturally, this inner product is non-degenerate and

$$(\lambda, \mu) = (H_\lambda, H_\mu) = \lambda(H_\mu) = \mu(H_\lambda).$$

For convenience, we sometimes identify $\lambda \in \mathfrak{h}^*$ with $H_\lambda \in \mathfrak{h}$ and write

$$(\lambda, \mu) = (\lambda, H_\mu) = (H_\mu, \lambda) = (H_\lambda, \mu) = (\mu, H_\lambda).$$

THEOREM 3. *Let $\mathfrak{h}_0^*$ denote the set of elements in the dual space of the Cartan subalgebra $\mathfrak{h}$ of a semisimple Lie algebra of the form*

$$\sum_{\varphi \in \Sigma} a_\varphi \varphi \qquad (a_\varphi\text{'s are real}),$$

then:

(1) $\mathfrak{h}_0^*$ *is a real vector space whose dimension equals* dim $\mathfrak{h}$.
(2) *The Cartan inner product of* $\mathfrak{g}$ *induces a Euclidean metric on* $\mathfrak{h}^*$.
(3) *If* $\alpha_1, \ldots, \alpha_n$ *are n independent roots, then any* $\varphi \in \Sigma$ *can be written as a linear combination of them with rational coefficients.*

Proof. If $H \in \mathfrak{h}$, then

$$(H, H) = \sum_{\varphi \in \Sigma} \varphi(H)^2 = \sum_{\varphi \in \Sigma} (H, H_\varphi)^2 = \sum_{\varphi \in \Sigma} (H, \varphi)^2. \tag{5.6}$$

We first prove that (φ, α) is a rational number. By Lemma 1,

$$(\varphi, \alpha) = -\tfrac{1}{2}(q_{\varphi,\alpha} - p_{\varphi,\alpha})(\alpha, \alpha), \tag{5.7}$$

where $q_{\varphi,\alpha}$ and $p_{\varphi,\alpha}$, are non-negative integers. From (5.6) and (5.7), it follows that

$$(\alpha, \alpha) = \sum_{\varphi \in \Sigma} (\alpha, \varphi)^2 = \tfrac{1}{4} \sum_{\varphi \in \Sigma} (q_{\varphi,\alpha} - p_{\varphi,\alpha})^2 (\alpha, \alpha)^2.$$

Now $(\alpha, \alpha) \neq 0$, thus

$$(\alpha, \alpha) = \frac{4}{\sum_{\varphi \in \Sigma} (q_{\varphi,\alpha} - p_{\varphi,\alpha})^2}. \tag{5.8}$$

By substituting (5.8) into (5.7), it follows that (φ, α) is rational.

Let $\mu = \sum_{\alpha \in \Sigma} a_\alpha \alpha$ (a_α are real), then from (5.6)

$$(\mu, \mu) = \sum_{\varphi \in \Sigma} \left(\sum_{\alpha \in \Sigma} a_\alpha \alpha, \varphi \right)^2 = \sum_{\varphi \in \Sigma} \left(\sum_{\alpha \in \Sigma} a_\alpha (\varphi, \alpha) \right)^2.$$

Thus $(\mu, \mu) \geqslant 0$. If $(\mu, \mu) = 0$, then $(\mu, \varphi) = (\sum_{\alpha \in \Sigma} a_\alpha \alpha, \varphi) = 0$ for all $\varphi \in \Sigma$. Since $\mathfrak{g}$ has n linearly independent roots over the complex numbers, hence $(\mu, \mathfrak{h}) = 0$ and $\mu = 0$. This proves that the Killing form induces a Euclidean metric on $\mathfrak{h}_0^*$.

Now let $\alpha_1, \ldots, \alpha_n$ be n linearly independent roots over the complex numbers and $\varphi \in \Sigma$, then φ can be uniquely written as

$$\varphi = \sum_{i=1}^{n} a_i \alpha_i, \tag{5.9}$$

where the a_i's are complex. We want to show that the a_i's are rational. Consider the equations

$$(\varphi, \alpha_k) = \sum_{i=1}^{n} a_i (\alpha_j, \alpha_k), \qquad (1 \leqslant k \leqslant n) \tag{5.10}$$

as a system of linear equations of the a_i's. The determinant $|(\alpha_i, \alpha_k)|$ of this system is non-zero, thus the values of the a_i's are uniquely determined by (5.10). Since the coefficients of (5.10) are rational, the a_i's are also rational. Thus the dimension of $\mathfrak{h}_0^*$ over the real numbers is also n. This completes the proof of Theorem 3.

If $\mathfrak{h}_0$ denotes the real linear space consisting of all real linear combinations of H_α ($\alpha \in \Sigma$), then $\mathfrak{h}_0$ is the dual space of $\mathfrak{h}_0^*$ and the restriction of the Killing form to $\mathfrak{h}_0$ induces a Euclidean metric on $\mathfrak{h}_0$.

From the previous discussion, we know that the root system Σ of a semisimple Lie algebra with respect to a Cartan subalgebra $\mathfrak{h}$ is a set of vectors in the n-dimensional space with the properties:

1° If $\alpha \in \Sigma$ then $-\alpha \in \Sigma$ and for $k \neq \pm 1$, $k\alpha \notin \Sigma$.

2° Let $\alpha, \beta \in \Sigma$ and $\alpha \neq \pm\beta$. If p and q are largest non-negative integers such that $\beta+k\alpha \in \Sigma$ $(-p \leqslant k \leqslant g)$, then

$$\frac{2(\beta, \alpha)}{(\alpha, \alpha)} = -(q-p).$$

In general, if a set Σ of non-zero vectors in a Euclidean space has the properties 1° and 2°, then it is said to be a σ system. Naturally, root systems are σ systems.

From the proof of Lemma 1, we can obtain the following properties of σ systems:

3° If $\alpha, \beta \in \Sigma$, then $\beta - \dfrac{2(\beta, \alpha)}{(\alpha, \alpha)}\alpha \in \Sigma$.

4° If $\alpha, \beta \in \Sigma$, $\alpha \neq \pm\beta$ and p, q are largest non-negative integers such that $\beta+k\alpha \in \Sigma$ $(-p \leqslant k \leqslant q)$, then $\beta+k\alpha \notin \Sigma$ for $k > q$ or $k < -p$.

THEOREM 4. *Let Σ be a σ system, $\alpha, \beta \in \Sigma$ and $\alpha \neq \pm\beta$. If $\langle\alpha, \beta\rangle$ denotes the angle between α and β, then*

$$\cos\langle\alpha, \beta\rangle = \frac{\varepsilon}{2}\sqrt{r}, \quad \varepsilon = \pm 1, \quad r = 0, 1, 2, 3.$$

Suppose that $(\alpha, \alpha) \leqslant (\beta, \beta)$, *then when* $r \neq 0$,

$$\frac{(\beta, \beta)}{(\alpha, \alpha)} = r,$$

$$\frac{2(\alpha, \beta)}{(\alpha, \alpha)} = \varepsilon r, \quad \frac{2(\alpha, \beta)}{(\beta, \beta)} = \varepsilon;$$

when $r = 0$,

$$\frac{2(\alpha, \beta)}{(\alpha, \alpha)} = \frac{2(\alpha, \beta)}{(\beta, \beta)} = 0.$$

Proof. We have

$$4\cos^2\langle\alpha, \beta\rangle = \frac{4(\alpha, \beta)^2}{(\alpha, \alpha)(\beta, \beta)} = \frac{2(\alpha, \beta)}{(\alpha, \alpha)} \cdot \frac{2(\alpha, \beta)}{(\beta, \beta)},$$

thus $4\cos^2\langle\alpha, \beta\rangle$ is an integer. Since $\cos^2\langle\alpha, \beta\rangle \leqslant 1$, hence

$$4\cos^2\langle\alpha, \beta\rangle = 0, 1, 2, 3.$$

Write

$$\cos\langle\alpha, \beta\rangle = \frac{\varepsilon}{2}\sqrt{r}, \qquad r = 0, 1, 2, 3.$$

From $(\alpha, \alpha) \leqslant (\beta, \beta)$, it follows that

$$\frac{2(\alpha, \beta)}{(\alpha, \alpha)} \geqslant \frac{2(\alpha, \beta)}{(\beta, \beta)}.$$

Thus when $r \neq 0$, we have

$$\frac{2(\alpha, \beta)}{(\alpha, \alpha)} = \varepsilon r, \qquad \frac{2(\alpha, \beta)}{(\beta, \beta)} = \varepsilon, \qquad \frac{(\beta, \beta)}{(\alpha, \alpha)} = r.$$

When $r = 0$,

$$\frac{2(\alpha, \beta)}{(\alpha, \alpha)} = \frac{2(\alpha, \beta)}{(\beta, \beta)} = 0.$$

Theorem 4 has the following important corollary.

COROLLARY. *Any σ system is finite.*

Proof. Let Σ be a σ system. Since the angle between any two elements only has a finite set of possible values, the set

$$\left\{ \frac{\alpha}{(\alpha, \alpha)^{1/2}} \,\middle|\, \alpha \in \Sigma \right\}$$

on the unit sphere cannot have limit points. Thus Σ is finite.

A non-empty subset of a σ system is said to be a σ subsystem if this subset also has properties 1° and 2°.

Let Σ be a σ system. If Σ_1, and Σ_2 are two non-empty subsets satisfying $\Sigma = \Sigma_1 \cup \Sigma_2$ and $(\alpha_1, \alpha_2) = 0$ for all $\alpha_1 \in \Sigma_1$ and $\alpha_2 \in \Sigma_2$, then Σ_1 and Σ_2 are σ subsystems. Σ is then said to be decomposed into the union of two orthogonal σ subsystems. If Σ is indecomposable, then it is said to be a simple σ system.

THEOREM 5. *Let $\mathfrak{g}$ be a semisimple Lie algebra, $\mathfrak{h}$ be a Cartan subalgebra and Σ be the root system of $\mathfrak{g}$ with respect to $\mathfrak{h}$. Suppose that $\mathfrak{g}$ is the direct sum of two non-zero semisimple ideals $\mathfrak{g}_1$ and $\mathfrak{g}_2$ and $\mathfrak{h}$ is written as $\mathfrak{h}_1 \dot{+} \mathfrak{h}_2$ where $\mathfrak{h}_i$ is a Cartan subalgebra of $\mathfrak{g}_i$ $(i = 1, 2)$. If Σ_i is the root system of $\mathfrak{g}_i$ with respect to $\mathfrak{h}_i$ $(i = 1, 2)$ then every root α_i of $\mathfrak{g}_i$ $(i = 1, 2)$ becomes a linear function on $\mathfrak{h}$ by defining*

$$\alpha_i(H) = \alpha_i(H_i) \quad \text{where} \quad H = H_1 + H_2 \quad \text{and} \quad H_i \in \mathfrak{h}_i, \quad i = 1, 2.$$

Moreover, every $\alpha_i \in \Sigma_i$ $(i = 1, 2)$ becomes a root of $\mathfrak{g}$ with respect to $\mathfrak{h}$ and Σ is decomposed into the union of the orthogonal σ subsystems Σ_1 and Σ_2.

Proof. We show first that every $\alpha_i \in \Sigma_i$ $(i = 1, 2)$ is a root of $\mathfrak{g}$ with respect to $\mathfrak{h}$. Let $E_{\alpha_i} \in \mathfrak{g}_i$ be a root vector corresponding to the root α_i $(i = 1, 2)$, i.e.

$$[H_i, E_{\alpha_i}] = \alpha_i(H_i) E_{\alpha_i}.$$

Let $H \in \mathfrak{h}$ and write $H = H_1 + H_2$, then

$$[H, E_{\alpha_i}] = [H_i, E_{\alpha_i}] = \alpha_i(H_i) E_{\alpha_i} = \alpha_i(H) E_{\alpha_i}.$$

Thus α_i is a root of $\mathfrak{g}$ with respect to $\mathfrak{h}$ and E_{α_i} is a corresponding root vector.

Now we show that if $\alpha_1 \in \Sigma_1$ and $\alpha_2 \in \Sigma_2$ then $(\alpha_1, \alpha_2) = 0$. Let $H_{\alpha_i} \in \mathfrak{h}_i$ $(i = 1, 2)$ satisfy

$$(H_i, H_{\alpha_i}) = \alpha_i(H_i), \qquad i = 1, 2.$$

If $H = H_1 + H_2 \in \mathfrak{h}$, then

$$(H, H_{\alpha_i}) = (H_i, H_{\alpha_i}) = \alpha_i(H) = \alpha_i(H_i), \qquad i = 1, 2.$$

Thus

$$(\alpha_1, \alpha_2) = (H_{\alpha_1}, H_{\alpha_2}) = \alpha_1(H_{\alpha_2}) = 0.$$

Finally, we show that $\Sigma = \Sigma_1 \cup \Sigma_2$. Let $\alpha \in \Sigma$ and E_α be a root vector of α. Write $E_\alpha = E_1 + E_2$ where $E_1 \in \mathfrak{g}_1$ and $E_2 \in \mathfrak{g}_2$. Then for any $H \in \mathfrak{h}$ we have

$$\begin{aligned}[H, E_\alpha] &= \alpha(H)E_\alpha = \alpha(H)E_1 + \alpha(H)E_2 \\ &= [H, E_1 + E_2] = [H, E_1] + [H, E_2].\end{aligned}$$

Since $\mathfrak{g} = \mathfrak{g}_1 \dot{+} \mathfrak{g}_2$, hence

$$[H, E_i] = \alpha(H)E_i, \qquad i = 1, 2.$$

Thus $E_i \in \mathfrak{g}^\alpha$ $(i = 1, 2)$. Now $\mathfrak{g}^\alpha$ is one-dimensional, hence E_1 and E_2 are linearly dependent. Since $\mathfrak{g} = \mathfrak{g}_1 \dot{+} \mathfrak{g}_2$, so either $E_1 = 0$ or $E_2 = 0$. Thus $E_\alpha \in \mathfrak{g}_1$ or $E_\alpha \in \mathfrak{g}_2$. Now $E_\alpha \in \mathfrak{g}_1$ implies $\alpha \in \Sigma_1$ and $E_\alpha \in \mathfrak{g}_2$ implies $\alpha \in \Sigma_2$. Hence $\Sigma = \Sigma_1 \cup \Sigma_2$.

THEOREM 5′. *Let $\mathfrak{g}$ be a semisimple Lie algebra, $\mathfrak{h}$ be a Cartan subalgebra and Σ be the root system of $\mathfrak{g}$ with respect to $\mathfrak{h}$. Suppose that Σ is decomposed into the union of two orthogonal subsets Σ_1 and Σ_2, then $\mathfrak{g}$ is correspondingly decomposed into the direct sum of two ideals $\mathfrak{g}_1$ and $\mathfrak{g}_2$ and $\mathfrak{h}$ is correspondingly decomposed into the direct sum of a Cartan subalgebra $\mathfrak{h}_1$ of $\mathfrak{g}_1$ and a Cartan subalgebra $\mathfrak{h}_2$ of $\mathfrak{g}_2$. Moreover, the restriction of Σ_i to $\mathfrak{h}_i$ is the root system of $\mathfrak{g}_i$ with respect to $\mathfrak{h}_i$ $(i = 1, 2)$.*

Proof. Let

$$\begin{aligned}\mathfrak{h}_1 &= \{H_1 \mid H_1 \in \mathfrak{h} \text{ such that } \alpha_2(H_1) = 0\ (\forall \alpha_2 \in \Sigma_2)\}. \\ \mathfrak{h}_2 &= \{H_2 \mid H_2 \in \mathfrak{h} \text{ such that } \alpha_1(H_2) = 0\ (\forall \alpha_1 \in \Sigma_1)\}.\end{aligned}$$

We claim that $\mathfrak{h} = \mathfrak{h}_1 \dot{+} \mathfrak{h}_2$. In fact, since $\mathfrak{h}$ has n independent roots, we naturally have $\mathfrak{h}_1 \cap \mathfrak{h}_2 = \{0\}$. Secondly, since Σ_1 and Σ_2 are orthogonal, hence $n = n_1 + n_2$, where n_i $(i = 1, 2)$ is the number of independent roots of Σ_i and $\dim \mathfrak{h}_i = n_i$ $(i = 1, 2)$. Thus $\mathfrak{h} = \mathfrak{h}_1 \dot{+} \mathfrak{h}_2$.

Now let

$$\mathfrak{g}_i = \left\{\mathfrak{h}_i + \sum_{\alpha_i \in \Sigma} a_{\alpha_i} E_{\alpha_i} \,\middle|\, a_{\alpha_i} \in C,\ E_{\alpha_i} \text{ is a root vector of } \alpha_i\right\}, \qquad i = 1, 2.$$

Obviously, as vector spaces, $\mathfrak{g}$ is the direct sum of $\mathfrak{g}_1$, $\mathfrak{g}_2$. We now show that $\mathfrak{g}_1$ and $\mathfrak{g}_2$ are ideals of $\mathfrak{g}$. In fact, let $H \in \mathfrak{h}$ and write $H = H_1 + H_2$ where $H_1 \in \mathfrak{h}_1$ and $H_2 \in \mathfrak{h}_2$, then

$$\left.\begin{aligned}&[H, \mathfrak{h}_1] = 0, \\ &[H, E_{\alpha_1}] = \alpha_1(H)E_{\alpha_1} = \alpha_1(H_1)E_{\alpha_1}, \quad \text{for all } \alpha_1 \in \Sigma\end{aligned}\right\}. \tag{5.11}$$

Thus $[\mathfrak{h}, \mathfrak{g}] \subseteq \mathfrak{g}_1$. If $\alpha_1, \beta_1 \in \Sigma$, then $\alpha_1+\beta_1 \notin \Sigma_2$; otherwise, from $\alpha_1+\beta_1 \in \Sigma_2$ it follows that

$$(\alpha_1, \alpha_1+\beta_1) = (\beta_1, \alpha_1+\beta_1) = 0$$

and thus

$$(\alpha_1+\beta_1, \alpha_1+\beta_1) = 0;$$

hence $\alpha_1+\beta_1 = 0 \notin \Sigma_2$. Thus if $\alpha_1, \beta_1 \in \Sigma_1$ and $\alpha_1+\beta_1 \in \Sigma$, then $\alpha_1+\beta_1 \in \Sigma_1$ and

$$[E_{\alpha_1}, E_{\beta_1}] = N_{\alpha_1\beta_1}E_{\alpha_1+\beta_1} \in \mathfrak{g}_1.$$

If $\alpha \in \Sigma_1$, then since Σ_1 and Σ_2 are orthogonal, we have $-\alpha \in \Sigma$. Let $[E_{\alpha_1}, E_{-\alpha_1}] = H_{\alpha_1}$, then

$$\alpha_2(H_{\alpha_1}) = (\alpha_2, \alpha_1) = 0, \quad \text{for all } \alpha_2 \in \Sigma_2. \text{ Thus } H_{\alpha_1} \in \mathfrak{h}_1.$$

Finally, if $\alpha \in \Sigma_1$, and $\alpha_2 \in \Sigma_2$, then $\alpha_1+\alpha_2 \notin \Sigma$. For otherwise, let $\alpha_1+\alpha_2 \in \Sigma_1$, then $(\alpha_1+\alpha_2, \alpha_2) = 0$; since $(\alpha_2, \alpha_2) \neq 0$, we have $(\alpha_1, \alpha_2) \neq 0$. Hence

$$[E_{\alpha_1}, E_{\alpha_2}] = 0.$$

This proves that $\mathfrak{g}_1$ is an ideal of $\mathfrak{g}$. Similarly, $\mathfrak{g}_2$ is an ideal of $\mathfrak{g}$. Thus we have $\mathfrak{g} = \mathfrak{g}_1 \dot{+} \mathfrak{g}_2$. Now from (5.11) it follows that $\mathfrak{h}_1$ is a Cartan subalgebra of $\mathfrak{g}_1$ and the restriction of Σ_1 to $\mathfrak{h}_1$ is the root system of $\mathfrak{g}_1$ with respect to $\mathfrak{h}_1$. Similarly, $\mathfrak{h}_2$ is a Cartan subalgebra of $\mathfrak{g}_2$ and the restriction of Σ_2 to $\mathfrak{h}_2$ is the root system of $\mathfrak{g}_2$ with respect to $\mathfrak{h}_2$. The proof of Theorem 5′ is now completed.

COROLLARY. *A semisimple Lie algebra* $\mathfrak{g}$ *is simple iff the root system* Σ *of* $\mathfrak{g}$ *is simple.*

5.3. Dependence of structure of semisimple Lie algebras on root systems

Let S and S' be sets of vectors in the Euclidean spaces R and R'. S and S' are said to be *equivalent* if there exists a one–one correspondence f from S onto S' which satisfies

$$(f(\alpha), f(\beta)) = (\alpha, \beta) \quad \text{for all } \alpha, \beta \in S.$$

We say that S and S' are *similar* if there exists a one–one correspondence f from S onto S' which satisfies

$$(f(\alpha), f(\beta)) = k^2(\alpha, \beta) \quad \text{for all } \alpha, \beta \in S,$$

where k is a positive real number. k is called the coefficient of similarity. Obviously, if two root systems are similar and $k = 1$, then they are equivalent.

LEMMA 2. *Let* Σ *and* Σ' *be the root systems of the semisimple Lie algebras* $\mathfrak{g}$ *and* $\mathfrak{g}'$ *with respect to their Cartan subalgebras* $\mathfrak{h}$ *and* $\mathfrak{h}'$. *If* Σ *and* Σ' *are similar then they are equivalent.*

Proof. We have

$$(\beta, \beta) = \sum_{\alpha\in\Sigma} (\alpha, \beta)^2 \quad \text{for all } \beta \in \Sigma$$

and

$$(f(\beta), f(\beta)) = \sum_{f(\alpha)\in\Sigma'} (f(\alpha), f(\beta))^2.$$

If k is the coefficient of similarity, then

$$(f(\beta), f(\beta)) = k^2(\beta, \beta) = k^2 \sum_{\alpha\in\Sigma} (\alpha, \beta)^2$$

and

$$\sum_{f(\alpha)\in\Sigma} (f(\alpha), f(\beta))^2 = \sum_{\alpha\in\Sigma} k^4(\alpha, \beta)^2 = k^4 \sum_{\alpha\in\Sigma} (\alpha, \beta)^2.$$

Thus $k^2 = 1$ and $k = 1$.

LEMMA 3. *Let $\mathfrak{g}$ be a semisimple Lie algebra, $\mathfrak{h}$ be a Cartan subalgebra and Σ be the root system of $\mathfrak{g}$ with respect to $\mathfrak{h}$. For any $\alpha \in \Sigma$, choose $E_\alpha \in \mathfrak{g}^\alpha$ such that $(E_\alpha, E_{-\alpha}) = 1$. If $\alpha+\beta \neq 0$, let*

$$[E_\alpha, E_\beta] = N_{\alpha\beta}E_{\alpha+\beta},$$

then

(i) *If $\alpha, \beta \in \Sigma$* and *$\alpha+\beta \neq 0$, then*

$$N_{\alpha\beta} = -N_{\beta\alpha}.$$

(ii) *If $\alpha, \beta, \gamma, \in \Sigma$ and $\alpha+\beta+\gamma = 0$ (we also say that α, β and γ form a triangle), then*

$$N_{\alpha\beta} = N_{\beta\gamma} = N_{\gamma\alpha}.$$

(iii) *If $\alpha, \beta, \gamma, \delta \in \Sigma$, $\alpha+\beta+\gamma+\delta = 0$ (we also say that α, β, γ and δ form a parallelogram) and the sum of any two of α, β, γ and δ is not zero, then*

$$N_{\alpha\beta}N_{\gamma\delta}+N_{\alpha\gamma}N_{\delta\beta}+N_{\alpha\delta}N_{\beta\gamma} = 0.$$

(iv) *Let $\alpha, \beta \in \Sigma$ and $\alpha+\beta \neq 0$. If p, q are largest non-negative integers such that $\beta+k\alpha$ $(-p \leqslant k \leqslant q)$ are roots, then*

$$N_{\alpha\beta}N_{-\alpha,-\beta} = -R_{\alpha\beta} = -\frac{q(p+1)}{2}(\alpha, \alpha).$$

Thus $R_{\alpha\beta} > 0$ and if Σ is given, then $R_{\alpha\beta}$ is uniquely determined.

Proof. (i) is obvious. We now prove (ii). Since $\alpha+\beta = -\gamma$, thus $\alpha+\beta$ is a root. Similarly, $\beta+\gamma$ and $\gamma+\alpha$ are in Σ. From

$$([E_\alpha, E_\beta], E_\gamma)+(E_\beta,[E_\alpha, E_\gamma]) = 0,$$

$$([E_\alpha, E_\beta], E_\gamma) = N_{\alpha\beta}(E_{-\gamma}, E_\gamma) = N_{\alpha\beta},$$

and

$$(E_\beta[E_\alpha, E_\gamma]) = N_{\alpha\gamma}(E_\beta, E_{-\beta}) = N_{\alpha\gamma},$$

it follows that $N_{\alpha\beta}+N_{\alpha\gamma} = 0$. By (i), we have $N_{\alpha\beta} = N_{\gamma\alpha}$. Similarly, it can be proved that $N_{\alpha\beta} = N_{\beta\gamma}$.

Now we prove (iii). By the Jacobi identity, we have

$$[E_\alpha, [E_\beta, E_\gamma]]+[E_\beta, [E_\gamma, E_\alpha]]+[E_\gamma, [E_\alpha, E_\beta]] = 0.$$

If $\beta+\gamma \in \Sigma$, then α, $\beta+\gamma$ and δ are in Σ. Thus

$$[E_\alpha, [E_\beta, E_\gamma]] = N_{\beta\gamma}[E_\alpha, E_{\beta+\gamma}] = N_{\beta\gamma}N_{\alpha,\beta+\gamma}E_{\alpha+\beta+\gamma}.$$

By (ii), we have

$$[E_\alpha, [E_\beta, E_\gamma]] = N_{p\gamma}N_{\delta\alpha}E_{-\delta}. \tag{5.12}$$

If $\beta+\gamma \notin \Sigma$, then $[E_\beta, E_\gamma] = 0$ and $N_{\beta\gamma} = 0$, hence (5.12) is still true. Thus

$$[E_\alpha, [E_\beta, E_\gamma]] = -N_{\alpha\delta}N_{\beta\gamma}E_{-\delta}.$$

Similarly, we have

$$[E_\beta, [E_\gamma, E_\alpha]] = -N_{\beta\delta}N_{\gamma\alpha}E_{-\delta}$$

and

$$[E_\gamma, [E_\alpha, E_\beta]] = -N_{\gamma\delta}N_{\alpha\beta}E_{-\delta}.$$

(iii) can now be derived by adding up these three equations and using the Jacobi identity.

Finally, we prove (iv). If $\alpha+\beta \notin \Sigma$, then $N_{\alpha\beta} = 0$ and $q = 0$. In this case, (iv) is true. Now suppose $\alpha+\beta \in \Sigma$ and let the α-string of roots containing β be

$$\beta-p\alpha, \ldots, \beta-\alpha, \beta, \beta+\alpha, \ldots, \beta+q\alpha.$$

If X_0 is a root vector corresponding to the root $\beta-p\alpha$ and

$$X_1 = (\operatorname{ad} E_\alpha)X_0, \quad X_2 = (\operatorname{ad} E_\alpha)X_1, \ldots,$$
$$X_k = (\operatorname{ad} E_\alpha)X_{k-1}, \ldots,$$

then X_k is a root vector corresponding to the root $\beta-p\alpha+k\alpha$. We have

$$(\operatorname{ad} E_{-\alpha})X_0 = 0.$$

It can be proved by induction that

$$(\operatorname{ad} E_{-\alpha})X_{k+1} = \frac{(k+1)(q+p-k)}{2}(\alpha, \alpha)X_k. \tag{5.13}$$

In fact, if $k = -1$, then (5.13) holds. Now suppose that (5.13) holds for $k \geqslant -1$, then

$$\begin{aligned}(\operatorname{ad} E_{-\alpha})X_{k+2} &= \operatorname{ad} E_{-\alpha} \operatorname{ad} E_\alpha X_{k+1}\\
&= -\operatorname{ad} H_\alpha X_{k+1} + \operatorname{ad} E_\alpha \operatorname{ad} E_{-\alpha} X_{k+1}\\
&= -(\beta-p\alpha+(k+1)\alpha, \alpha)X_{k+1} + \operatorname{ad} E_\alpha \frac{(k+1)(q+p-k)}{2}(\alpha, \alpha)X_k\\
&= \left\{-(\beta, \alpha)+(p-k-1)(\alpha, \alpha)+\frac{(k+1)(q+p-k)}{2}(\alpha, \alpha)\right\}X_{k+1}\\
&= \left\{\frac{q-p}{2}+p-k-1+\frac{(k+1)(q+p-k)}{2}\right\}(\alpha, \alpha)X_{k+1}\\
&= \frac{(k+2)(q+p-k-1)}{2}(\alpha, \alpha)X_{k+1}.\end{aligned}$$

Thus (5.13) also holds for $k+1$. (5.13) can be rewritten as

$$\operatorname{ad} E_{-\alpha} \operatorname{ad} E_\alpha X_k = \frac{(k+1)(q+p-k)}{2}(\alpha, \alpha) X_k.$$

In particular,

$$\operatorname{ad} E_{-\alpha} \operatorname{ad} E_\alpha X_p = \frac{(p+1)q}{2}(\alpha,\alpha) X_k.$$

Since X_p and E_β are dependent, thus

$$[E_{-\alpha}, [E_\alpha, E_\beta]] = \frac{q(p+1)}{2}(\alpha, \alpha) E_\beta.$$

Now

$$[E_{-\alpha}, [E_\alpha, E_\beta]] = N_{\alpha\beta} N_{-\alpha, \alpha+\beta} E_\beta.$$

and by (ii),

$$[E_{-\alpha}, [E_\alpha, E_\beta]] = - N_{\alpha\beta} N_{-\alpha, -\beta} E_\beta.$$

(iv) now follows from comparing (5.14) and (5.15). The proof of Lemma 3 is now completed.

Let $\mathfrak{g}$ be a semisimple Lie algebra, $\mathfrak{h}$ be a Cartan subalgebra and Σ be the root system of $\mathfrak{g}$ with respect to $\mathfrak{h}$. For each $\alpha \in \Sigma$, choose $E_\alpha \in \mathfrak{g}^\alpha$ such that

$$(E_\alpha, E_{-\alpha}) = 1$$

and $H_2 \in \mathfrak{h}$ such that

$$(H, H_\alpha) = \alpha(H) \quad \text{for all } H \in \mathfrak{h}.$$

The structure formulas of $\mathfrak{g}$ can then be written as:

$$\begin{aligned} [\mathfrak{h}, \mathfrak{h}] &= 0, \\ [H, E_\alpha] &= \alpha(H) E_\alpha, \\ [E_\alpha, E_{-\alpha}] &= H_\alpha, \\ [E_\alpha, E_\beta] &= N_{\alpha\beta} E_{\alpha+\beta}, \quad \text{if } \alpha+\beta \neq 0. \end{aligned}$$

Now H_α $(\forall \alpha \in \Sigma)$ is uniquely determined but E_α is not. Let

$$E'_\alpha = \mu_\alpha E_\alpha$$

where $\mu_\alpha \neq 0$ and

$$\mu_\alpha \mu_{-\alpha} = 1,$$

then we also have

$$(E'_\alpha, E'_{-\alpha}) = 1.$$

If $\alpha+\beta \neq 0$, let

$$[E'_\alpha, E'_\beta] = N'_{\alpha\beta} E'_{\alpha+\beta},$$

then

$$N'_{\alpha\beta} = \frac{\mu_\alpha \mu_\beta}{\mu_{\alpha+\beta}} N_{\alpha\beta} \quad \text{for all } \alpha, \beta \in \Sigma \text{ satisfying } \alpha+\beta \in \Sigma.$$

If these equations are true for all $\alpha, \beta \in \Sigma$ satisfying $\alpha+\beta \in \Sigma$ and $\mu_\alpha \mu_{-\alpha} = 1$, then the two sets of numbers $\{N_{\alpha\beta}\}$ and $\{N'_{\alpha\beta}\}$ are said to be *equivalent*.

THEOREM 6.† *Let* $\mathfrak{g}$ *and* $\mathfrak{g}'$ *be two semisimple Lie algebras and* Σ *and* Σ' *be root systems with respect to the Cartan subalgebras* $\mathfrak{h}$ *and* $\mathfrak{h}'$ *respectively. If* Σ *and* Σ' *are similar, then* $\mathfrak{g}$ *and* $\mathfrak{g}'$ *are isomorphic.*

Proof. For simplicity, let $\Sigma = \Sigma'$. Since the dimensions of $\mathfrak{h}$ and $\mathfrak{h}'$ are equal to the number of linearly independent vectors of Σ, we have $\dim \mathfrak{h} = \dim \mathfrak{h}'$.

For any $\alpha \in \Sigma$, there exist $H_\alpha \in \mathfrak{h}$ and $H'_\alpha \in \mathfrak{h}'$ such that

$$(H, H_\alpha) = \alpha(H) \qquad \text{for all } H \in \mathfrak{h},$$

and

$$(H', H'_\alpha) = \alpha(H') \qquad \text{for all } H' \in \mathfrak{h}'.$$

For every $\alpha \in \Sigma$, choose $E_\alpha \in \mathfrak{g}$ and $E'_\alpha \in \mathfrak{g}'$ such that

$$(E_\alpha, E_{-\alpha}) = 1 \quad \text{and} \quad (E'_\alpha, E'_{-\alpha}) = 1.$$

The structure formulas of $\mathfrak{g}$ and $\mathfrak{g}'$ are respectively

$$\begin{array}{ll|ll} [\mathfrak{h}, \mathfrak{h}] = 0 & & [\mathfrak{h}', \mathfrak{h}'] = 0 & \\ [H, E_\alpha] = \alpha(H)E_\alpha & & [H', E'_\alpha] = \alpha(H')E'_\alpha & \\ [E_\alpha, E_{-\alpha}] = H_\alpha & & [E'_\alpha, E'_{-\alpha}] = H'_\alpha & \\ [E_\alpha, E_\beta] = N_{\alpha\beta}E_{\alpha+\beta} & (\alpha+\beta \neq 0) & [E'_\alpha, E'_\beta] = N'_{\alpha\beta}E'_{\alpha+\beta} & (\alpha+\beta \neq 0) \end{array}$$

If $N_{\alpha\beta} = N'_{\alpha\beta}$ for all $\alpha, \beta \in \Sigma$ satisfying $\alpha+\beta \neq 0$, then the mapping

$$\sum_{\alpha\in\Sigma} a_\alpha H_\alpha + \sum_{\alpha\in\Sigma} b_\alpha E_\alpha \to \sum_{\alpha\in\Sigma} a_\alpha H'_\alpha + \sum_{\alpha\in\Sigma} b_\alpha E'_\alpha$$

is an isomorphism from $\mathfrak{g}$ to $\mathfrak{g}'$; it also follows that this isomorphism induces an equivalent mapping from Σ to itself.

We now show that E'_α ($\alpha \in \Sigma$) can be chosen such that $N_{\alpha\beta} = N'_{\alpha\beta}$. For this purpose, select a basis $e_1 \ldots, e_n$ of the Euclidean space containing Σ. Introduce an ordering into this space by defining

$$x = x_1e_1 + \ldots + x_ne_n > y = y_1e_1 + \ldots + y_ne_n$$

if there exists some k such that $x_k > y_k$ and $x_i = y_i$ ($1 \leq i \leq k-1$). Obviously, this ordering satisfies:

1° For any x, y in the space, one of $x > y$, $x = y$, $x < y$ is true.
2° If $x > y$ and $y > z$, then $x > z$.
3° If $x > y$, $\lambda > 0$, then $\lambda x > \lambda y$ and $x+z > y+z$.

Thus the vectors in Σ can be arranged in increasing order with respect to this ordering. If $\alpha \in \Sigma$ and $\alpha > 0$, then it is said to be a positive root, otherwise, it is said to be a negative root. We will use induction on the positive roots with respect to this ordering to choose E'_α such that $N'_{\alpha\beta} = N_{\alpha\beta}$.

†H. Weyl, Theorie der Darstellung kontinuierlicher halbeinfach Gruppen durch lineare Transformationen. I, II, III, *Math. Zeit.* **23** (1925), 271–309; **24** (1926), 328–76; **24** (1926), 377–95.

Let ϱ be a fixed positive root and suppose that for roots satisfying $-\varrho < \alpha < \varrho$ we have chosen $E'_\alpha \in \mathfrak{g}'$ such that $(E'_\alpha, E'_{-\alpha}) = 1$ and

$$N_{\alpha\beta} = N'_{\alpha\beta} \quad \text{for any } \alpha, \beta, \alpha+\beta \in \Sigma \text{ satisfying}$$
$$-\varrho < \alpha, \beta, \alpha+\beta < \varrho.$$

If ϱ cannot be written as $\varrho = \gamma+\delta$ where $-\varrho < \gamma, \delta < \varrho$, then choose any roots E'_ϱ and $E'_{-\varrho}$ such that $(E'_\varrho, E'_{-\varrho}) = 1$. If $\varrho = \gamma+\delta$ for some γ and δ satisfying $-\varrho < \gamma, \delta < \varrho$, then use

$$[E'_\gamma, E'_\delta] = N_{\gamma,\delta} E'_\varrho \tag{5.16}$$

to choose E'_ϱ and then determine $E'_{-\varrho}$ by $(E'_\varrho, E'_{-\varrho}) = 1$. We must show that

$$N_{\alpha\beta} = N'_{\alpha\beta} \quad \text{for any } \alpha, \beta, \alpha+\beta \in \Sigma \text{ satisfying} -\varrho \leqslant \alpha, \beta, \alpha+\beta \leqslant \varrho.$$

Consider the following cases separately.

(1) $-\varrho < \alpha, \beta, \alpha+\beta < \varrho$. By the induction assumption, $N_{\alpha\beta} = N'_{\alpha\beta}$.

(2) $-\varrho < \alpha, \beta < \varrho$ and $\alpha+\beta = \varrho$. If the decomposition $\varrho = \alpha+\beta$ is the same as the decomposition $\varrho = \gamma+\delta$, then by the choice of E'_ϱ and (5.16), we have $N'_{\alpha\beta} = N_{\alpha\beta}$. If these decompositions are not the same, then $\alpha+\beta+(-\gamma)+(-\delta) = 0$, i.e. α, β, $-\gamma$ and $-\delta$ form a parallelogram. By (iii) of Lemma 3, we have

$$N_{\alpha\beta}N_{-\gamma,-\delta} = -N_{\alpha,-\gamma}N_{-\delta,\beta} - N_{\alpha,-\delta}N_{\beta,-\gamma},$$

and

$$N'_{\alpha\beta}N'_{-\gamma,-\delta} = -N'_{\alpha,-\gamma}N'_{-\delta,\beta} - N'_{\alpha,-\delta}N'_{\beta,-\gamma}.$$

Now from $0 < \alpha, \gamma < \varrho$ it follows that $-\varrho < \alpha-\gamma < \varrho$. Similarly, we have $-\varrho < -\delta$, $\beta, \beta-\delta < \varrho$, $-\varrho < \alpha, -\delta, \alpha-\delta < \varrho$ and $-\varrho < \beta, -\gamma, \beta-\gamma < \varrho$. According to the induction assumption, we have

$$N_{\alpha,-\gamma} = N'_{\alpha,-\gamma}, \quad N_{-\delta,\beta} = N'_{-\delta,\beta},$$

and

$$N_{\alpha,-\delta} = N'_{\alpha,-\delta}, \quad N_{\beta,-\gamma} = N'_{\beta,-\gamma}.$$

Thus

$$N_{\alpha\beta}N_{-\gamma,-\delta} = N'_{\alpha\beta}N'_{-\gamma,-\delta}. \tag{5.17}$$

From $N_{\gamma\delta} = N'_{\gamma\delta}$ and (iv) of Lemma 3, we get $N_{-\gamma,-\delta} = N'_{-\gamma,-\delta}$.
Thus by (5.17), we have $N_{\alpha\beta} = N'_{\alpha\beta}$.

(3) $-\varrho < \alpha, \beta < \varrho$ and $\alpha+\beta = -\varrho$. In this case, $\varrho = (-\alpha)+(-\beta)$ and $-\varrho < -\alpha, -\beta < \varrho$. Thus by (2), we get $N_{-\alpha,-\beta} = N'_{-\alpha,-\beta}$. Now by (iv) of Lemma 3, we have $N_{\alpha\beta} = N'_{\alpha\beta}$.

(4) $-\varrho \leqslant \alpha, \beta, \alpha+\beta \leqslant \varrho$. In this case, $\alpha+\beta+(-\alpha-\beta) = 0$ and at most one of α, β and $-\alpha-\beta$ is equal to $\pm\varrho$. By (ii) of Lemma 3, this case can be reduced to one of the cases discussed above. The proof of Theorem 5 is now completed.

The following theorem is the converse of Theorem 6 and is a direct conclusion of Theorem 6 in Chapter 3 (conjugacy of Cartan subalgebras).

Theorem 7. *Let $\mathfrak{g}$ be a semisimple Lie algebra, $\mathfrak{h}$ and $\mathfrak{h}'$ be two Cartan subalgebras and Σ, Σ' be root systems of $\mathfrak{g}$ with respect to $\mathfrak{h}$ and $\mathfrak{h}'$, then Σ and Σ' are equivalent. In other words, root systems are determined up to equivalence by $\mathfrak{g}$ but not the choice of a Cartan subalgebra; or equivalently, semisimple Lie algebras with non-equivalent root systems are not isomorphic.*

Proof. According to Theorem 6 of Chapter 3, there is an automorphism σ such that $\sigma(\mathfrak{h}) = \mathfrak{h}'$. Let α be a root of $\mathfrak{g}$ with respect to $\mathfrak{h}$, i.e. there exists E_α such that

$$[H, E_\alpha] = \alpha(H)E_\alpha, \quad \text{for all } H \in \mathfrak{h}.$$

Applying σ to both sides of this equation, we get

$$[\sigma(H), \sigma(E_\alpha)] = \alpha(H)\,\sigma(E_\alpha) \quad \text{for all } H \in \mathfrak{h}.$$

Define a linear function on $\mathfrak{h}'$ by

$$\alpha'(H') = \alpha(\sigma^{-1}(H')) \quad \text{for all } H' \in \mathfrak{h}';$$

then

$$[H', \sigma(E_\alpha)] = \alpha'(H')\,\sigma(E_\alpha) \quad \text{for all } H' \in \mathfrak{h}'.$$

This proves that α' is a root of $\mathfrak{g}$ with respect to $\mathfrak{h}'$, thus σ induces a mapping from Σ to Σ'. Denote this mapping also by σ, then

$$\alpha \xrightarrow{\sigma} \alpha', \quad \alpha \in \Sigma, \quad \alpha' \in \Sigma'.$$

Since the cardinalities of Σ and Σ' are equal, this mapping is one–one.

Embed α into $\mathfrak{h}$ and obtain H_α satisfying

$$(H, H_\alpha) = \alpha(H) \quad \text{for all } H \in \mathfrak{h}.$$

Since σ preserves the Killing form, we have

$$(\sigma(H), \sigma(H_\alpha)) = \alpha(H) \quad \text{for all } H \in \mathfrak{h}.$$

Let $\sigma(H) = H'$, then

$$(H', \sigma(H_\alpha)) = \alpha'(H') \quad \text{for all } H' \in \mathfrak{h}.$$

Thus if $\alpha' \in \Sigma'$ is embedded into $\mathfrak{h}'$, then $H'_{\alpha'} = \sigma(H_\alpha)$. Hence for all $\alpha, \beta \in \Sigma$, we have

$$\begin{aligned}(\sigma(\alpha), \sigma(\beta) &= (H'_{\sigma(\alpha)}, H'_{\sigma(\beta)}) = (\sigma(H_\alpha), \sigma(H_\beta)) \\ &= (H_\alpha, H_\beta) = (\alpha, \beta).\end{aligned}$$

Thus σ is an equivalence from Σ onto Σ'.

The following theorem will be used in later part of this book.

Theorem 8.† *In a semisimple Lie algebra $\mathfrak{g}$, there exist root vectors E_α ($\alpha \in \Sigma$) such that*

$$(E_\alpha, E_{-\alpha}) = 1$$

† See footnote of Theorem 6.

and all structure constants $N_{\alpha\beta}$ $(\alpha+\beta \in \Sigma)$ *are non-zero real numbers satisfying*

$$N_{\alpha\beta} = -N_{-\alpha,\,-\beta},$$

thus

$$N_{\alpha\beta}^2 = R_{\alpha\beta} = \frac{q(p+1)}{2}(\alpha, \alpha) > 0.$$

Hence there exist root vectors E_α $(\forall \alpha \in \Sigma)$ *such that all* $N_{\alpha,\beta}$ *are determined by* Σ.

Proof. The mapping

$$\alpha \to \alpha' = -\alpha$$

is an equivalence of the root system Σ of $\mathfrak{g}$. For any α', choose $Y_{\alpha'} = -E_{-\alpha}$ such that

$$(Y_{\alpha'}, Y_{-\alpha'}) = (E_{-\alpha}, E_\alpha) = 1.$$

Let

$$[Y_{\alpha'}, Y_{\beta'}] = N'_{\alpha'+\beta'} Y_{\alpha'+\beta'} \quad \text{for all } \alpha+\beta \neq 0,$$

then

$$N'_{\alpha'\beta'} = -N_{-\alpha,\,-\beta} \quad \text{for all } \alpha+\beta \neq 0.$$

By Theorem 6, we can choose μ_α $(\forall\, \alpha \in \Sigma)$ such that $\mu_\alpha \mu_{-\alpha} = 1$ and if $Z_\alpha = \mu_\alpha E_\alpha$, then

$$[Z_\alpha, Z_\beta] = -N_{-\alpha,\,-\beta} Z_{\alpha+\beta} \quad \text{for all } \alpha+\beta \neq 0.$$

Then

$$-N_{-\alpha,\,-\beta} = \frac{\mu_\alpha \mu_\beta}{\mu_{\alpha+\beta}} N_{\alpha,\beta} \quad \text{for all } \alpha+\beta \in \Sigma.$$

Let $\tilde{\mu}_\alpha = \sqrt{\mu_\alpha}\,(\forall \alpha \in \Sigma)$ and choose suitable signs so that $\tilde{\mu}_\alpha \tilde{\mu}_{-\alpha} = -1$. If $F_\alpha = \tilde{\mu}_\alpha E_\alpha$ and $[F_\alpha, F_\beta] = \tilde{N}_{\alpha\beta} F_{\alpha+\beta}$, then for $\alpha+\beta \in \Sigma$, we have

$$\tilde{N}_{\alpha\beta} = \frac{\tilde{\mu}_\alpha \tilde{\mu}_\beta}{\tilde{\mu}_{\alpha+\beta}} N_{\alpha\beta} = -\frac{\tilde{\mu}_\alpha \tilde{\mu}_\beta}{\tilde{\mu}_{\alpha+\beta}} \frac{\mu_{\alpha+\beta}}{\mu_\alpha \mu_\beta} N_{-\alpha,\,-\beta}$$

$$= -\frac{\tilde{\mu}_{-\alpha} \tilde{\mu}_{-\beta}}{\tilde{\mu}_{-\alpha,\,-\beta}} N_{-\alpha,\,-\beta} = -\tilde{N}_{-\alpha,\,-\beta}.$$

Thus

$$\tilde{N}_{\alpha\beta}^2 = -\tilde{N}_{\alpha\beta} \tilde{N}_{-\alpha,\,-\beta} = R_{\alpha\beta} = \frac{q(p+1)}{2}(\alpha, \alpha).$$

This proves Theorem 8.

Finally, we give the following definition:

Definition. Let $H_1, \ldots, H_n$ be an arbitrary basis of the Cartan subalgebra $\mathfrak{h}$ of a semi-simple Lie algebra $\mathfrak{g}$ and Σ be the root system of $\mathfrak{g}$ with respect to $\mathfrak{h}$. If E_α $(\forall \alpha \in \Sigma)$ is a root vector of α, $(E_\alpha, E_{-\alpha}) = 1$ and $N_{\alpha\beta} = -N_{-\alpha,\,-\beta}$, then $\{H_1, \ldots, H_n, E_\alpha \,|\, \alpha \in \Sigma\}$ is said to be a *Weyl basis* of $\mathfrak{g}$.

Theorem 8 has the following

COROLLARY. *Any semisimple Lie algebra has a Weyl basis.*

Weyl bases are important for the study of compact real forms of complex semisimple Lie algebras.

5.4. Root systems of the classical Lie algebras

(A_n) Let $n = m+1$. We know that

$$\mathfrak{h} = \left\{ H_{\lambda_1, \ldots, \lambda_m} \middle| \sum_{i=1}^{m} \lambda_i = 0 \right\}$$

is a Cartan subalgebra of A_n. The root system $\Sigma(A_n)$ of A_n with respect to $\mathfrak{h}$ consists of al roots

$$\lambda_i - \lambda_k, \quad i \neq k \quad \text{and} \quad 1 \leqslant i, k \leqslant m.$$

There are $m(m-1) = m^2 - m$ roots. A root vector corresponding to $\lambda_i - \lambda_k$ is E_{ik}. We now compute the restriction of the Killing form to $\mathfrak{h}$. Choose a basis of A_n consisting of all E_{ik} ($i \neq k$, $1 \leqslant i, k \leqslant m$) and a basis of $\mathfrak{h}$, then

$$\begin{aligned}
(H_{\lambda_1, \ldots, \lambda_m}, H_{\mu_1, \ldots, \mu_m}) &= \sum_{\substack{i,k=1 \\ i \neq k}}^{m} (\lambda_i - \lambda_k)(\mu_i - \mu_k) \\
&= \sum_{\substack{i,k=1 \\ i \neq k}}^{m} (\lambda_i \mu_i + \lambda_k \mu_k - \lambda_i \mu_k - \lambda_k \mu_i) \\
&= 2(m-1) \sum_{i=1}^{m} \lambda_i \mu_i - 2 \sum_{\substack{i,k=1 \\ i \neq k}}^{m} \lambda_i \mu_k \\
&= 2(m-1) \sum_{i=1}^{m} \lambda_i \mu_i - 2 \sum_{i=1}^{m} \lambda_i \sum_{k=1}^{m} \mu_k + 2 \sum_{i=1}^{m} \lambda_i \mu_i . \\
&= 2m \sum_{i=1}^{m} \lambda_i \mu_i . \qquad (5.18)
\end{aligned}$$

In particular,

$$(H_{\lambda_1, \ldots, \lambda_m}, H_{\lambda_1, \ldots, \lambda_m}) = 2m \sum_{i=1}^{m} \lambda_i^2 .$$

To embed $\lambda_i - \lambda_k$ into $\mathfrak{h}$, we want to find an element $H_{\mu_1, \ldots, \mu_m}$ such that

$$(H_{\lambda_1, \ldots, \lambda_m}, H_{\mu_1, \ldots, \mu_m}) = \lambda_i - \lambda_k, \qquad (5.19)$$

for all $H_{\lambda_1, \ldots, \lambda_m} \in \mathfrak{h}$, i.e. for all $\lambda_1, \ldots, \lambda_m$ satisfying $\sum_1^m \lambda_i = 0$. From (5.18) and (5.19), we get

$$2m \sum_{s=1}^{m} \lambda_s \mu_s = \lambda_i - \lambda_k,$$

for all $\lambda_1, \ldots, \lambda_m$ satisfying $\sum_1^m \lambda_i = 0$. From this equation, it follows that

$$\mu_s = \begin{cases} c+\dfrac{1}{2m}, & s = i, \\ c-\dfrac{1}{2m}, & s = k \\ c, & s \neq i, k, \end{cases}$$

where c is a constant. Since $\sum_1^m \mu_s = 0$, hence $c = 0$. Thus

$$\mu_s = \begin{cases} \dfrac{1}{2m}, & s = i, \\ -\dfrac{1}{2m}, & s \neq k, \\ 0, & s \neq i, k. \end{cases}$$

Now by identifying $\mathfrak{h}_0$ and $\mathfrak{h}_0^*$, we get

$$\lambda_i - \lambda_k = \frac{1}{2m}(E_{ii}-E_{kk}).$$

The square of the length of the root $\lambda_i - \lambda_k$ is

$$(\lambda_i-\lambda_k, \lambda_i-\lambda_k) = \left(\frac{1}{2m}\right)^2 (E_{ii}-E_{kk}, E_{ii}-E_{kk})$$
$$= \left(\frac{1}{2m}\right)^2 2m\cdot 2 = \frac{1}{m}.$$

If $e_i = (1/2m)E_{ii}$ $(i = 1, \ldots, m)$ form an orthogonal basis of the $(n+1)$-dimensional Euclidean space and

$$(e_i, e_i) = \frac{1}{2m},$$

then $\mathfrak{h}_0$ consists of all vectors

$$\sum_i^m \mu_i e_i \left(\mu_i's \text{ are real and } \sum_i^m \mu_i = 0\right)$$

and the root system $\Sigma(A_n)$ can be denoted by

$$\{e_i - e_k \mid i \neq k,\ i, k = 1, \ldots, m\}.$$

(B_n) Let $m = 2n+1$. We know that.

$$\mathfrak{h} = \left\{ H_{\lambda_1 \ldots \lambda_n} = \left[\begin{array}{c|ccc|ccc} 0 & & 0 & & & 0 & \\ \hline & \lambda_1 & & 0 & & & \\ 0 & & \ddots & & & & 0 \\ & 0 & & \lambda_n & & & \\ \hline & & & & -\lambda_1 & & 0 \\ & & & & & \ddots & \\ 0 & & 0 & & 0 & & -\lambda_n \end{array}\right] \right\}$$

is a Cartan subalgebra $\mathfrak{h}$. The corresponding root system $\Sigma(B_n)$ consists of all roots

$$\pm\lambda_i\pm\lambda_k \; (i<k) \quad \text{and} \quad \pm\lambda_i \; (i, k = 1, \ldots, n).$$

We now compute

$$\begin{aligned}(H_{\lambda_1,\ldots,\lambda_n}, H_{\mu_1,\ldots,\mu_n}) &= \sum_{i<k} (\pm\lambda_i\pm\lambda_k)(\pm\mu_i\pm\mu_k) + \sum_{i=1}^{n} (\pm\lambda_i)(\pm\mu_i)\\
&= \sum_{i<k} \{(\lambda_i+\lambda_k)(\mu_i+\mu_k)+(\lambda_i-\lambda_k)(\mu_i-\mu_k)+(-\lambda_i+\lambda_k)(-\mu_i+\mu_k)\\
&\quad +(-\lambda_i-\lambda_k)(-\mu_i-\mu_k)\} + \sum_{i}^{n} \{\lambda_i\mu_i+(-\lambda_i)(-\mu_i)\}\\
&= 2\sum_{i<k} \{(\lambda_i+\lambda_k)(\mu_i+\mu_k)+(\lambda_i-\lambda_k)(\mu_i-\mu_k)\}+2\sum_{1}^{n} \lambda_i\mu_i\\
&= 2\sum_{i<k} (\lambda_i\mu_i+\lambda_i\mu_k+\lambda_k\mu_i+\lambda_k\mu_k+\lambda_i\mu_i-\lambda_i\mu_k-\lambda_k\mu_i+\lambda_k\mu_k)\\
&\quad +2\sum_{i}^{n} \lambda_i\mu_i\\
&= 4\sum_{i<k} (\lambda_i\mu_i+\lambda_k\mu_k)+2\sum_{i}^{n} \lambda_i\mu_i\\
&= 4(n-1)\sum_{1}^{n} \lambda_i\mu_i+2\sum_{1}^{n} \lambda_i\mu_i\\
&= (4n-2)\sum_{1}^{n} \lambda_i\mu_i.\end{aligned}$$

In particular,

$$(H_{\lambda_1,\ldots\lambda_n}, H_{\lambda_1,\ldots\lambda_n}) = (4n-2)\sum_{1}^{n} \lambda_i^2.$$

We want to embed the roots $\pm\lambda_i\pm\lambda_k$ and $\pm\lambda_i$ into $\mathfrak{h}$. For this purpose, we first find an element $H_{\mu_1,\ldots,\mu_n}$ in $\mathfrak{h}$ satisfying

$$(H_{\lambda_1,\ldots\lambda_n}, H_{\mu_1,\ldots,\mu_n}) = \lambda_i+\lambda_k;$$

this is equivalent to

$$(4n-2)\sum_{i}^{n} \lambda_s\mu_s = \lambda_i+\lambda_k \quad \text{for any} \quad \lambda_1, \ldots, \lambda_n.$$

Thus

$$\mu_s = \begin{cases} \dfrac{1}{4n-2}, & s = i, k\\ 0, & s \neq i, k.\end{cases}$$

Now identify $\mathfrak{h}_0^*$ and $\mathfrak{h}_0$ and let

$$H_i = \begin{bmatrix} 0 & 0 & 0\\ 0 & E_{ii} & 0\\ 0 & 0 & -E_{ii}\end{bmatrix};$$

then

$$\lambda_i + \lambda_k = \frac{1}{4n-2}\,(H_i + H_k).$$

Similarly,

$$\pm\lambda_i \pm \lambda_k = \frac{1}{4n-2}\,(\pm H_i \pm H_k)$$

and

$$\pm\lambda_i = \frac{1}{4n-2}\,(\pm H_i).$$

The squares of the lengths of the roots $\pm\lambda_i \pm\lambda_k$ and $\pm\lambda_i$ are

$$\begin{aligned}(\pm\lambda_i \pm\lambda_k, \pm\lambda_i \pm\lambda_k) &= \left(\frac{1}{4n-2}\right)^2 (\pm H_i \pm H_k,\ \pm H_i \pm H_k)\\ &= \left(\frac{1}{4n-2}\right)^2 (4n-2)\,2 = \frac{1}{2n-1}\end{aligned}$$

and

$$\begin{aligned}(\pm\lambda_i,\ \pm\lambda_i) &= \left(\frac{1}{4n-2}\right)^2 (\pm H_i, \pm H_i)\\ &= \left(\frac{1}{4n-2}\right)^2 (4n-2) = \frac{1}{4n-2}\,.\end{aligned}$$

If $e_i = (1/(4n-2))\,H_i$ $(1 \leqslant i \leqslant n)$ form an orthogonal basis of the n-dimensional Euclidean space, then

$$(e_i, e_i) = \frac{1}{4n-2}\,.$$

Thus $\mathfrak{h}_0$ consists of all vectors

$$\sum_1^n \mu_i e_i \quad (\mu_i\text{'s are real}),$$

and the root system $\Sigma(B_n)$ can be denoted by

$$\{\pm e_i \pm e_k, \pm e_i \,|\, i \neq k, k = 1, \ldots, n\}.$$

Notice that the root system of B_1 is $\{\pm e_1\}$ and

$$(e_1, e_1) = (-e_1, -e_1) = \frac{1}{4n-2} = \frac{1}{2},$$

$$(e_1, -e = \quad 1 \quad = -\frac{1}{2}\,.$$

Recall that the root system of A_1 is $\{e_1 - e_2, e_2 - e_1\}$ and

$$(e_1 - e_2, e_1 - e_2) = (e_2 - e_1, e_2 - e_1) = \tfrac{1}{2},$$
$$(e_1 - e_2, e_2 - e_1) = -\tfrac{1}{2},$$

thus the root systems of A_1 and B_1 are similar. Hence A_1 and B_1 are isomorphic.

(D_n) $(n \geq 2)$. Let $m = 2n$. We know that

$$\mathfrak{h}_0 = \left\{ H_{\lambda_1, \dots, \lambda_n} = \left[\begin{array}{ccc|ccc} \lambda_1 & & & & & \\ & \ddots & & & 0 & \\ & & \lambda_n & & & \\ \hline & & & -\lambda_1 & & \\ & 0 & & & \ddots & \\ & & & & & -\lambda_n \end{array}\right] \right\}$$

is a Cartan subalgebra of D_n. The corresponding root system $\Sigma(D_n)$ consists of all roots

$$\pm\lambda_i \pm \lambda_k \qquad (i < k,\ i, k = 1, \dots, n).$$

We now compute

$$\begin{aligned}
(H_{\lambda_1, \dots, \lambda_n}, H_{\mu_1, \dots, \mu_n}) &= \sum_{i<k} (\pm\lambda_i \pm \lambda_k)(\pm\mu_i \pm \mu_k) \\
&= 2 \sum_{i<k} \{(\lambda_i + \lambda_k)(\mu_i + \mu_k) + (\lambda_i - \lambda_k)(\mu_i - \mu_k)\} \\
&= 4 \sum_{i<k} (\lambda_i \mu_i + \lambda_k \mu_k) \\
&= (4n-4) \sum_1^n \lambda_i \mu_i .
\end{aligned}$$

In particular,

$$(H_{\lambda_1, \dots, \lambda_n}, H_{\lambda_1, \dots, \lambda_n}) = (4n-4) \sum_1^n \lambda_i^2 .$$

We want to embed the roots $\pm\lambda_i \pm \lambda_k$ into $\mathfrak{h}$. For this purpose, we first find elements $H_{\mu_1, \dots, \mu_n}$ in $\mathfrak{h}$ satisfying

$$(H_{\lambda_1, \dots, \lambda_n}, H_{\mu_1, \dots, \mu_n}) = \pm\lambda_i \pm \lambda_k ,$$

or equivalently,

$$(4n-4) \sum_1^n \lambda_s \mu_s = \pm\lambda_i \pm \lambda_k .$$

Thus

$$\mu_s = \begin{cases} \pm \dfrac{1}{4n-4}, & s = i, k \\ 0, & s \neq i, k. \end{cases}$$

Now identify $\mathfrak{h}_0^*$ and $\mathfrak{h}_0$ and let

$$H_i = \begin{bmatrix} E_{ii} & 0 \\ 0 & -E_{ii} \end{bmatrix},$$

then

$$\pm\lambda_i\pm\lambda_k = \frac{1}{4n-4}(\pm H_i\pm H_k).$$

The squares of the lengths of the roots are

$$(\pm\lambda_i\pm\lambda_k,\ \pm\lambda_i\pm\lambda_k) = \left(\frac{1}{4n-4}\right)^2(\pm H_i\pm H_k,\ \pm H_i\pm H_k)$$
$$= \left(\frac{1}{4n-4}\right)^2(4n-4)2 = \frac{1}{2n-2}.$$

If $e_i = (1/(4n-4))\,H_i\,(1\leqslant i\leqslant n)$ form an orthogonal basis of the n-dimensional Euclidean space, then

$$(e_i,\ e_i) = \frac{1}{4n-4}.$$

Thus $\mathfrak{h}_0$ consists of all vectors

$$\sum_1^n \mu_i e_i \qquad (\mu_i\text{'s are real}),$$

and the root system $\Sigma(D_n)$ can be denoted by

$$\{\pm e_i\pm e_k\,|\,i<k,\ i,\ k = 1,\ \ldots,\ n\}.$$

Notice that $\Sigma(D_n) = \{\pm e_1\pm e_2\}$ and

$$(\pm e_1\pm e_2,\ \pm e_1\pm e_2) = \frac{1}{2n-2} = \frac{1}{2},$$
$$(e_1+e_2,\ -e_1-e_2) = -\tfrac{1}{2},$$
$$(e_1-e_2,\ -e_1+e_2) = -\tfrac{1}{2},$$
$$(\pm e_1\pm e_2,\ \pm e_1\mp e_2) = 0.$$

Thus $\Sigma(D_2)$ is the union of two orthogonal subsets:

$$\Sigma(D_2) = \{e_1+e_2,\ -e_1-e_2\}\cup\{e_1-e_2,\ -e_1+e_2\},$$

where each σ subsystem is equivalent to $\Sigma(A_1)$, thus $D_2 \simeq A_1 \dotplus A_1$.

(C_n) We know that

$$\mathfrak{h}_0 = \left\{ H_{\lambda_1,\ldots,\lambda_n} = \left[\begin{array}{ccc|ccc} \lambda_1 & & & & & \\ & \ddots & & & 0 & \\ & & \lambda_n & & & \\ \hline & & & -\lambda_1 & & \\ & 0 & & & \ddots & \\ & & & & & -\lambda_n \end{array}\right]\right\}$$

is a Cartan subalgebra. The corresponding root system $\Sigma(C_n)$ consists of all roots

$$\pm\lambda_i\pm\lambda_k,\ \pm 2\lambda_i \qquad (i < k,\ \ i, k = 1, \ldots, n).$$

We now compute

$$\begin{aligned}(H_{\lambda_1, \ldots, \lambda_1}, H_{\mu_1, \ldots, \mu_n}) &= \sum_{i<k} (\pm\lambda_i\pm\lambda_k)(\pm\mu_i\pm\mu_k)+\sum_i (\pm 2\lambda_i)(\pm 2\mu_i)\\ &= 2\sum_{i<k} \{(\lambda_i+\lambda_k)(\mu_i+\mu_k)+(\lambda_i-\lambda_k)(\mu_i-\mu_k)\}+8\sum_i \lambda_i\mu_i\\ &= 4\sum_{i<k} (\lambda_i\mu_i+\lambda_k\mu_k)+8\sum_i \lambda_i\mu_i\\ &= 4(n-1)\sum_{i=1}^{n} \lambda_i\mu_i+8\sum_1^n \lambda_i\mu_i\\ &= 4(n+1)\sum_1^n \lambda_i\mu_i.\end{aligned}$$

In particular,

$$(H_{\lambda_1, \ldots, \lambda_n}, H_{\lambda_1, \ldots, \lambda_n}) = (4n+4)\sum_1^n \lambda_i^2.$$

We want to embed the roots $\pm\lambda_i\pm\lambda_k$ and $\pm 2\lambda_i$ into $\mathfrak{h}$. From

$$(H_{\lambda_1, \ldots, \lambda_n}, H_{\mu_1, \ldots, \mu_n}) = \pm\lambda_i\pm\lambda_k \quad \text{or} \quad \pm 2\lambda_i,$$

we get

$$(4n+4)\sum_1^n \lambda_s\mu_s = \pm\lambda_i\pm\lambda_k \quad \text{or } \pm 2\lambda_i, \quad \text{for any } \lambda_1, \ldots, \lambda_n.$$

Thus for $\pm\lambda_i\pm\lambda_k$

$$\mu_s = \begin{cases} \pm\dfrac{1}{4n+4}, & s = i, k,\\ 0, & s \neq i, k,\end{cases}$$

and for $\pm 2\lambda_i$

$$\mu_s = \begin{cases} \pm\dfrac{1}{2n+2}, & s = i,\\ 0, & s \neq i.\end{cases}$$

Now identify $\mathfrak{h}_0^*$ and $\mathfrak{h}_0$ and let

$$H_i = \begin{bmatrix} E_{ii} & 0\\ 0 & -E_{ii}\end{bmatrix},$$

then

$$)\pm H_i\pm H_k),$$

$$\pm 2\lambda_i = \frac{1}{4n+4}(\pm 2H_i).$$

The squares of the lengths of roots are

$$(\pm\lambda_i\pm\lambda_k,\ \pm\lambda_i\pm\lambda_k) = \left(\frac{1}{4n+4}\right)^2 (\pm H_i\pm H_k,\ \pm H_i\pm H_k)$$

$$= \left(\frac{1}{4n+4}\right)^2 (4n+4)\cdot 2 = \frac{2}{4n+4},$$

$$(\pm 2\lambda_i,\ \pm 2\lambda_i) = \left(\frac{1}{4n+4}\right)^2 (\pm 2H_i,\ \pm 2H_i)$$

$$= \left(\frac{1}{4n+4}\right)^2 (4n+4)4 = \frac{4}{4n+4}.$$

If $e_i = (1/(4n+4))\, H_i$, then

$$(e_i,\ e_i) = \frac{1}{4n+4}$$

and $e_1, \ldots, e_n$ generate an n-dimensional Euclidean space. The root system $\Sigma(C_n)$ can be denoted by

$$\{\pm e_i \pm e_k,\ \pm 2e_i \mid i < k,\quad i,\ k = 1,\ \ldots,\ n\}.$$

Recall that $A_1 \approx B_1$ because their root systems are equivalent. Similarly, it can be shown that $\Sigma(A)_1$ and $\Sigma(C_1)$ are equivalent, thus $A_1 \approx C_1$.

CHAPTER 6

FUNDAMENTAL SYSTEMS OF ROOTS OF SEMISIMPLE LIE ALGEBRAS AND WEYL GROUPS

6.1. Fundamental systems of roots and prime roots

Let Σ be a σ system. A subset $\Pi = \{\alpha_1, \ldots, \alpha_n\}$ of Σ is said to be a fundamental system of vectors, if the following two conditions are satisfied:

1° $\alpha_1, \ldots, \alpha_n$ are linearly independent.

2° Any $\alpha \in \Sigma$ can be written as

$$\alpha = \varepsilon(m_1\alpha_1 + \ldots + m_n\alpha_n),$$

where $\varepsilon = \pm 1$ and $m_1, \ldots, m_n$ are non-negative integers. In particular, a fundamental system of vectors of the root system of a semisimple Lie algebra is said to be a fundamental system of roots of the Lie algebra.

If Π is a fundamental system of vectors of a σ system Σ, then Π has the following properties:

3° If $\alpha, \beta \in \Pi$ and $\alpha \neq \beta$, then

$$(\alpha, \beta) \leqslant 0$$

and $2(\beta, \alpha)/(\alpha, \alpha)$, $2(\alpha, \beta)/(\beta, \beta)$ are integers not greater than zero.

Proof. According to property 2°, $\beta - \alpha \in \Sigma$. If p and q are largest non-negative integers such that $\beta + k\alpha \in \Sigma$ $(-p \leqslant k \leqslant q)$, then $p = 0$. Since Σ is a σ system, we have

$$\frac{2(\beta, \alpha)}{(\alpha, \alpha)} = -(q-p) = -q \leqslant 0.$$

Thus

$$(\beta, \alpha) \leqslant 0 \quad \text{and} \quad \frac{2(\alpha, \beta)}{(\beta, \beta)} \leqslant 0;$$

The following theorem assures the existence of fundamental systems of vectors.

THEOREM 1.[†] *If Σ is a σ system, then there exists a linearly independent subset $\Pi = \{\alpha_1, \ldots, \alpha_n\}$ satisfying*

(i) *If $\alpha, \beta \in \Pi$ and $\alpha \neq \beta$, then*

$$(\alpha, \beta) \leqslant 0. \tag{6.1}$$

Thus $2(\alpha, \beta)/(\alpha, \alpha)$ and $2(\alpha, \beta)/(\beta, \beta)$ are integers not greater than zero.

(ii) *Any $\alpha \in \Sigma$ can be written as*

$$\alpha = \varepsilon(m_1\alpha_1 + \ldots + m_n\alpha_n), \tag{6.2}$$

where $\varepsilon = \pm 1$ and $m_1, \ldots, m_n$ are non-negative integers.

Proof. Let Σ be contained in the m-dimensional Euclidean space R^m and $\{e_1, \ldots, e_m\}$ be a fixed basis of R^m. Introduce an ordering into R^m by defining

$$x = a_1e_1 + \ldots + a_me_m > y = b_1e_1 + \ldots + b_me_m$$

iff there exists k such that

$$a_1 = b_1, \ldots, a_k = b_k \quad \text{and} \quad a_{k+1} > b_{k+1}.$$

If $x > 0$, then x is said to be a positive vector; if $x < 0$, then it is said to be a negative vector.

We first prove the following:

LEMMA 1. *Positive vectors $x_1, \ldots, x_p$ of a σ system Σ satisfying*

$$(x_j, x_k) \leqslant 0, \quad \text{for} \quad j \neq k \tag{6.3}$$

are linearly independent.

Proof. Suppose that $x_1, \ldots, x_p$ are linearly dependent. Without loss of generality, assume that

$$x_p = \sum_{i=1}^{p-1} \lambda_i x_i$$

Now rewrite this equation as

$$x_p = \Sigma'\lambda_i x_i + \Sigma''\lambda_i x_i,$$

where all λ_i in $\Sigma'\lambda_i x_i$ are positive and all λ_i in $\Sigma''\lambda_i x_i$ are negative. Let

$$y = \Sigma'\lambda_i x_i \quad \text{and} \quad z = \Sigma''\lambda_i x_i,$$

then $x_p = y + z$. Since $x_p > 0$ and $z \leqslant 0$, thus $y \neq 0$ and $y > 0$. From (6.3) we have $(y, z) \geqslant 0$. Thus

$$(x_p, y) = (y, y) + (z, y) > 0.$$

On the other hand, by (6.3), we have

$$(x_p, y) = \Sigma'\lambda_i(x_p, x_i) \leqslant 0$$

—contradiction. Thus $x_1, \ldots, x_p$ must be linearly independent.

[†] E. B. Dynkin, The structure of semisimple Lie algebras, *Uspehi Mat. Nauk* (N.S.) **2** (1947), 59–127. *Amer. Math. Soc. Transl.* No. 17 (1950).

We now prove Theorem 1. If a positive vector in Σ is not the sum of two other positive vectors in Σ, then we call it a prime vector. Let $\Pi = \{\alpha_1, \ldots, \alpha_n\}$ be the collection of all prime vectors of Σ and call Π a system of prime vectors. We will show that Π has properties (i) and (ii) in Theorem 1.

Let $\alpha, \beta \in \Pi$ and $\alpha \neq \beta$. If $(\alpha, \beta) > 0$, then $2(\alpha, \beta)/(\alpha, \alpha)$ and $2(\alpha, \beta)/(\beta, \beta)$ are positive integers. By property 2° of σ systems (see Section 5.2), $\beta-\alpha \in \Sigma$ and $\alpha-\beta \in \Sigma$. Now let $\varphi = \beta-\alpha$ be a positive vector, then $\beta = \varphi+\alpha$; this is a contradiction to the assumption that β is prime. Similarly, $\alpha-\beta$ cannot be positive. Thus $(\alpha, \beta) \leqslant 0$. By Lemma 1, Π is a linearly independent set.

We now show that Π has property (ii). Since Σ is a finite set, so is the set of all positive vectors. Order the positive vectors in Σ increasingly with respect to the ordering. We will show by induction that (6.2) holds for positive vectors with $\varepsilon = 1$. Let α be a positive vector. If α is a prime vector, then $\alpha = \alpha_i$ (for some i). If α is not a prime vector, then $\alpha = \beta+\gamma$ where $\beta, \gamma \in \Sigma$ and $\beta > 0$, $\gamma > 0$. Now $\alpha > \beta$ and $\alpha > \gamma$, thus by the induction assumption, both β and γ can be written as linear combinations of prime vectors with positive integral coefficients. Hence α can be written in the form of (6.2) with $\varepsilon = 1$. If α is a negative vector, then $-\alpha > 0$ and $-\alpha$ can be written in the form of (6.2). This completes the proof of Theorem 1.

COROLLARY. *If an ordering is defined on the Euclidean space containing a σ system Σ by means of a basis, then the system of prime roots is a fundamental system of vectors of Σ.*

Naturally, different orderings of the Euclidean space may give rise to different fundamental systems of roots or even the same fundamental system of roots. Conversely, we have the following:

THEOREM 1′. *If Π is a fundamental system of roots of a σ system Σ, then there exists an ordering of the Euclidean space containing Σ such that Π is the system of prime vectors determined by this ordering.*

Proof. Let $\Pi = \{\alpha_1, \ldots, \alpha_n\}$, then Σ is contained in the n-dimensional Euclidean space spanned by Π. Introduce an ordering on this space by the following definition: $x = a_1\alpha_1+ \ldots +a_n\alpha_n > y = b_1\alpha_1+ \ldots +b_n\alpha_n$ if there exists an integer k such that

$$a_1 = b_1, \ldots, a_k = b_k \quad \text{and} \quad a_{k+1} > b_{k+1}.$$

With respect to this ordering, a vector $\beta = \sum_1^n m_i\alpha_i \in \Sigma$ is positive iff all m_i $(i = 1, \ldots, n)$ are non-negative integers. Let $\alpha_i = \beta+\gamma$, where $\beta = \sum_1^n m_i\alpha_i \geqslant 0$ and $\gamma = \sum_1^n n_i\alpha_i \geqslant 0$, then $\alpha_i = \sum_1^n (m_j+n_j)\alpha_j$. Now $\alpha_1, \ldots, \alpha_n$ are linearly independent, thus $m_j+n_j = 0$ when $i \neq j$ and $m_i+n_i = 1$. Hence $m_j = n_j = 0$ for $i \neq j$ and $m_i = 1, n_i = 0$ or $m_i = 0, n_i = 1$. In either case, α_i is a prime vector, thus Π is a system of prime vectors. This completes the proof of Theorem 1′.

THEOREM 2. *A fundamental system of roots Π of a σ system Σ uniquely determines Σ, i.e. linear combinations of the form* (6.2) *that are contained in Σ are determined by Π. Therefore, σ systems with equivalent fundamental systems of roots are equivalent.*

Proof. Let $\Pi = \{\alpha_1, \ldots, \alpha_n\}$ be a fundamental system of vectors of Σ, then Σ is contained in the n-dimensional Euclidean space R^n spanned by $\alpha_1, \ldots, \alpha_n$. As in the proof of Theorem 1′, define an ordering on R^n by means of $\alpha_1, \ldots, \alpha_n$. Let Σ_+ be the collection of positive roots and

$$\Sigma_m = \left\{\sum_1^n m_i\alpha_i \,\middle|\, \sum_1^n m_i\alpha_i \in \Sigma_+ \quad \text{and} \quad \sum_1^n m_i \leqslant m\right\},$$

then

$$\Sigma_1 \subseteq \Sigma_2 \ldots \subseteq \Sigma_m \subseteq \Sigma_{m+1} \ldots \subseteq \Sigma_+.$$

We will prove by induction that each Σ_m $(m = 1, 2, \ldots)$ is uniquely determined by Π.

First observe that $\Sigma_1 = \Pi$, thus Σ_1 is determined by Π. Suppose that Σ_m has already been determined; we will show that every $\gamma \in \Sigma_{m+1} - \Sigma_m$ can be expressed as $\gamma = \beta + \alpha$ where $\beta \in \Sigma_m$ and $\alpha \in \Pi$. Since $\{\alpha_1, \ldots, \alpha_n, \gamma\}$ is a linearly dependent set, by Lemma 1, $(\gamma, \alpha_i) > 0$ for some i. By property 2° of σ systems, $\beta = \gamma - \alpha_i \in \Sigma$. β cannot be negative, for otherwise $\alpha_i = \gamma + (-\beta)$ is the sum of two positive vectors. Thus $\beta \in \Sigma_m$.

Since Σ_m has already been constructed, thus for any $\beta \in \Sigma_m$ and $\alpha \in \Pi$ all elements that are of the form $\beta - j\alpha$ $(j = 1, \ldots)$ and in Σ are known. Let p be the largest non-negative integer such that $\beta - j\alpha \in \Sigma$ $(j = 0, 1, 2, \ldots, p)$ and $q = p - 2(\beta, \alpha)/(\alpha, \alpha)$. If $q > 0$, then $\beta + \alpha \in \Sigma_{m+1}$, otherwise, $\beta + \alpha \notin \Sigma_{m+1}$. In this way, we can determine Σ_{m+1}.

Therefore, starting with Π, we can determine Σ_+, thus Σ is also determined.

COROLLARY. *If the fundamental systems of roots of two semisimple Lie algebras are equivalent, then the Lie algebras are isomorphic.*

Proof. It is a direct consequence of Theorem 2 and Theorem 6 of Chapter 5.

By a proof similar to the one of Theorem 6 of Chapter 5, the following stronger result can be obtained.

THEOREM 3. *Let $\mathfrak{g}$ and $\mathfrak{g}'$ be two semisimple Lie algebras, $\mathfrak{h}$ and $\mathfrak{h}'$ be their Cartan subalgebras and Σ and Σ' be their root systems with fundamental system of roots $\Pi = \{\alpha_1, \ldots, \alpha_n\}$ and $\Pi' = \{\alpha_1', \ldots, \alpha_n'\}$ respectively. Suppose that*

$$\alpha_i \to \alpha_i', \qquad i = 1, \ldots, n$$

is an equivalence between Π and Π', i.e.

$$(\alpha_i, \alpha_k) = (\alpha_i', \alpha_k'), \qquad i, k = 1, \ldots, n.$$

If two sets of root vectors $E_{\pm\alpha_i} \in \mathfrak{g}^{\pm\alpha_i}$ and $E'_{\pm\alpha_i'} \in \mathfrak{g}^{\pm\alpha_i'}$ $(i = 1, \ldots, n)$ satisfy

$$(E_{\alpha_i}, E_{-\alpha_i}) = 1 \quad \text{and} \quad (E'_{\alpha_i'}, E'_{-\alpha_i'}) = 1, \qquad i = 1, \ldots, n,$$

then there exists one and only one isomorphism f from $\mathfrak{g}$ onto $\mathfrak{g}'$ such that

$$f(E_{\pm\alpha_i}) = E'_{\pm\alpha_i}, \qquad i = 1, 2, \ldots, n.$$

Proof. We first show the uniqueness of f. We have

$$[E_{\alpha_i}, E_{-\alpha_i}] = H_{\alpha_i} \quad \text{and} \quad [E'_{\alpha_i'}, E'_{-\alpha_i'}] = H'_{\alpha_i'}, \qquad i = 1, 2, \ldots, n.$$

From (6.4), it follows that $f(H_{\alpha_i}) = H'_{\alpha'_i}$. Now H_{α_i} $(i = 1, \ldots, n)$ generate $\mathfrak{h}$, thus the image of any element in $\mathfrak{h}$ under f is determined.

Define an ordering on $\mathfrak{h}_0^*$ so that $\alpha_1, \ldots, \alpha_n$ is a system of prime roots with respect to this ordering. Then the roots in Σ can be arranged in ascending order. We will prove by induction on the positive roots that for any $\alpha \in \Sigma_+$, the image of any element in $\mathfrak{g}^{\pm\alpha}$ is uniquely determined by (6.4). Suppose that this is true for all $\varrho \in \Sigma_+$ such that $\varrho < \alpha$. If α is a prime root, then obviously the image of any element in $\mathfrak{g}^{\pm\alpha}$ is uniquely determined by (6.4). If α is not a prime root, let $\alpha = \beta+\gamma$ where $\alpha > \beta$, $\gamma > 0$. By the induction assumption, the image of any element in $\mathfrak{g}^{\pm\beta}$ or $\mathfrak{g}^{\pm\gamma}$ is uniquely determined by (6.4). Now $\mathfrak{g}^{\pm\alpha} = [\mathfrak{g}^{\pm\beta}, \mathfrak{g}^{\pm\gamma}]$, thus the image of any element in $\mathfrak{g}^{\pm\alpha}$ is uniquely determined. This proves the uniqueness of f.

Secondly, we show the existence of f by a method similar to the method of proof of Theorem 6 in Chapter 5. Notice that if Π and Π' are equivalent, then by Theorem 2, Σ and Σ' are also equivalent. Let α' be the image of $\alpha \in \Sigma$. For any $\alpha \in \Sigma - \Pi$, choose root vectors $E_{\pm\alpha}$ such that

$$(E_\alpha, E_{-\alpha}) = 1.$$

If $\alpha, \beta, \in \Sigma$ and $\alpha \neq \pm\beta$, let

$$[E_\alpha, E_\beta] = N_{\alpha,\beta} E_{\alpha+\beta}.$$

We want to show that for any $\alpha' \in \Sigma' - \Pi'$, root vectors $E'_{\pm\alpha'} \in \mathfrak{g}^{\pm\alpha'}$ can be chosen that satisfy

$$(E'_{\alpha'}, E'_{-\alpha'}) = 1$$

and if $\alpha', \beta' \in \Sigma'$ and $\alpha' \neq \pm\beta'$, then

$$[E'_{\alpha'}, E'_{\beta'}] = N_{\alpha\beta} E'_{\alpha'+\beta'}.$$

If these root vectors exist, then as in the proof of Theorem 6 in Chapter 5, the mapping

$$\sum_1^n a_i H_{\alpha_i} + \sum_{\alpha\in\Sigma} b_\alpha E_\alpha \to \sum_1^n a_i H'_{\alpha'_i} + \sum_{\alpha\in\Sigma} b_\alpha E'_{\alpha'}$$

is an isomorphism from $\mathfrak{g}$ onto $\mathfrak{g}'$ satisfying (6.4).

Let Π be the system of prime roots with respect to a certain ordering on $\mathfrak{h}_0^*$, then Π' is the system of prime roots with respect to a corresponding ordering on $\mathfrak{h}'^*_0$.

Suppose that ϱ' is a fixed positive root of Σ'. Assume that for all roots α' satisfying $-\varrho' < \alpha' < \varrho$ we have chosen root vectors $E'_{\alpha'} \in \mathfrak{g}'^{\alpha'}$ such that $(E'_{\alpha'}, E'_{-\alpha'}) = 1$ and

$$[E'_{\alpha'}, E'_{\beta'}] = N_{\alpha\beta} E'_{\alpha'+\beta'}, \quad \text{where} \quad \alpha', \beta', \alpha'+\beta' \in \Sigma' \quad \text{and}$$
$$-\varrho' < \alpha', \beta', \alpha'+\beta' < \varrho'.$$

If $\varrho = \alpha'_i$ is a prime root, then choose $E'_{\pm\varrho'} = E'_{\pm\alpha'_i}$. If $\varrho' = \gamma' + \delta'$ is not a prime root, then use

$$[E'_{\gamma'}, E'_{\delta'}] = N_{\gamma\delta} E'_{\gamma'+\delta'},$$

to choose $E'_{\varrho'}$ and use $(E'_{\varrho'}, E'_{-\varrho'}) = 1$ to determine $E'_{-\varrho'}$. In either case, it can be proved by a method similar to the method of proof of Theorem 6 in Chapter 5 that

$$[E'_{\alpha'}, E'_{\beta'}] = N_{\alpha\beta}E'_{\alpha'+\beta'}, \quad \text{where} \quad \alpha', \beta', \alpha'+\beta' \in \Sigma' \quad \text{and}$$
$$-\varrho' \prec \alpha', \beta', \alpha'+\beta' \prec \varrho'.$$

The proof of Theorem 3 is now completed.

We know that a fundamental system Π of vectors of a σ system Σ has the properties:

1° Π is a linearly independent set.

2° If $\alpha, \beta \in \Pi$ and $\alpha \neq \beta$, then $2(\alpha, \beta)/(\alpha, \alpha)$ is an integer not greater than zero.

In general, a set Π of vectors in an Euclidean space is said to be a π system if it has properties 1° and 2°. Thus fundamental systems of vectors of σ systems are π systems, in particular, fundamental systems of roots of semisimple Lie algebras are π systems. A π system is said to be simple if it cannot be decomposed into the union of two non-empty mutually orthogonal subsets.

THEOREM 4. *Let Σ be a σ system and Π be a fundamental system of vectors; then Σ is simple iff Π is simple.*

Proof. Suppose that $\Sigma = \Sigma_1 \cup \Sigma_2$ is the union of two mutually orthogonal σ subsystems. Let $\Pi_1 = \Sigma_1 \cap \Pi$ and $\Pi_2 = \Sigma_2 \cap \Pi$, then Π_1 and Π_2 are mutually orthogonal and non-empty. Thus Π_1 and Π_2 are π systems.

Conversely, let $\Pi = \Pi_1 \cup \Pi_2$ be the union of two mutually orthogonal non-empty subsets, then Π_1 and Π_2 are π systems. Denote the set of vectors in Σ_m that are linearly dependent with Π_i by Σ_{m_i} $(i = 1, 2)$, then $(\Sigma_{m1}, \Sigma_{m2}) = 0$ $(m = 1, \ldots)$. Suppose that $\Sigma_m = \Sigma_{m1} \cup \Sigma_{m2}$, we will show that $\Sigma_{m+1} = \Sigma_{m+1,1} \cup \Sigma_{m+1,2}$. Let $\gamma \in \Sigma_{m+1}$, then $\gamma = \beta+\alpha$, where $\beta \in \Sigma_m$ and $\alpha \in \Pi$. By the induction assumption, $\beta \in \Sigma_{m,1}$ or $\beta \in \Sigma_{m,2}$. For convenience, let $\beta \in \Sigma_{m,1}$. Assume that $\alpha \in \Pi_2$, then when $\beta-\alpha$ is written as a linear combination of $\alpha_1, \ldots, \alpha_n$, there are positive and negative coefficients. Hence $\beta-\alpha \notin \Sigma$. Now if p is the largest non-negative integer such that $\beta-j\alpha \in \Sigma$ $(j = 0, 1, 2, \ldots, p)$, then $p = 0$. If q is the largest non-negative integer such that $\beta+j\alpha \in \Sigma$ $(j = 0, 1, 2, \ldots, q)$, then

$$q = p - \frac{2(\beta, \alpha)}{(\alpha, \alpha)} = 0, \quad \text{because} \quad \frac{2(\beta, \alpha)}{(\alpha, \alpha)} = 0.$$

This contradicts $\beta+\alpha \in \Sigma_{m+1}$. Thus $\alpha \in \Pi_1$ and $\gamma = \beta+\alpha \in \Sigma_{m+1,1}$. This proves that $\Sigma_{m+1} = \Sigma_{m+1,1} \cup \Sigma_{m+1,2}$. Let $\Sigma_{+1} = \Sigma_{11} \cup \Sigma_{21} \cup \ldots$ and $\Sigma_{+2} = \Sigma_{12} \cup \Sigma_{22} \cup \ldots$, then $\Sigma_+ = \Sigma_{+1} \cup \Sigma_{+2}$ and Σ_{+1}, Σ_{+2} are mutually orthogonal.

COROLLARY. *Let $\mathfrak{g}$ be a semisimple Lie algebra with fundamental systems of roots Π, then $\mathfrak{g}$ is simple iff Π is simple.*

6.2. Fundamental systems of roots of the classical Lie algebras

From Section 5.4, we know that the root systems of the classical Lie algebras are

$$\Sigma(A_n) = \{e_i - e_k,\ i \neq k,\ i, k = 1, 2, \ldots, n+1\},$$
$$\Sigma(B_n) = \{\pm e_i \pm e_k,\ i < k;\ \pm e_i;\ i, k = 1, 2, \ldots, n\},$$
$$\Sigma(C_n) = \{\pm e_i \pm e_k,\ i < k;\ \pm 2e_i;\ i, k = 1, 2, \ldots, n\},$$
$$\Sigma(D_n) = \{\pm e_i \pm e_k,\ i < k,\ i, k = 1, 2, \ldots, n\} \quad (n \geqslant 2).$$

Thus fundamental systems of roots of A_n, B_n, C_n and D_n are

$$\Pi(A_n) = \{e_1 - e_2,\ e_2 - e_3,\ \ldots,\ e_n - e_{n+1}\},$$
$$\Pi(B_n) = \{e_1 - e_2,\ e_2 - e_3,\ \ldots,\ e_{n-1} - e_n,\ e_n\},$$
$$\Pi(C_n) = \{e_1 - e_2,\ e_2 - e_3,\ \ldots,\ e_{n-1} - e_n,\ 2e_n\},$$
$$\Pi(D_n) = \{e_1 - e_2,\ e_2 - e_3,\ \ldots,\ e_{n-1} - e_n,\ e_{n-1} + e_n\} \quad (n \geqslant 2).$$

Denote the roots of any of these systems by $\alpha_1, \ldots, \alpha_n$; we now compute the inner products of the fundamental roots.

(A_n)

$$(\alpha_i, \alpha_k) = \begin{cases} \dfrac{1}{m}, & i = k, \\ -\dfrac{1}{2m}, & |i-k| = 1 \\ 0, & |i-k| > 1. \end{cases}$$

(B_n)

$$(\alpha_i, \alpha_k) = \begin{cases} \dfrac{1}{2n-1}, & i = k \\ -\dfrac{1}{4n-2}, & |i-k| = 1, \\ 0, & |i-k| > 1, \end{cases} \quad (1 \leqslant i, k \leqslant n-1).$$

$$(\alpha_i, \alpha_n) = \begin{cases} 0, & i < n-1, \\ -\dfrac{1}{4n-2}, & i = n-1, \\ \dfrac{1}{4n-2}, & i = n. \end{cases}$$

(C_n)

$$(\alpha_i, \alpha_k) = \begin{cases} \dfrac{1}{2n+2}, & i = k, \\ -\dfrac{1}{4n-4}, & |i-k| = 1, \quad (1 \leqslant i, k \leqslant n-1) \\ 0, & |i-k| > 1, \end{cases}$$

$$(\alpha_i, \alpha_n) = \begin{cases} 0, & i < n-1, \\ -\dfrac{1}{4n+4}, & i = n-1, \\ \dfrac{1}{n+1}, & i = n. \end{cases}$$

(D_n)

$$(\alpha_i, \alpha_k) = \begin{cases} \dfrac{1}{2n-2}, & i = k, \\ -\dfrac{1}{4n-4}, & |i-k| = 1, \ (1 \leqslant i, k \leqslant n-1) \\ 0, & |i-k| > 1, \end{cases}$$

$$(\alpha_i, \alpha_n) = \begin{cases} \dfrac{1}{2n-2}, & i = n, \\ -\dfrac{1}{4n-4}, & i = n-2, \\ 0, & i \neq n-2, n. \end{cases}$$

We now compute the angles between vectors of these fundamental systems of roots:

(A_n)

$$(\alpha_i, \alpha_k) = \begin{cases} 90^\circ, & |i-k| \neq 1, \\ 120^\circ, & |i-k| = 1. \end{cases}$$

(B_n)

$$(\alpha_i, \alpha_k) = \begin{cases} 90^\circ, & |i-k| \neq 1 \\ 120^\circ, & |i-k| = 1 \end{cases} \quad (1 \leqslant i, k \leqslant n-1),$$

$$(\alpha_i, \alpha_n) = \begin{cases} 90^\circ, & i < n-1, \\ 135^\circ, & i = n-1. \end{cases}$$

(C_n)

$$(\alpha_i, \alpha_k) = \begin{cases} 90^\circ, & |i-k| \neq 1 \\ 120^\circ, & |i-k| = 1. \end{cases} \quad (1 \leqslant i, k \leqslant n-1),$$

$$(\alpha_i, \alpha_n) = \begin{cases} 90^\circ, & i < n-1, \\ 135^\circ, & i = n-1. \end{cases}$$

(D_n)

$$(\alpha_i, \alpha_k) = \begin{cases} 90°, & |i-k| \neq 1, \\ 120°, & |i-k| = 1, \end{cases} \quad (1 \leqslant i, k \leqslant n-1)$$

$$(\alpha_i, \alpha_n) = \begin{cases} 90°, & i \neq n-2, \\ 120°, & i = n-2. \end{cases}$$

As an application of the Corollary to Theorem 2, we prove $A_3 \approx D_3$. We have

$$\Pi(A_3) = \{e_1-e_2, e_2-e_3, e_3-e_4\},$$

$$(e_i, e_j) = \delta_{ij}\frac{1}{2m} = \delta_{ij}\frac{1}{8},$$

Therefore,

$$(e_1-e_2, e_1-e_2) = (e_2-e_3, e_2-e_3) = (e_3-e_4, e_3-e_4) = \tfrac{1}{4},$$
$$(e_1-e_2, e_2-e_3) = (e_2-e_3, e_3-e_4) = -\tfrac{1}{8},$$
$$(e_1-e_2, e_3-e_4) = 0.$$

Now

$$\Pi(D_3) = \{e_1-e_2, e_2-e_3, e_2+e_3\}.$$

$$(e_i, e_j) = \delta_{ij}\frac{1}{4n-4} = \delta_{ij}\frac{1}{8},$$

thus

$$\begin{aligned}(e_1-e_2, e_1-e_2) &= (e_2-e_3, e_2-e_3) \\ &= (e_2+e_3, e_2+e_3) = \tfrac{1}{4}, \\ (e_1-e_2, e_2-e_3) &= (e_1-e_2, e_2+e_3) = -\tfrac{1}{8}, \\ (e_2-e_3, e_2+e_3) &= 0.\end{aligned}$$

An equivalent mapping from $\Pi(A_3)$ to $\Pi(D_3)$ can be defined by:

$$e_1-e_2 \to e_2-e_3,$$
$$e_2-e_3 \to e_1-e_2,$$
$$e_3-e_4 \to e_2+e_3.$$

Thus $A_3 \approx D_3$. Similarly, it can be shown that $B_2 \approx C_2$.

6.3. Weyl groups

Let Σ be a σ system, $\alpha \in \Sigma$, then the mapping

$$\xi \to \xi - \frac{2(\xi, \alpha)}{(\alpha, \alpha)}\alpha, \quad (\xi \in R^m) \tag{6.5}$$

is a reflection of the Euclidean space R^m containing Σ with respect to the hyperplane

$$P_\alpha^* : (\xi, \alpha) = 0;$$

it is said to be the reflection determined by α and is denoted by S_α. If $\Pi = \{\alpha_1, \ldots, \alpha_n\}$ is a fundamental system of vectors of Σ, then Σ is contained in the n-dimensional Euclidean space R^n spanned by Π. Let R^{m-n} be the orthogonal complement of R^n in R^m, then $R^{m-n} \subset P_\alpha^*$. Thus elements in R^{m-n} are fixed under S_α and S_α can be considered as a reflection in R^n. In what follows, S_α will always be considered as a reflection in R^n, thus S_α is an orthogonal transformation in R^n and det $S_\alpha = -1$. All S_α ($\forall \alpha \in \Sigma$) generate a group W called the Weyl group† of the σ system Σ. Naturally, all elements of W are orthogonal transformations.

THEOREM 5. *The Weyl group of the σ system Σ can be considered as a group of permutations on Σ, thus it is a finite group.*

Proof. Let $\xi = \beta \in \Sigma$ in (6.5); then

$$S_\alpha = \beta - \frac{2(\beta, \alpha)}{(\alpha, \alpha)}\alpha.$$

If $\beta = \pm\alpha$, then $S_\alpha(\alpha) = -\alpha$. If $\beta \neq \pm\alpha$, let p and q be largest non-negative integers such that $\beta + k\alpha$ $(-p \leqslant k \leqslant q)$ all belong to Σ, then

$$\frac{2(\beta, \alpha)}{(\alpha, \alpha)} = -(q-p).$$

Thus $S_\alpha(\beta) = \beta + (q-p)\alpha \in \Sigma$. This proves that every S_α induces a permutation on the vectors in Σ. Since W is generated by S_α ($\alpha \in W$), thus every element in W induces a permutation on Σ.

If S and T induce the same permutation on Σ, then $S^{-1}T$ leaves every vector in Σ fixed; in particular, it leaves $\alpha_1, \ldots, \alpha_n$ fixed. Now $\alpha_1, \ldots, \alpha_n$ are linearly independent, thus $S^{-1}T = I$ (identity mapping) on R^n and $S = T$. This proves that W can be considered as a group of permutations on Σ, thus it is a finite group.

The Weyl group of the root system of a semisimple Lie algebra $\mathfrak{g}$ is said to be the Weyl group of the Lie algebra $\mathfrak{g}$. If $\mathfrak{h}$ is a Cartan subalgebra of $\mathfrak{g}$, then the Weyl group of $\mathfrak{g}$ is generated by reflections S_α ($\alpha \in \Sigma$) in $\mathfrak{h}_0^*$ with respect to the hyperplanes P_α^* ($\alpha \in \Sigma$) perpendicular to α ($\alpha \in \Sigma$) and

$$S_\alpha : \xi \to \xi - \frac{2(\xi, \alpha)}{(\alpha, \alpha)}\alpha \quad (\xi \in \mathfrak{h}_0^*).$$

W can also be considered as the group generated by reflections in $\mathfrak{h}_0$ with respect to the hyperplanes

$$P_\alpha : (H_\alpha, H) = \alpha(H) = 0 \quad (\alpha \in \Sigma)$$

perpendicular to the dual element H_α of α; then

$$S_\alpha : H \to H - \frac{2(H, H_\alpha)}{(H_\alpha, H_\alpha)}H_\alpha \quad (H \in \mathfrak{h}_0).$$

† See footnote of Theorem 6 in Section 5.3.

We will sometimes consider W as the group of reflections in $\mathfrak{h}^*$ (or $\mathfrak{h}$) determined by all roots α and defined by

$$\xi \to \xi - \frac{2(\xi, \alpha)}{(\alpha, \alpha)}\alpha \left(\text{or } H \to H - \frac{2(H, H_\alpha)}{(H_\alpha, H_\alpha)} H_\alpha\right).$$

Obviously, then every element in W maps $\mathfrak{h}_0^*$ (or $\mathfrak{h}_0$) onto itself.

We now study the Weyl groups of the classical Lie algebras and consider them as groups of permutations on some $\mathfrak{h}_0^*$ of these algebras.

(A_n) Recall that the root system of A_n is

$$\Sigma(A_n) = \{\lambda_i - \lambda_k \,|\, i \neq k,\ i, k = 1, \ldots, m\} \qquad (m = n+1),$$

where $\lambda_1, \ldots, \lambda_m$ form an orthogonal basis of the m-dimensional Euclidean space with lengths

$$(\lambda_i, \lambda_i) = \frac{1}{2m}, \quad i = 1, \ldots, m.$$

Thus

$$\mathfrak{h}_0^*(A_n) = \left\{\sum_1^m x_i\lambda_i \,|\, x_i \quad \text{real and } \sum_1^m x_i = 0\right\}.$$

Let

$$x = x_1\lambda_1 + \ldots + x_i\lambda_i + \ldots + x_k\lambda_k + \ldots + x_m\lambda_m$$

be a vector in $\mathfrak{h}_0^*$. The reflection determined by the root $\lambda_i - \lambda_k$ is

$$x \to x - \frac{2(x, \lambda_i - \lambda_k)}{(\lambda_i - \lambda_k, \lambda_i - \lambda_k)}(\lambda_i - \lambda_k).$$

We have

$$\frac{(x, \lambda_i - \lambda_k)}{(\lambda_i - \lambda_k, \lambda_i - \lambda_k)} = \frac{\dfrac{x_i - x_k}{2m}}{\dfrac{1}{m}} = \frac{x_i - x_k}{2},$$

thus the reflection determined by $\lambda_i - \lambda_k$ is

$$\begin{aligned} x &= x_1\lambda_1 + \ldots + x_i\lambda_i + \ldots + x_k\lambda_k + \ldots + x_m\lambda_m \\ &\to x_1\lambda_1 + \ldots + x_k\lambda_i + \ldots + x_i\lambda_k + \ldots + x_m\lambda_m, \end{aligned}$$

which transposes the coefficients λ_i and λ_k of x. Thus the Weyl group of A_n can be considered as the symmetric group on m variables.

(B_n) We have

$$\Sigma(B_n) = \{\pm\lambda_i \pm \lambda_k,\ \pm\lambda_i \,|\, i < k, \quad i, k = 1, 2, \ldots, n\},$$

where $\lambda_1, \ldots, \lambda_n$ form an orthogonal basis of $\mathfrak{h}_0^*(B_n)$ with lengths

$$(\lambda_i, \lambda_i) = \frac{1}{4n-2}, \quad i = 1, \ldots, n.$$

Let

$$x = x_1\lambda_1 + \dots + x_n\lambda_n \quad (x_i\text{'s are real})$$

be an arbitrary vector of $\mathfrak{h}_0^*(B_n)$ and compute

$$\frac{(x, \pm\lambda_i\pm\lambda_k)}{(\pm\lambda_i\pm\lambda_k, \pm\lambda_i\pm\lambda_k)} = \frac{\dfrac{\pm x_i\pm x_k}{4n-2}}{\dfrac{2}{4n-2}} = \frac{\pm x_i\pm x_k}{2},$$

$$\frac{(x, \pm\lambda_i)}{(\pm\lambda_i, \pm\lambda_i)} = \frac{\dfrac{\pm x_i}{4n-2}}{\dfrac{1}{4n-2}} = \pm x_i.$$

Thus the reflections determined by $\pm\lambda_i\pm\lambda_k$ are

$$x = x_1\lambda_1 + \dots + x_n\lambda_n \to x_1\lambda_1 + \dots + x_k\lambda_i + \dots + x_i\lambda_k + \dots + x_n\lambda_n$$
$$\text{or } x_1\lambda_1 + \dots - x_k\lambda_i + \dots - x_i\lambda_k + \dots + x_n\lambda_n.$$

and the reflections determined by $\pm\lambda_i$ are

$$x = x_1\lambda_1 + \dots + x_n\lambda_n \to x_1\lambda_1 + \dots - x_i\lambda_i + \dots + x_n\lambda_n.$$

Thus the Weyl group of B_n can be considered as the group of permutations on n variables that change the signs of arbitrarily many variables.

(C_n) We have

$$\Sigma(C_n) = \{\pm\lambda_i\pm\lambda_k, \pm 2\lambda_i \mid i < k, \quad i, k = 1, 2, \dots, n\},$$

where $\lambda_1, \dots, \lambda_n$ form an orthogonal basis of $\mathfrak{h}_0^*(C_n)$ with lengths

$$(\lambda_i, \lambda_i) = \frac{1}{4n+4}, \quad i = 1, 2, \dots, n.$$

Let

$$x = x_1 y_1 + \dots + x_n y_n \quad (x_i\text{'s are real})$$

be an arbitrary vector of $\mathfrak{h}_0^*(C_n)$ and compute

$$\frac{(x, \pm\lambda_i\pm\lambda_k)}{(\pm\lambda_i\pm\lambda_k, \pm\lambda_i\pm\lambda_k)} = \frac{\pm x_i\pm x_k}{2},$$

$$\frac{(x, \pm 2x_i)}{(\pm 2\lambda_i, \pm 2\lambda_i)} = \frac{\pm x_i}{2}.$$

Thus the reflections determined by $\pm\lambda_i\pm\lambda_k$ are

$$x = x_1\lambda_1 + \dots + x_n\lambda_n \to x_1\lambda_1 + \dots + x_k\lambda_i + \dots + x_i\lambda_k + \dots + x_n\lambda_n,$$
$$\text{or } \quad x_1\lambda_1 + \dots - x_k\lambda_i + \dots - x_i\lambda_k + \dots + x_n\lambda_n,$$

and the reflections determined by $\pm 2\lambda_i$ are

$$x = x_1\lambda_1 + \ldots + x_n\lambda_n \to x_1\lambda_1 + \ldots - x_i\lambda_i + \ldots + x_n\lambda_n.$$

Thus C_n has the same Weyl group as B_n.

(D_n) We have

$$\Sigma(D_n) = \{\pm\lambda_i \pm \lambda_k \mid i < k, \quad i, k = 1, 2, \ldots, n\}$$

where $\lambda_1, \ldots, \lambda_n$ form an orthogonal basis of $\mathfrak{h}_0^*(D_n)$ with lengths

$$(\lambda_i, \lambda_k) = \frac{1}{4n-4}.$$

Let

$$x = x_1\lambda_1 + \ldots + x_n\lambda_n \quad (x_i\text{'s are real})$$

be an arbitrary vector of $\mathfrak{h}_0^*(D_n)$ and compute

$$\frac{(x, \pm\lambda_i \pm \lambda_k)}{(\pm\lambda_i \pm \lambda_k, \pm\lambda_i \pm \lambda_k)} = \frac{\pm x_i \pm x_k}{2}.$$

Thus the reflections determined by $\pm\lambda_i \pm \lambda_k$ are

$$x = x_1\lambda_1 + \ldots + x_n\lambda_n \to x_1\lambda_1 + \ldots + x_k\lambda_i + \ldots + x_i\lambda_k + \ldots + x_n\lambda_n$$
$$\text{or} \quad x_1\lambda_1 + \ldots - x_k\lambda_i + \ldots - x_i\lambda_k + \ldots + x_n\lambda_n.$$

Thus the Weyl group of D_n can be considered as the group of permutations on n variables that change the signs of an even number of variables.

6.4. Properties of Weyl groups

Let Σ be a σ system with fundamental system of vectors $\Pi = \{\alpha_1, \ldots, \alpha_n\}$, then Σ is contained in the space R^n spanned by $\alpha_1, \ldots, \alpha_n$. Thus the Weyl group W of Σ is a group of orthogonal transformations in R^n. W can also be considered as a group of permutations on Σ, hence it is finite.

Let $\alpha \in \Sigma$ and P_α^* be the hyperplane defined by

$$(\alpha, \xi) = 0, \quad \xi \in R^n.$$

Let

$$K = R^n - \bigcup_{\alpha \in \Sigma} P_\alpha^*,$$

i.e. K is the complement of the union of all p_α^* ($\alpha \in \Sigma$). A component of K is said to be a Weyl chamber of R^n with respect to Σ.

LEMMA 2. *If $\alpha_1, \ldots \alpha_n$ is a fundamental system of vectors, then the set*

$$C_0 = \{\xi \in R^n \mid (\alpha_i, \xi) > 0, i = 1, \ldots, n\}$$

is a Weyl chamber and for any $\xi \in C_0$, $\alpha \in \Sigma$, either $(\alpha, \xi) > 0$ or $(\alpha, \xi) < 0$.

Proof. Define an ordering on R^n by means of $\alpha_1, \ldots, \alpha_n$ so that $\{\alpha_1, \ldots, \alpha_n\}$ is the set of prime vectors with respect to this ordering. Let $\alpha = \sum_1^n m_i\alpha_i \in \Sigma_+$, then $m_1, \ldots, m_n$ are non-negative integers. Thus for any $\xi \in C_0$,

$$(\alpha, \xi) = \sum_1^n m_i(\alpha, \xi) > 0.$$

Similarly, if $\alpha \in \Sigma_-$, then $(\alpha, \xi) < 0$ for all $\xi \in C_0$. This proves the second statement of Lemma 2. Thus $C_0 \subset K$.

We now show that C_0 is connected. Let ξ_1 and ξ_2 be any two points in C_0. The line segment connecting ξ_1 and ξ_2 is $\lambda\xi_1+(1-\lambda)\xi_2$ $(0 \leqslant \lambda \leqslant 1)$. Obviously, when $0 \leqslant \lambda \leqslant 1$, we have

$$(\alpha_i, \lambda\xi_1+(1-\lambda)\xi_2) = \lambda(\alpha_i, \xi_1)+(1-\lambda)(\alpha_i, \xi) > 0.$$

Thus $\lambda\xi_1+(1-\lambda)\xi_2 \in C_0\,(0 \leqslant \lambda \leqslant 1)$ and C_0 is connected.

Finally, we prove that C_0 is a maximal connected set. Suppose that $\xi' \in K-C_0$ and ξ' is connected to a point ξ_0 by a continuous curve $l(t)$ $(0 \leqslant t \leqslant 1)$ in K and $l(0) = \xi'$, $l(1) = \xi_0$. Since $\xi' \notin C_0$, we may suppose that $(\alpha_1, \xi') < 0$. Now $\big(\alpha_1, l(t)\big)$ is a continuous function of t and

$$\big(\alpha_1, l(0)\big) = (\alpha_1, \xi') < 0, \quad (\alpha_1, l(1)) = (\alpha_1, \xi_0) > 0;$$

thus there exists $t = t_0\,(0 < t_0 < 1)$ such that $\big(\alpha_1, l(t_0)\big) = 0$. Hence $l(t_0) \notin K$. This contradicts with the assumption that $l(t)$ is a continuous curve in K. Thus C_0 is a maximal connected set, so it is a Weyl chamber.

For any fundamental vector α_i, let $S_i = S_{\alpha_i}$ and denote the subgroup of W generated by $S_1, \ldots, S_n$ by W', then we have:

LEMMA 3. *W' acts transitively on the Weyl chambers, i.e. if C_1 is any Weyl chamber, then there exists an $S \in W'$ such that $C_0 = S(C_1)$.*

Proof. We first prove that elements in W' map a Weyl chamber to a Weyl chamber. It suffices to show that: if ξ_1 and ξ_2 are any two points of K that can be connected by a continuous curve, then $S(\xi_1)$ and $S(\xi_2)$ can also be so connected. In fact, $S(l(t))$ $(0 \leqslant t \leqslant 1)$ connects $S(\xi_1)$ and $S(\xi_2)$. Now S is an orthogonal transformation and it permutes the vectors in Σ, thus

$$\big(\alpha, S(l(t))\big) = (S^{-1}(\alpha), l(t)) \neq 0 \quad \text{for all } \alpha \in \Sigma.$$

Thus $S(l(t)) \in K\,(0 \leqslant t \leqslant 1)$.

Now to prove Lemma 3, it suffices to show that for any $\xi_1 \in C_1$, there exists $S \in W'$ such that $S(\xi_1) \in C_0$.

Consider all $S(\xi_1)$ $(S \in W')$. If ξ_0 is a fixed point in C_0, then one of the points $S(\xi_1)$ $(S \in W')$ is nearest to ξ_0; let it be $S_0(\xi_1)$, then

$$||S_0(\xi_1)-\xi_0|| \leqslant ||S(\xi_1)-\xi_0|| \quad \text{for all } S \in W',$$

where

$$||\xi-\eta|| = \sqrt{[(\xi-\eta, \xi-\eta)]}, \quad \xi, \eta \in R^n.$$

We claim that $S_0(\xi_1) \in C_0$. Otherwise, there exists a prime root α_i such that

$$(\alpha_i, S_0(\xi_1)) < 0.$$

Thus

$$S_i(S_0(\xi_1)) = S_0(\xi_1) - \frac{2(\alpha_i, S_0(\xi_1))}{(\alpha_i, \alpha_i)}\alpha_i,$$

and

$$\begin{aligned} \|S_i S_0(\xi_1) - \xi_0\| &= \left\| S_0(\xi_1) - \xi_0 - \frac{2(\alpha_i, S_0(\xi_1))}{(\alpha_i, \alpha_i)}\alpha_i \right\| \\ &= \left\{ \|S_0(\xi_1) - \xi_0\|^2 + \left\| \frac{2(\alpha_i, S_0(\xi_1))}{(\alpha_i, \alpha_i)}\alpha_i \right\|^2 - 2\left(S_0(\xi_1) - \xi_0, \frac{2(\alpha_i, S_0(\xi_1))}{(\alpha_i, \alpha_i)}\alpha_i\right) \right\}^{1/2} \\ &= \left\{ \|S_0(\xi_1) - \xi_0\|^2 + \left\| \frac{2(\alpha_i, S_0(\xi_1))}{(\alpha_i, \alpha_i)}\alpha_i \right\|^2 - \frac{4(\alpha_i, S_0(\xi_1))}{(\alpha_i, \alpha_i)}(S_0(\xi_1) - \xi_0, \alpha_i) \right\}^{1/2} \\ &= \left\{ \|S_0(\xi_1) - \xi_0\|^2 + \frac{4(\alpha_i, S_0(\xi_1))^2}{(\alpha_i, \alpha_i)} - \frac{4(\alpha_i, S_0(\xi_1))^2}{(\alpha_i, \alpha_i)} \right. \\ &\quad \left. + \frac{4(\alpha_i, S_0(\xi_1)(\alpha_0, \xi_0))}{(\alpha_i, \alpha_i)} \right\}^{1/2} < \|S_0(\xi_1) - \xi_0\|. \end{aligned}$$

This is because $(\alpha_i, S_0(\xi_1)) < 0$, $(\alpha_i, \xi_0) > 0$. In other words, $S_i(S_0(\xi_1))$ is closer to ξ_0 than $S_0(\xi_1)$. This is a contradiction, thus $S_0(\xi_1) \in C_0$.

LEMMA 4. *Let C_1 be any Weyl chamber, then there are n linearly independent vectors $\beta_1, \ldots, \beta_n$ in Σ such that*

$$C_1 = \{\xi \in R^n | (\beta_i, \xi) > 0 \quad \textit{for } i = 1, \ldots, n\}$$

and $\{\beta_1, \ldots, \beta_n\}$ is uniquely determined by C_1. Moreover, $\beta_1, \ldots, \beta_n$ form a fundamental system of vectors.

Proof. By Lemma 3, there exists $S \in W'$ such that $C_0 = S(C_1)$. We claim that the vectors $S^{-1}(\alpha_1), \ldots, S^{-1}(\alpha_n)$ have all the properties stated in the lemma. In fact,

$$\begin{aligned} C_1 = S^{-1}(C_0) &= S^{-1}(\{\xi \in R^n | (\alpha_i, \xi) > 0, i = 1, \ldots, n\}) \\ &= \{S^{-1}(\xi) \in R^n | (\alpha_i, \xi) > 0, i = 1, \ldots, n\} \\ &= \{S^{-1}(\xi) \in R^n | (S^{-1}(\alpha_i), S^{-1}(\xi)) > 0, i = 1, \ldots, n\} \\ &= \{\xi \in R^n | (S^{-1}(\alpha_i), \xi) > 0, i = 1, \ldots, n\}. \end{aligned}$$

Since S induces a permutation on the vectors in Σ, $S^{-1}(\alpha_1), \ldots, S^{-1}(\alpha_n)$ also form a fundamental system of vectors.

To show that $\{S^{-1}(\alpha_1), \ldots, S^{-1}(\alpha_n)\}$ is uniquely determined by C_1, it suffices to show that C_0 is uniquely determined by $\{\alpha_1, \ldots, \alpha_n\}$. We know that

$$C_0 = \{\xi \in R^n | (\alpha, \xi) < 0 \quad \text{for all } \alpha \in \Sigma_+\}.$$

Since $\alpha_1, \ldots, \alpha_n$ are linearly independent, any proper subset of $\{\alpha_1, \ldots, \alpha_n\}$ cannot determine C_0. If $\beta_1, \ldots, \beta_m$ are linearly independent roots and

$$C_0 = \{\xi \in R^n | (\beta_i, \xi) > 0, \quad i = 1, \ldots, m\},$$

then $\beta_1, \ldots, \beta_m$ are all positive and any proper subset of $\{\beta_1, \ldots, \beta_m\}$ cannot determine C_0. Thus there exists $\xi_1 \in R^n$ such that $(\beta_i, \xi_1) > 0$ $(i = 2, \ldots, m)$ and $(\beta_1, \xi_1) \leqslant 0$; clearly, $\xi_1 \notin C_0$. Choose any $\xi_0 \in C_0$, then the line segment connecting ξ_0 and ξ_1 passes through $P^*_{\beta_1}: (\beta_1, \xi) = 0$. Let the point of intersection of this line segment with $P^*_{\beta_1}$ be ξ_2, then $(\beta_i, \xi_2) > 0$ $(i = 2, \ldots, m)$ and $(\beta_1, \xi_2) = 0$. Choose ξ_3 such that $(\beta_1, \xi_3) = 0$ and $(\alpha, \xi_3) \neq 0$ for all $\alpha \in \Sigma_+ - \{\beta_1\}$. Thus there exists a real number t such that $(\beta_i, \xi_2 + t\xi_3) > 0$ $(i = 2, \ldots, m)$, $(\beta_1, \xi_2 + t\xi_3) = 0$ and $(\alpha, \xi_2 + t\xi_3) \neq 0$ for all $\alpha \in \Sigma_+ - \{\beta_1\}$. Let $\xi_4 = \xi_2 + t\xi_3$, then the line segment connecting ξ_0 and ξ_4 is contained in C_0; since $(\xi_4, \alpha) \neq 0$ for all $\alpha \in \Sigma_+ - \{\beta_1\}$, $(\alpha, \xi_4) > 0$ for all $\alpha \in \Sigma_+ - \{\beta_1\}$. Now $\xi_4 \notin C_0$, thus β_1 is equal to one of $\alpha_1, \ldots, \alpha_n$. This proves that C_0 is uniquely determined by $\{\alpha_1, \ldots, \alpha_n\}$.

The hyperplanes $P^*_{\alpha_1}, \ldots, P^*_{\alpha_n}$ are said to be walls of the Weyl chamber C_0. In general, if a Weyl chamber C_1 is determined by the linearly independent roots $\beta_1, \ldots, \beta_n$, i.e.

$$C_1 = \{\xi \in R^n | (\beta_i, \xi) > 0 \quad \text{for } i = 1, \ldots, n\},$$

then $P^*_{\beta_1}, \ldots, P^*_{\beta_n}$ are said to be walls of C_1.

LEMMA 5. *If $\alpha \in \Sigma$, then P^*_α is a wall of some Weyl chamber.*

Proof. Let $\xi_1 \in P^*_\alpha$ and $(\beta, \xi_1) \neq 0$ for all $\beta \in \Sigma - \{\pm\alpha\}$. If $\xi_2 \notin P^*_\alpha$, then there exists a real number t such that $(\beta, \xi_1)(\beta, \xi_1, +t\xi_2) > 0$ for all $\beta \in \Sigma - \{\pm\alpha\}$. Thus P^*_α is a wall of the Weyl chamber C_1 containing $\xi_1 + t\xi_2$. In fact, let $\beta_1, \ldots, \beta_n$ be linearly independent roots such that

$$C_1 = \{\xi \in R^n | (\beta_i, \xi) > 0 \quad \text{for } i = 1, \ldots, n\}.$$

If $\beta_1, \ldots, \beta_n \neq \pm\alpha$, then it follows from $(\beta_i, \xi_1 + t\xi_2) > 0$ $(i = 1, \ldots, n)$ that $(\beta_i, \xi_1) > 0$ $(i = 1, \ldots, n)$; thus $(\alpha, \xi_1) \neq 0$. This is a contradiction.

LEMMA 6. *Suppose $\{\alpha_1, \ldots, \alpha_n\}$ is a system of fundamental vectors of Σ and the system of prime roots with respect to some ordering defined on R^n. If $\alpha \in \Sigma_+$ and $\alpha \neq \alpha_i$, then $S_{\alpha_i}(\alpha) > 0$ and $S_{\alpha_i}(\alpha_i) = -\alpha_i$.*

Proof. If $\alpha = \sum_1^n m_i \alpha_i \in \Sigma_+$ then $m_i \geqslant 0$ $(i = 1, \ldots, n)$ and

$$S_{\alpha_i}(\alpha) = \alpha - \frac{2(\alpha, \alpha_i)}{(\alpha_i, \alpha_i)}\alpha_i$$

$$= \sum_{j \neq 1} m_j \alpha_j + \left(m_i - \frac{2(\alpha, \alpha_i)}{(\alpha_i, \alpha_i)}\right)\alpha_i.$$

If $\alpha \neq \alpha_i$, then there exists some $j \neq i$ such that $m_j > 0$, thus $S_{\alpha_i}(\alpha) > 0$. Obviously, $S_{\alpha_i}(\alpha_i) = -\alpha_i$.

THEOREM 6. *Let $\{\alpha_1, \ldots, \alpha_n\}$ be a fundamental system of vectors of a σ system Σ, then the reflections $S_i = S_{\alpha_i}$ $(i = 1, \ldots, n)$ generate the Weyl group W of Σ and if $\alpha \in \Sigma$, then there* *exist $\alpha_i \in \{\alpha_1, \ldots, \alpha_n\}$ and $S \in W$ such that $S(\alpha_i) = \alpha$. Moreover, W is a regular transitive permutation group on the Weyl chambers (regularity means that if $S \in W$ and $S(C) = C$ for some Weyl chamber C, then $S = I$).*

Proof. Let α be any root, then $\pm\alpha$ are the only two roots perpendicular to P^*_α and P^*_α is a wall of some Weyl chamber. If $S \in W'$ and $S(C_0) = C_1$, then for some fundamental root α_i, $S(P^*_{\alpha_i}) = P^*_\alpha$. Thus $S(\alpha_i) = \pm\alpha$. If $S(\alpha_i) = -\alpha$, then $(SS_{\alpha_i})(\alpha_i) = S(-\alpha_i) = \alpha$. So there always exists $S \in W'$ such that $S(\alpha_i) = \alpha$. Thus $S_\alpha = SS_iS^{-1}$ and $S_\alpha \in W'$. This proves that $W = W'$. By Lemma 3', $W(= W')$ acts transitively on the Weyl chambers.

We now show that W is regular on the Weyl chambers. Let $S \in W$ and $S(C) = C$ for some Weyl chamber C; we want to show that $S = I$. Since W is transitive on the Weyl chambers, we may assume that

$$C = C_0 = \{\xi \in R^n \mid (\alpha_i, \xi) > 0 \quad \text{for } i = 1, 2, \ldots, n\}.$$

By Lemma 4, $S(C_0) = C_0$ iff S induces a permutation on the fundamental vectors, or equivalently, S induces a permutation on the positive roots. Since W is generated by $S_1, \ldots, S_n$, we may let $S = S_{i_1} \ldots . S_{i_p}$ $(1 \leq i_1, \ldots, i_p \leq n)$. We use induction on p to show that $S = I$. Suppose $S = I$ if $S = S_{j_1} \ldots S_{j_{p'}}$ $(1 \leq j_1, \ldots, j_{p'} \leq n)$ and $p' < p$.

Let $S' = S_{i_2} \ldots . S_{i_p}$, then $S = S_{i_1} S'$. If $S' = I$, then $S(\alpha_i) = S_{i_1}S'(\alpha_{i_1}) = S_{i_1}(\alpha_{i_1}) = -\alpha_{i_1} < 0$—contradiction. Thus $S' \neq I$. By the induction assumption, there exists a positive vector α such that $S'(\alpha) < 0$. Now $S(\alpha) = S_{i_1}S'(\alpha) > 0$, thus by Lemma 6, $S'(\alpha) = -\alpha_{i_1}$. If β is a positive root and $S'(\beta) < 0$, then from $S(\beta) = S_iS'(\beta) > 0$, we have $S'(\beta) = -\alpha_i$, thus $\beta = \alpha$. We have proved that if β is a positive root and $\beta \neq \alpha$, then $S'(\beta) > 0$. We now prove that α is a fundamental vector. Let $\alpha = \sum_1^n m_i\alpha_i$; if α is not one of α_i $(i = 1, \ldots, n)$, then $S'(\alpha) = \sum_1^n m_iS'(\alpha_i) > 0$—contradiction. So $\alpha = \alpha_j$ for some j. Thus $S'(\alpha_j) = -\alpha_{i_1} < 0$ and for any positive root $\beta \neq \alpha_j$, $S'(\beta) > 0$. Let $\beta_t = S_{i_t}S_{i_{t+1}} \cdots . S_{i_p}(\alpha_j)$ $(t = 2, 3, \ldots, p+1)$ and $\beta_{p+1} = \alpha_j$, then $\beta_{p+1} = \alpha_j > 0$ and $\beta_2 = S'(\alpha_j) < 0$. Hence for some k $(2 < k \leq p+1)$, $\beta_k > 0$ and $\beta_{k-1} < 0$. Let $S_{i_k} \ldots S_{i_p} = T$, $S_{i_2} \ldots S_{i_{k-2}} = T'$ and if $k = p+1$, let $T = I$. Then $S'(\alpha_j) = T'S_{{k-1}}T(\alpha_j)$ and $\beta_k = T(\alpha_j) > 0$, $S_{i_{k-1}}T(\alpha_j) = \beta_{k-1} < 0$. By Lemma 6, $T(\alpha_j) = \alpha_{i_{k-1}}$. Thus $T^{-1}S_{i_{k-1}}T = S_j$ and $S' = T'S_{i_{k-1}}T = T'TS_j$. Since $S_j^2 = 1$, $T'T(\alpha) = T'TS_j(S_j(\alpha)) = S'(S_j(\alpha))$. If α is positive and $\alpha \neq \alpha_j$, then $S_j(\alpha) > 0$ and $S_j(\alpha) \neq \alpha_j$, so $S'(S_j(\alpha)) > 0$. If $\alpha = \alpha_j$, then $S_j(\alpha) = -\alpha_j$ and $S'(S_j(\alpha_j)) = -S'(\alpha_j) = \alpha_{i_1} > 0$. Hence $T'T$ transforms positive vectors to positive vectors. Now $T'T$ is the product of $p-2$ elements (possibly with repetitions) from S_i $(i = 1, \ldots, n)$, thus $T'T = I$. So $S' = T'TS_j = S_j$. Now we have $S'(\alpha_j) = -\alpha_{i_1}$ and $S'(\alpha_j) = S_j(\alpha_j) = -\alpha_j$, thus $j = i_1$ and $S = S_{i_1} S' = S_{i_1}^2 = I$. This proves that W is regular on the Weyl chambers.

Corollary. *If $S \in W$ and S transforms positive roots to positive roots or fundamental roots to fundamental roots, then $S = 1$.*

With these results on Weyl groups of σ systems, we can prove the following two theorems on σ systems and semisimple Lie algebras.

Theorem 7. *Let Σ be a σ system, then any two fundamental systems of vectors are equivalent and any such equivalent mapping can be extended to an equivalent mapping of Σ onto itself. Therefore, fundamental systems of vectors of equivalent σ systems are equivalent.*

Proof. It suffices to prove the first part of the theorem. Let $\{\alpha_1, \ldots, \alpha_n\}$ and $\{\beta_1, \ldots, \beta_n\}$ be two fundamental systems of vectors of Σ, then they determine two Weyl chambers

$$C_\alpha = \{\xi \in R^n \,|\, (\alpha_i, \xi) > 0 \text{ for } i = 1, \ldots, n\}$$
$$C_\beta = \{\xi \in R^n \,|\, (\beta_i, \xi) > \text{ for } i = 1, \ldots, n\}.$$

By Theorem 6, there exists $S \in W$ such that $S(C_\alpha) = C_\alpha$. By Lemma 4, we may assume that $\beta_i = S(\alpha_i)$ $(i = 1, \ldots, n)$. Since S is an orthogonal transformation, it is an equivalent mapping from $\{\alpha_1, \ldots, \alpha_n\}$ to $\{\beta_1, \ldots, \beta_n\}$. Thus it is also an equivalent mapping from Σ onto itself.

THEOREM 8. *Any two fundamental systems of roots of a semisimple Lie algebra are equivalent. Two semisimple Lie algebras are isomorphic iff they have equivalent fundamental systems of roots.*

Proof. This is a direct consequence of Theorem 7 in Chapter 5, and Theorem 3 and Theorem 7 in this Chapter.

From Theorem 8 and Corollary to Theorem 4, it follows that for the determinations of all non-isomorphic simple Lie algebras it suffices to determine all simple π systems that are not equivalent and then determine a simple Lie algebra for each of these π systems. This is the purpose of the following chapter.

Finally, we make the following remark: when the Weyl group of a semisimple Lie algebra is considered to be the group generated by reflections with respect to the hyperplanes

$$\varrho_\alpha : (\alpha, H) = \alpha(H) = 0 \qquad (\alpha \in \Sigma)$$

in $\mathfrak{h}_0$, then the components of the set

$$\mathfrak{h}_0 - \bigcup_{\alpha \in \Sigma} P_\alpha$$

are also called Weyl chambers of $\mathfrak{g}$. The Weyl group then is a regular transitive group of permutations on the Weyl chambers. Any Weyl chamber is determined by a unique set of fundamental root systems $\{\alpha_1, \ldots, \alpha_n\}$ by

$$C = \{H \in \mathfrak{h}_0 \,|\, \alpha_i(H) > 0 \quad \text{for } i = 1, \ldots, n\}$$

and the Weyl group is generated by $S_1, \ldots, S_n$ defined by

$$S_i(H) = H - \frac{2(H, H_{\alpha_i})}{(H_{\alpha_i}, H_{\alpha_i})} H_{\alpha_i}, \quad \forall H \in \mathfrak{h}_0.$$

CHAPTER 7

CLASSIFICATION OF SIMPLE LIE ALGEBRAS

7.1. Diagrams of π systems

Recall that a set of vectors Π in an Euclidean space is said to be a π system if

1° Π consists of linearly independent vectors.

2° If $\alpha, \beta \in \Pi$ and $\alpha \neq \beta$, then $2(\beta, \alpha)/(\alpha, \alpha)$ is zero or a negative integer.

Π is said to be a simple π system if it cannot be decomposed into the union of two mutually orthogonal π subsystems.

Let Π be a π system, $\alpha, \beta \in \Pi$ and $\alpha \neq \beta$. If $\langle \alpha, \beta \rangle$ denotes the angle between α and β, then by Theorem 4 of Chapter 5

$$\cos \langle \alpha, \beta \rangle = \frac{\varepsilon}{2} \sqrt{r}, \quad \varepsilon = \pm 1, \quad r = 0, 1, 2, 3. \tag{7.1}$$

Now

$$\cos \langle \alpha, \beta \rangle = \frac{(\alpha, \beta)}{\sqrt{[(\alpha, \alpha)(\beta, \beta)]}} \leqslant 0,$$

hence $\varepsilon = -1$ in (7.1). Then $\langle \alpha, \beta \rangle$ can only have one of the values 90°, 120°, 135°, 150°. Assume $(\alpha, \alpha) \leqslant (\beta, \beta)$, then by Theorem 4 of Chapter 5, we have the following cases:

If $\langle \alpha, \beta \rangle = 120°$, then $(\beta, \beta) = (\alpha, \alpha)$,

$$\frac{2(\beta, \alpha)}{(\alpha, \alpha)} = -1 \quad \text{and} \quad \frac{2(\beta, \alpha)}{(\beta, \beta)} = -1.$$

If $\langle \alpha, \beta \rangle = 135°$, then $(\beta, \beta) = 2(\alpha, \alpha)$,

$$\frac{2(\beta, \alpha)}{(\alpha, \alpha)} = -2 \quad \text{and} \quad \frac{2(\beta, \alpha)}{(\beta, \beta)} = -1.$$

If $\langle \alpha, \beta \rangle = 150°$, then $(\beta, \beta) = 3(\alpha, \alpha)$,

$$\frac{2(\beta, \alpha)}{(\alpha, \alpha)} = -3 \quad \text{and} \quad \frac{2(\beta, \alpha)}{(\beta, \beta)} = -1.$$

For each vector in a π system Π, construct a point on the plane. If the angle between two vectors is 120°, 135° or 150°, then connect their corresponding points by 1, 2 or 3 line segments, respectively. The diagram obtained in this way is said to be the angle diagram of the π system Π. If to each point of the angle diagram we attach the square of the length (α, α) of the corresponding vector α, then we obtain the diagram of the π system Π. Equivalent π systems have the same diagram.

Obviously, we have the following:

LEMMA 1. *The diagram of a π system Π is connected iff Π is a simple π system.*

It will be proved that lengths of vectors in a simple system can at most take two values and that the ratio of the squares of these two different lengths is either 2 : 1 or 3 : 1. If in an angle diagram the longer vectors correspond to dots of the form ●, the shorter vectors correspond to small circles, then we obtain the *Dynkin diagram $\Gamma(\Pi)$* of the simple π system. Two simple systems with the same Dynkin diagram are equivalent. The Dynkin diagram of fundamental system of roots of a simple Lie algebra is also said to be the Dynkin diagram of the algebra.

Diagrams and Dynkin diagrams of the classical Lie algebras A_n, B_n, C_n and D_n are:

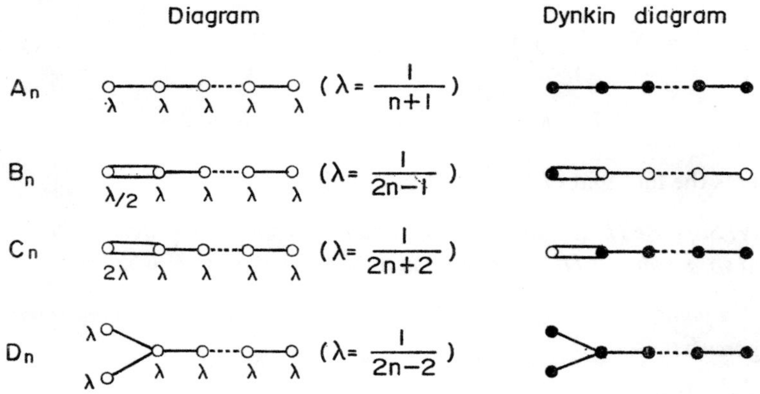

7.2. Classification of simple π systems†

We study first the forms of angle diagrams of simple π systems.

Let $\Pi = \{\alpha_1, \ldots, \alpha_n\}$. If $\lambda_1, \ldots, \lambda_n$ are n real variables, then

$$\left(\sum_1^n \lambda_i \alpha_i, \sum_1^n \lambda_i \alpha_i\right) = \sum_{i,j=1}^n (\alpha_i, \alpha_j) \lambda_i \lambda_j \tag{7.2}$$

is a positive definite bilinear form, since it is the square of the length of a vector in a Euclidean space.

In the following, a vector in Π and the corresponding dot in the angle diagram will be denoted by the same letter.

† E. B. Dynkin, The structure of semisimple Lie algebras, *Amer. Math. Soc. Transl.* (1) **9** (1955), 328–469, Séminaire Sophus Lie, Paris, 1955.

LEMMA 2. *The angle diagram*

is the only angle diagram of simple π systems that contains two dots connected by three line segments.

Proof. Let α_1 and α_2 in the angle diagram of Π be connected by three line segments. If there is another point α_3 in the diagram, then it is connected to α_1 or α_2. Now α_1, α_2 and α_3 are linearly independent, thus from geometric considerations, we have

$$\langle\alpha_1, \alpha_2\rangle + \langle\alpha_2, \alpha_3\rangle + \langle\alpha_3, \alpha_1\rangle < 360°.$$

This is impossible because $\langle\alpha_1, \alpha_2\rangle = 150°$, $\langle\alpha_2, \alpha_3\rangle \geqslant 120°$, $\langle\alpha_3, \alpha_1\rangle \geqslant 90°$.

In Lemma 3 to 6, we assume that the simple π system Π under consideration has no more than two line segments between any two dots.

LEMMA 3. *The angle diagram of a simple π system Π cannot contain any closed path.*

Proof. Assume that $\alpha_1, \ldots, \alpha_k \in \Pi$ form a closed path in the angle diagram, then $(\alpha_1, \alpha_2) \neq 0, (\alpha_2, \alpha_3) \neq 0, \ldots, (\alpha_{k-1}, \alpha_k) \neq 0, (\alpha_k, \alpha_1) \neq 0$. Let $||\alpha_i|| = \sqrt{\{(\alpha_i, \alpha_i)\}}$ and $\alpha_{k+1} = \alpha_1$ then

$$\left(\sum_1^k \frac{1}{||\alpha_i||}\alpha_i, \sum_1^k \frac{1}{||\alpha_i||}\alpha_i\right) \leqslant \sum_1^k \frac{1}{||\alpha_i||^2}(\alpha_i, \alpha_i) + 2\sum_1^k \frac{1}{||\alpha_i||\,||\alpha_{i+1}||}(\alpha_i, \alpha_{i+1})$$

$$= k + 2\sum_1^k \cos\langle\alpha_i, \alpha_{i+1}\rangle \leqslant k - k = 0.$$

This contradicts the fact that (7.1) is positive definite.

COROLLARY. *Suppose Π' is a simple π subsystem of a simple π system Π. If $\beta \in \Pi - \Pi'$ and β is connected to a point in Π', then there is only one point in Π' which is connected to β.*

LEMMA 4. *No point in the angle diagram of a simple π system Π is connected to more than three line segments.*

Proof. Let $\alpha \in \Pi$ be connected to k (> 3) line segments. If $\alpha_1, \ldots, \alpha_l$ ($l \leqslant k$) are all dots connected to α, then by the Corollary to Lemma 3, $\alpha_1, \ldots, \alpha_l$ are mutually orthogonal. Let V be the space spanned by α and $\alpha_1, \ldots, \alpha_l$. Choose $\gamma \neq 0$ in V such that γ is orthogonal to α_i ($i = 1, \ldots, l$). Thus

$$\cos^2\langle\gamma, \alpha\rangle + \sum_1^l \cos^2\langle\alpha_i, \alpha\rangle = 1.$$

Now $\alpha, \alpha_1, \ldots, \alpha_l$ are linearly independent, so

$$\sum_1^l 4\cos^2\langle\alpha_i, \alpha\rangle < 4. \tag{7.3}$$

If α_i is connected to α by t line segments, then $4\cos^2\langle\alpha_i, \alpha\rangle = t$, thus

$$\sum_1^l 4\cos^2\langle\alpha_i, \alpha\rangle = k.$$

This contradicts with (7.3), thus $k \leqslant 3$.

The points $\alpha_1, \ldots, \alpha_k$ in the angle diagram of Π are said to form a chain if α_1 is connected to α_2, α_2 is connected to $\alpha_3, \ldots, \alpha_{k-1}$ is connected to α_k and α_i is not connected to α_j if $|i-j| > 1$. A chain $C = \{\alpha_1, \ldots, \alpha_k\}$ is said to be simple if α_i is connected to α_{i+1} $(i = 1, \ldots, k-1)$ by one line segment.

LEMMA 5. *If $C = \{\alpha_1, \ldots, \alpha_k\}$ is a simple chain of a simple π system Π and*

$$\Pi' = (\Pi - C) \cup \left\{\sum_1^k \alpha_i\right\},$$

then Π' is also a simple π system. The angle diagram of Π' can be obtained from the angle diagram of Π by shrinking the chain $C = \{\alpha_1, \ldots, \alpha_k\}$ to a point α and connecting any $\beta \in \Pi - C$ to α by t line segments if it is connected to some α_i $(i = 1, \ldots, k)$ by t line segments in the diagram of Π.

Proof. The vectors in Π' are linearly independent. Secondly,

$$\begin{aligned}(\alpha, \alpha) &= \left(\sum_1^k \alpha_i, \sum_1^k \alpha_i\right) \\ &= \sum_1^{k-1} \{(\alpha_i, \alpha_i) + 2(\alpha_i, \alpha_{i+1})\} + (\alpha_k, \alpha_k) \\ &= (\alpha_k, \alpha_k) = (\alpha_l, \alpha_l).\end{aligned}$$

If $\beta \in \Pi - C$ is connected to α_l by t line segments, then by Corollary to Lemma 3, β is not connected to any other α_i. Thus

$$(\beta, \alpha) = \left(\beta, \sum_1^k \alpha_i\right) = (\beta, \alpha_l).$$

If γ is not connected to C in Π, then

$$(\gamma, \alpha) = \left(\gamma, \sum_1^k \alpha_i\right) = 0.$$

Therefore, Π' is a π system and the angle diagram of Π' can be obtained from that of Π by shrinking the chain C to a point α and connecting any $\beta \in \Pi - C$ to α by t line segments if it is connected to some α_i $(i = 1, \ldots, k)$ by t line segments.

COROLLARY. *The following diagrams are not subdiagrams of angle diagrams of simple π systems:*

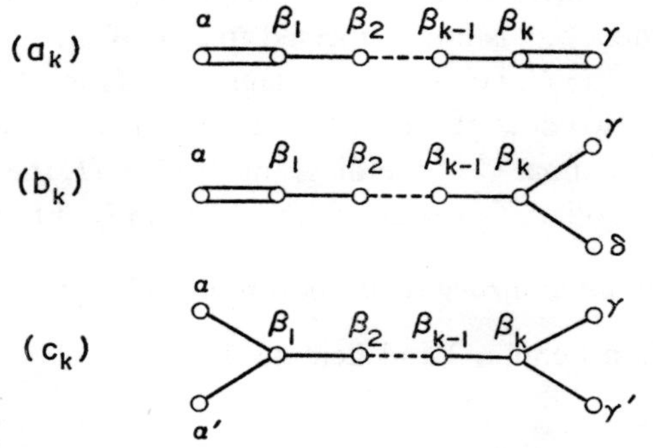

Proof. If (a_k), (b_k) and (c_k) are subdiagrams of angle diagrams of some simple π systems, then by Lemma 5, the diagrams:

(a_1)

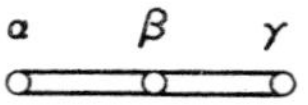

(b_1)

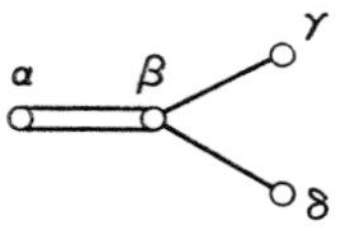

(c_1)

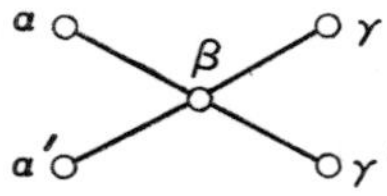

are subdiagrams of angle diagrams of simple π systems. This contradicts with Lemma 4.

LEMMA 6. *An angle diagram of a simple π system must have one of the following forms:*

$(a'_{p,q})$

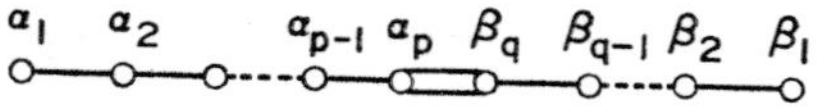

, $(p, q \geq 1)$.

(b'_n)

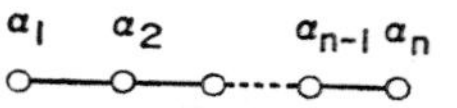

, $(n \geq 1)$.

$(c'_{p,q,r})$

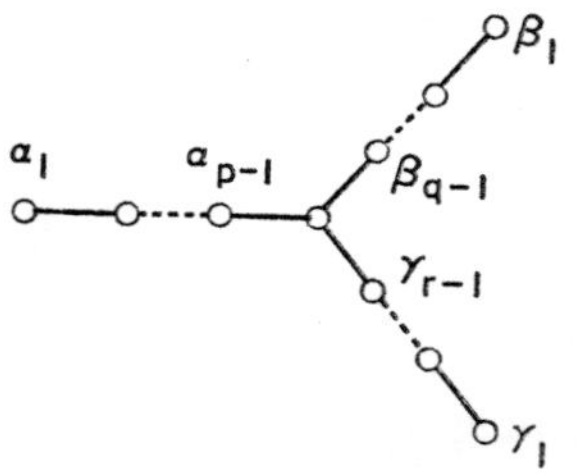

, $(p \geq q \geq r \geq 2)$.

Proof. Let  be a subdiagram of the angle diagram of Π and extend this subdiagram to a maximal chain C. By the Corollary to Lemma 5, C can only contain one subdiagram of the form 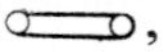, thus C is a diagram of type $(a'_{p,q})$. If $C \neq \Pi$, then there exists $\gamma \in \Pi - C$ which is connected to a point of C. By maximality of C, γ is not connected to α_1, or β_1; by the Corollary to Lemma 5, γ is not connected to any of $\alpha_1, \ldots, \alpha_p$ or $\beta_1, \ldots, \beta_q$. This is a contradiction. Thus $\Pi = C$, i.e. the angle diagram of Π is of type $(a'_{p,q})$.

Now suppose that any two dots of the angle diagram of Π are connected by at most one line segment. Choose a maximal C in the diagram. If $C = \Pi$, then the diagram is of type (b'_n). Otherwise, by the Corollary to Lemma 3, the diagram is of type $(c'_{p,q,r})$.

LEMMA 7. *The length of an arbitrary vector in a simple Π system can only take two values.*

Proof. This follows from Lemma 2 and Lemma 6.

THEOREM 1. *A Dynkin diagram of a simple π system must be of one of the following types:*

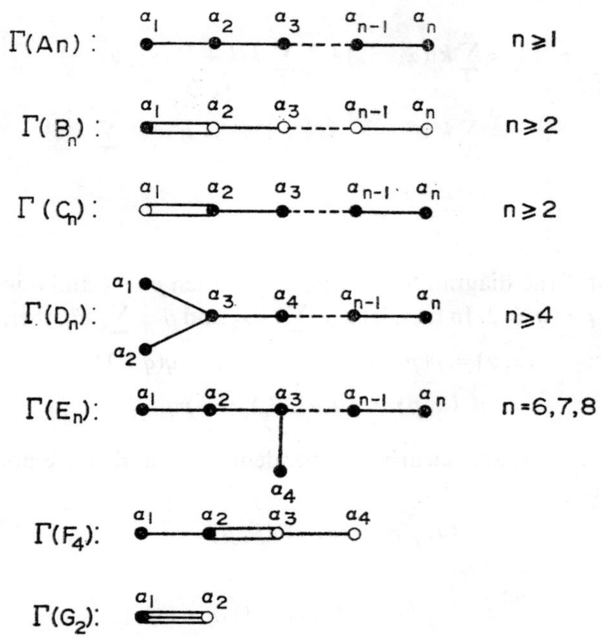

Proof. By Lemma 2, $\Gamma(G_2)$ is the only Dynkin diagram that has two points connected by three line segments. In what follows, we do not consider this diagram. Without loss of generality, let the shortest length of vectors in a simple π system Π be one. By Lemma 6, the Dynkin diagram of Π must be one of the types:

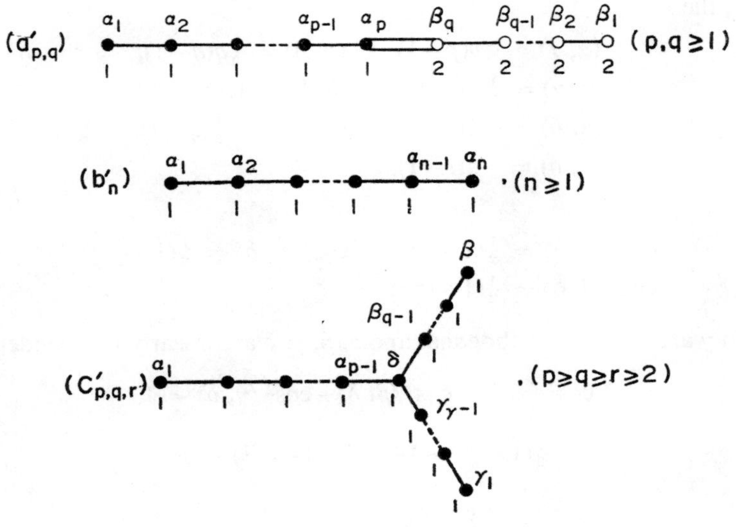

If every vector in a chain $C = \{\alpha_1, \ldots, \alpha_n\}$ has the same length, i.e. $\sqrt{\{(\alpha_i, \alpha_i)\}} = a$ $(i = 1, \ldots, n)$ let $\alpha = \sum_{k=1}^{n} k\alpha_k$, then

$$\begin{aligned}(\alpha, \alpha) &= \sum_1^n k^2(\alpha_k, \alpha_k) + 2\sum_1^{n-1} k(k+1)(\alpha_k, \alpha_{k+1})\\ &= \sum_1^n k^2a^2 - \sum_1^{n-1} k(k+1)a^2 = n^2a^2 - \sum_1^{n-1} ka^2\\ &= \frac{n(n+1)}{2}a^2.\end{aligned}$$

We show first that if the diagram is of type $(a'_{p,q})$, then $p = 1$ and q is arbitrary or $q = 1$ and p is arbitrary or $p = q = 2$. In fact, let $\alpha = \sum_1^p k\alpha_k$ and $\beta = \sum_1^q j\beta_j$, then

$$(\alpha, \alpha) = p(p+1)/2, \qquad (\beta, \beta) = q(q+1)$$

and

$$(\alpha, \beta) = pq(\alpha_p, \beta_q) = -pq.$$

Now $\alpha_1, \ldots, \alpha_p, \beta_1, \ldots, \beta_q$ are linearly independent, so α and β are not dependent. From Schwartz's inequality,

$$(\alpha, \beta)^2 < (\alpha, \alpha)^2 (\beta, \beta)^2,$$

hence

$$p^2q^2 < \tfrac{1}{2}pq(p+1)(q+1),$$

which is equivalent to

$$(p-1)(q-1) < 2.$$

Thus $p = 1$ and q is arbitrary or $q = 1$ and p is arbitrary or $p = q = 2$. The corresponding Dynkin diagrams are $\Gamma(B_{q+1})$, $\Gamma(C_{p+1})$ and $\Gamma(F_4)$.

If the diagram is of type (b'_n), then obviously, it is one of $\Gamma(A_n)$ $(n = 1, 2, \ldots)$.

Finally, suppose the diagram is of type $(C'_{p,q,r})$. Let $\alpha = \sum_1^{p-1}\alpha_i$, $\beta = \sum_1^{q-1} j\beta_j$, and $\gamma = \sum_1^{r-1} k\gamma_k$, then

$$\begin{aligned}&(\alpha, \alpha) = \tfrac{1}{2}p(p-1), && (\beta, \beta) = \tfrac{1}{2}q(q-1),\\ &(\gamma, \gamma) = \tfrac{1}{2}r(r-1), && (\delta, \delta) = 1.\\ &(\alpha, \delta) = -\tfrac{1}{2}(p-1), && (\beta, \delta) = -\tfrac{1}{2}(q-1),\\ &(\gamma, \delta) = -\tfrac{1}{2}(r-1).\end{aligned}$$

Thus

$$\begin{aligned}&\cos^2\langle\alpha, \delta\rangle = \tfrac{1}{2}(1-p^{-1}), \quad \cos^2\langle\beta, \delta\rangle = \tfrac{1}{2}(1-q^{-1}),\\ &\cos^2\langle\gamma, \delta\rangle = \tfrac{1}{2}(1-r^{-1}).\end{aligned}$$

Since α, β and γ are mutually orthogonal and α, β, γ, δ are linearly independent

$$\cos^2\langle\alpha, \delta\rangle + \cos^2\langle\beta, \delta\rangle + \cos^2\langle\gamma, \delta\rangle < 1,$$

i.e.

$$\tfrac{1}{2}(1-p^{-1}+1-q^{-1}+1-r^{-1}) < 1$$

or

$$p^{-1}+q^{-1}+r^{-1} < 1.$$

Now $p \geqslant q \geqslant r$, so $p^{-1} \leqslant q^{-1} \leqslant r^{-1}$. Thus $3r^{-1} > 1$. Since $r \geqslant 2$, r must be 2. Thus $p^{-1}+q^{-1} > \frac{1}{2}$, $2q^{-1} > \frac{1}{2}$ and $q < 4$; so $q = 2$ or $q = 3$. If $q = 2$, then $p^{-1} > 0$ and $p \geqslant 2$. Thus p is any integer $\geqslant 2$ and the Dynkin diagram is $\Gamma(D_{p+2})$. If $q = 3$, then $p^{-1} > \frac{1}{6}$ and $p \geqslant 3$, i.e. $3 \leqslant p \leqslant 5$; in this case, the Dynkin diagram is $\Gamma(E_{p+3})$.

THEOREM 2. *Each diagram in Theorem 1 is the Dynkin diagram of some simple π system.*

Proof. Let $e_1, \ldots, e_{n+1}$ form an orthonormal basis of the $(n+1)$-dimensional Euclidean space R^{n+1}.

If $\alpha_i = e_i - e_{i+1}$ $(1 \leqslant i \leqslant n)$,

then
$$(\alpha_i, \alpha_j) = \begin{cases} 2, & |i-j| = 0 \\ -1, & |i-j| = 1 \\ 0, & |i-j| \geqslant 2 \end{cases} \qquad i, j = 1, \ldots, n.$$

Thus $\Pi(A_n) = \{\alpha_1, \ldots, \alpha_n\}$ is a simple π system with Dynkin diagram $\Gamma(A_n)$.

Now consider the vectors α_i $(1 \leqslant i \leqslant n-1)$ and e_n; we have

$$(\alpha_i, e_n) = \begin{cases} -1, & i = n-1 \\ 0, & i < n-1 \end{cases}$$
$$(e_n, e_n) = 1.$$

Thus $\Pi(B_n) = \{\alpha_1, \ldots, \alpha_{n-1}, e_n\}$ is a simple π system with Dynkin diagram $\Gamma(B_n)$.

In R^n, the set $\Pi(C_n) = \{\alpha_1, \ldots, \alpha_{n-1}, 2e_n\}$ is a simple π system with Dynkin diagram $\Gamma(C_n)$.

The set $\Pi(D_n) = \{\alpha_1, \ldots, \alpha_{n-1}, e_{n-1}+e_n\}$ is also a simple π system. The Dynkin diagram of $\Pi(D_n)$ is $\Gamma(D_n)$. Notice that

$$(e_{n-1}+e_n, e_{n-1}+e_n) = 2,$$

$$(\alpha_i, e_{n-1}+e_n) = \begin{cases} 0, & i \neq n-2, \\ -1, & i = n-2. \end{cases}$$

We now consider the five exceptional cases. Let $e^{(n)} = \sum_1^n e_k$.

For $\Gamma(G_2)$, in R^3, let

$$\Pi(G_2) = \{e_1 - e_2, 3e_2 - e^{(3)}\}.$$

For $\Gamma(F_4)$, in R^4, let

$$\Pi(F_4) = \{e_1 - e_2, e_2 - e_3, e_3, \tfrac{1}{2}(e_4 - e_1 - e_2 - e_3)\}.$$

For $\Gamma(E_6)$, in R^7, let

$$\Pi(E_6) = \left\{e_1 - e_2, e_2 - e_3, e_3 - e_4, e_4 - e_5, e_5 - e_6, -e_1 - e_2 - e_3 + \frac{1}{2} e^{(6)} + \frac{\sqrt{2}}{2} e_7\right\}$$

For $\Gamma(E_7)$, in R^8, let

$$\Pi(E_7) = \{e_1 - e_2, e_2 - e_3, e_3 - e_4, e_4 - e_5, e_5 - e_6, e_6 - e_7, \tfrac{1}{2}e^{(8)} - e_1 - e_2 - e_3 - e_4\}.$$

For $\Gamma(E_8)$, in R^3, let

$$\Pi(E_8) = \{e_1-e_2,\ e_2-e_3,\ e_3-e_4,\ e_4-e_5,\ e_5-e_6,\ e_6-e_7,\ e_6+e_7,\ e_8-\tfrac{1}{2}e^{(8)}\}.$$

This completes the proof of Theorem 2.

Theorem 1 and Theorem 2 classify all simple π systems, i.e. all simple π systems that are not similar to each other are determined by these two theorems. $\Pi(A_n)$, $\Pi(B_n)$, $\Pi(C_n)$ and $\Pi(D_n)$ are fundamental systems of roots of the classical Lie algebras A_n, B_n, C_n and D_n, respectively. $\Pi(G_2)$, $\Pi(F_4)$ and $\Pi(E_n)$ $(n = 6, 7, 8)$ are called exceptional simple π systems. Using Theorem 1 and Theorem 2, we can now prove the following:

THEOREM 3. *Any simple σ system is similar to one of the root systems of A_n $(n \geq 1)$, B_n $(n \geq 2)$, C_n $(n \geq 3)$ and D_n $(n \geq 4)$ or the five exceptional simple σ systems $\Sigma(G_2)$, $\Sigma(F_4)$, $\Sigma(E_n)$ $(n = 6, 7, 8)$ that have fundamental systems of vectors $\Pi(G_2)$, $\Pi(F_4)$, $\Pi(E_n)$ $(n = 6, 7, 8)$, respectively.*

Proof. Let $\{e_1, \ldots, e_8\}$ be an orthonormal basis of R^8. It can be verified that

$$\Sigma(G_2) = \{e_i-e_j,\ \pm e^{(3)}-3e_i;\ i, j = 1, 2, 3\},$$
$$\Sigma(F_4) = \{\pm e,\ \pm e\pm e,\ \tfrac{1}{2}(\pm e_1\pm e_2\pm e_3\pm e_4);\ i, j = 1, 2, 3, 4\},$$
$$\Sigma(E_6) = \left\{e_i-e_j,\ \pm\sqrt{2}\,e_7,\ \pm\left(\frac{\sqrt{2}}{2}e_7+\frac{1}{2}e^{(6)}-e_i-e_j-e_k\right);\ i, j, k = 1, \ldots, 6\right\},$$
$$\Sigma(E_7) = \{e_i-e_j,\ (\tfrac{1}{2}e^{(8)}-e_i-e_j-e_k-e_m);\ i, j, k, m = 1, \ldots, 8\},$$
$$\Sigma(E_8) = \{\pm e_i\pm e_j, \pm(\tfrac{1}{2}e^{(8)}-e_i);\ \pm\tfrac{1}{2}e^{(8)}-e_i-e_j-e_k);\ i, j, k = 1, \ldots, 8\},$$

(where all i, j, k, m appearing in the same expression are distinct) are σ systems with fundamental systems of vectors $\Pi(G_2)$, $\Pi(F_4)$, $\Pi(E_6)$, $\Pi(E_7)$ and $\Pi(E_8)$, respectively. We will prove that there exist simple Lie algebras G_2, F_4, E_6, E_7 and E_8 that have fundamental systems of vectors $\Pi(G_2)$, $\Pi(F_4)$, $\Pi(E_6)$, $\Pi(E_7)$ and $\Pi(E_8)$ respectively.

7.3. The Lie algebra G_2

We know that

$$\mathfrak{h} = \{H_{\lambda_1\lambda_2\lambda_3} \mid \lambda_1, \lambda_2, \lambda_3 \in C\}$$

is a Cartan subalgebra of B_3. Let

$$\mathfrak{h}' = \left\{H_{\lambda_1\lambda_2\lambda_3} \mid \sum_1^3 \lambda_i = 0\right\},$$

then $\mathfrak{h}'$ is a subalgebra of $\mathfrak{h}$. Let

$$G_{\lambda_i} = \sqrt{2}\,E_{\lambda_i}+E_{-\lambda_j-\lambda_k},$$
$$G_{-\lambda_i} = \sqrt{2}\,E_{-\lambda_i}+E_{\lambda_j+\lambda_k},$$

where (i, j, k) are even permutations of 1, 2, 3 and

$$G_{\lambda_i-\lambda_k} = E_{\lambda_i-\lambda_k}, \quad i \neq k.$$

For any $H_{\lambda_1\lambda_2\lambda_3} \in \mathfrak{h}'$, we have

$$[H_{\lambda_1\lambda_2\lambda_3}, G_\alpha] = \alpha G_\alpha,$$

where $\alpha = \pm\lambda_i$ $(i = 1, 2, 3)$ or $\alpha = \lambda_i - \lambda_k$ $(i, k = 1, 2, 3, i \neq k)$.
If

$$H_1 = H_{1,0,0}, \quad H_2 = H_{0,1,0}, \quad H_3 = H_{0,0,1},$$

then

$$\begin{aligned}
[G_{\lambda_i}, G_{-\lambda_i}] &= 3H_i - (H_1 + H_2 + H_3),\\
[G_{\lambda_i - \lambda_k}, G_{\lambda_k - \lambda_i}] &= H_i - H_k,\\
\left.\begin{aligned}[G_{\lambda_i}, G_{\lambda_j}] &= -2G_{-\lambda_k}\\ [G_{-\lambda_i}, G_{-\lambda_j}] &= 2G_{\lambda_k}\end{aligned}\right\} &\ (i, j, k) \text{ is an even permutation},\\
[G_{\lambda_i}, G_{-\lambda_j}] &= 3G_{\lambda_i - \lambda_j} \quad (i \neq j),\\
[G_{\lambda_i - \lambda_j}, G_{\lambda_k}] &= \delta_{jk} G_{\lambda_i},\\
[G_{\lambda_i - \lambda_j}, G_{-\lambda_k}] &= -\delta_{ik} G_{-\lambda_j},\\
[G_{\lambda_i - \lambda_j}, G_{\lambda_k - \lambda_l}] &= \delta_{jk} G_{\lambda_i - \lambda_l} - \delta_{il} G_{\lambda_k - \lambda_j}\\
&(i \neq l \quad \text{and} \quad j \neq k).
\end{aligned}$$

Therefore, $\mathfrak{h}'$, $G_{\pm\lambda_i}$ $(i = 1, 2, 3)$ and $G_{\lambda_i - \lambda_k}$ $(i, k = 1, 2, 3, i \neq k)$ generate a 14-dimensional Lie algebra. Denote this Lie algebra by G_2. Obviously, $\mathfrak{h}'$ is a Cartan subalgebra of G_2.

We now show that G_2 is semisimple. Consider G_2 as a linear Lie algebra acting on the 7-dimensional space V_7 and let

$$v_0 = \begin{bmatrix}1\\0\\0\end{bmatrix}, \quad v_i = \begin{bmatrix}0\\e_i\\0\end{bmatrix}, \quad v_{i'} = \begin{bmatrix}0\\0\\e_i\end{bmatrix}, \quad i = 1, 2, 3,$$

where e_i is the 3-dimensional column vector with one as the ith component and zeros elsewhere. Then v_0, v_i and $v_{i'}$ $(i = 1, 2, 3)$ form a basis of V_7. We show first that V_7 is invariant under G_2. In fact, if V' is a non-zero invariant subspace, then V' is invariant under $\mathfrak{h}'$. Notice that

$$H_{\lambda_1\lambda_2\lambda_3} v_0 = 0, \quad H_{\lambda_1\lambda_2\lambda_3} v_i = \lambda_i v_i \quad \text{and} \quad H_{\lambda_1\lambda_2\lambda_3} v_{i'} = -\lambda_i v_{i'};$$

thus V' contains one of $v_0, v_1, v_2, v_3, v_{1'}, v_{2'}$ and $v_{3'}$. From

$$\begin{aligned}
G_{\lambda_i} v_0 &= -\sqrt{2} v_i, & G_{\lambda_i} v_j &= \varepsilon v_{k'}, & G_{\lambda_i - \lambda_j} v_{i'} &= -v_{j'},\\
G_{-\lambda_i} v_0 &= \sqrt{2} v_{i'}, & G_{-\lambda_i} v_{j'} &= -\varepsilon v_k, & G_{\lambda_i - \lambda_j} v_j &= v_i,\\
G_{\lambda_i} v_{i'} &= \sqrt{2} v_0,
\end{aligned}$$

where $\varepsilon = \operatorname{sgn}(i, j, k)$, it follows that $V' = V_7$. This proves that G_2 is irreducible. It is easy to see that the center of G_2 is $\{0\}$; by Corollary 3 to Theorem 2 of Chapter 2, G_2 is semisimple.

We have already seen that the roots of G_2 with respect to $\mathfrak{h}'$ are $\pm\lambda_i$ $(i = 1, 2, 3)$ and $\lambda_i - \lambda_k$ $(i, k = 1, 2, 3, i \neq k)$. We now show that $\lambda_1 - \lambda_2$ and λ_2 form a fundamental system of roots. Since $\lambda_1 + \lambda_2 + \lambda_3 = 0$, the roots of G_2 can be written as

$$\pm\lambda_1, \pm\lambda_2, \pm(\lambda_1+\lambda_2), \pm(\lambda_1-\lambda_2), \pm(2\lambda_1+\lambda_2), \pm(\lambda_1+2\lambda_2);$$

now $\lambda_1 = (\lambda_1-\lambda_2)+\lambda_2$, thus $\lambda_1-\lambda_2$ and λ_2 form a fundamental system of roots of G_2. Since $(\lambda_1-\lambda_2)+k\lambda_2$ $(0 \leqslant k \leqslant 3)$ are roots and $(\lambda_1-\lambda_2)+k\lambda_3$ $(k > 3)$ are not roots, thus

$$\frac{2(\lambda_1-\lambda_2, \lambda_2)}{(\lambda_2, \lambda_2)} = -3.$$

This proves that the Dynkin diagram of G_2 is

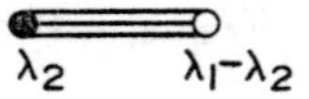

Thus G_2 is a simple Lie algebra with Dynkin diagram $\Gamma(G_2)$.

7.4. Classification of simple Lie algebras

THEOREM 4. *(W. Killing–E. Cartan). Any simple Lie algebra is isomorphic to one of the classical Lie algebras* A_n $(n \geqslant 1)$, B_n $(n \geqslant 1)$, C_n $(n \geqslant 1)$ *and* D_n $(n \geqslant 3)$ *or the five exceptional Lie algebras* G_2, F_4, E_6, E_7, *and* E_8. *Moreover,*

$$A_1 \approx B_1 \approx C_1, B_2 \approx C_2 \quad \text{and} \quad A_3 \approx D_3$$

and these are all the isomorphic relations among the classical Lie algebras and the five exceptional ones.

Proof. By Theorem 8 of Chapter 6, two simple Lie algebras are isomorphic iff they have similar fundamental systems of roots. By Corollary to Theorem 4 of Chapter 6, a semisimple Lie algebra is simple iff the fundamental system of roots is a simple π system. In Section 7.2 all simple π systems are determined (up to similarity), their Dynkin diagrams are $\Gamma(A_n)$ $(n \geqslant 1)$, $\Gamma(B_n)$ $(n \geqslant 2)$, $\Gamma(C_n)$ $(n \geqslant 3)$, $\Gamma(D_n)$ $(n \geqslant 4)$, $\Gamma(E_n)$ $(n = 6, 7, 8)$, $\Gamma(F_n)$ and $\Gamma(G_2)$. Therefore, it suffices to show that each of these diagrams is the Dynkin diagram of a simple Lie algebra.

By the results in Section 7.1, the classical Lie algebras A_n $(n \geqslant 1)$, B_n $(n \geqslant 1)$, C_n $(n \geqslant 1)$ and D_n $(n \geqslant 3)$ have Dynkin diagrams $\Gamma(A_n)$, $\Gamma(B_n)$, $\Gamma(C_n)$ and $\Gamma(D_n)$, respectively, and $A_1 \approx B_1 \approx C_1$, $B_2 \approx C_2$, $A_3 \approx D_3$. By the results in Section 7.3, the Lie algebra G_2 is simple and has Dynkin diagram $\Gamma(G_2)$.

It now remains to show the existence of simple Lie algebras F_4, E_6, E_7 and E_8 that have Dynkin diagrams $\Gamma(F_4)$, $\Gamma(E_6)$, $\Gamma(E_7)$ and $\Gamma(E_8)$, respectively. The existence of F_4 and E_8 will be shown in Chapter 12. Assume that E_8 exists, then the existence of E_6 and E_7 follow immediately from the following theorem, because $\Pi(E_6)$ and $\Pi(E_7)$ are subsystems of $\Pi(E_8)$.

THEOREM 5. *Let $\mathfrak{g}$ be a semisimple Lie algebra, $\mathfrak{h}$ be a Cartan subalgebra and Σ be the root system. If Σ' is a subset of Σ with the properties:*

(i) *If $\alpha, \beta \in \Sigma'$ and $\alpha+\beta \in \Sigma$, then $\alpha+\beta \in \Sigma'$.*
(ii) *If $\alpha \in \Sigma'$, then $-\alpha \in \Sigma'$.*

Then all H_α, E_α ($\alpha \in \Sigma'$) generate a semisimple subalgebra $\mathfrak{g}'$; all H_α ($\alpha \in \Sigma'$) span a subalgebra $\mathfrak{h}'$ of $\mathfrak{g}'$ and the root system of $\mathfrak{g}'$ with respect to $\mathfrak{h}'$ is similar to Σ'.

Proof. From (i) of Σ', it follows that $\mathfrak{g}'$ is a subalgebra. The structure formulas of $\mathfrak{g}'$ are

$$\begin{aligned} [H, H'] &= 0, && H, H' \in \mathfrak{h}, \\ [H, E_\alpha] &= \alpha(H)E_\alpha, && H \in \mathfrak{h}', \alpha \in \Sigma', \\ [E_\alpha, E_{-\alpha}] &= H_\alpha, && \alpha \in \Sigma', \\ [E_\alpha, E_\beta] &= N_{\alpha\beta}E_{\alpha+\beta}, && \alpha, \beta \in \Sigma'. \end{aligned}$$

To show that $\mathfrak{g}'$ is semisimple, we show that the Killing form $(X, Y)'$ of $\mathfrak{g}'$ is non-degenerate. Suppose $Z \in \mathfrak{g}'$ and $(Z, X)' = 0$ for all $X \in \mathfrak{g}'$. Choose a basis $H_{\gamma_1}, \ldots, H_{\gamma_{n'}}$ of $\mathfrak{h}'$ where $\{\gamma_1, \ldots, \gamma_{n'}\}$ is a maximal linearly independent subset of Σ', then $H_{\gamma_1}, \ldots, H_{\gamma_{n'}}$ and E_α ($\alpha \in \Sigma'$) form a basis of $\mathfrak{g}'$. Let

$$Z = \sum_1^{n'} a_i H_{\gamma_i} + \sum_{\alpha\in\Sigma'} b_\alpha E_\alpha.$$

For any $\beta \in \Sigma'$, we have

$$(H, E_{-\beta})' = \operatorname{Tr} \operatorname{ad}_{\mathfrak{g}'} H \operatorname{ad}_{\mathfrak{g}'} E_{-\beta} = 0, \quad \text{for all } H \in \mathfrak{h},$$

and

$$(E_\alpha, E_{-\beta})' = \operatorname{Tr} \operatorname{ad}_{\mathfrak{g}'} E_\alpha \operatorname{ad}_{\mathfrak{g}'} E_{-\beta} = 0, \quad \text{for all } \alpha \in \Sigma' - \{\beta\}.$$

Thus

$$(Z, E_{-\beta})' = b_\beta (E_\beta, E_{-\beta})'. \tag{7.4}$$

Now

$$\begin{aligned} \operatorname{ad} E_\beta \operatorname{ad} E_{-\beta} H &= \beta(H) H_\beta, \\ \operatorname{ad} E_\beta \operatorname{ad} E_{-\beta} E_\beta &= \beta(H) E_\beta, \\ \operatorname{ad} E_\beta \operatorname{ad} E_{-\beta} E_\gamma &= N_{-\beta,\gamma} N_{\beta,-\beta+\gamma} E_\gamma, \quad \text{if } \gamma \neq \beta. \end{aligned}$$

By Lemma 3 of Chapter 5,

$$N_{\beta,-\beta+\gamma} = N_{-\gamma,\beta} = -N_{\beta,-\gamma}$$

and

$$N_{-\beta,\gamma} N_{\beta,-\beta+\gamma} = -N_{-\beta,\gamma} N_{\beta,-\gamma} = \frac{q}{2}(p+1)(\beta, \beta),$$

where p, q are largest non-negative integers such that $-\gamma-k\beta$ ($-p \leqslant k \leqslant q$) are roots.

Therefore, when $\gamma \neq \beta$,

$$\operatorname{ad} E_\beta \operatorname{ad} E_{-\beta} E_\gamma = \tfrac{1}{2}q(p+1)(\beta, \beta)E_\gamma.$$

Thus

$$(E_\beta, E_{-\beta})' \geqslant 2(\beta, \beta) > 0,$$

and it follows from (7.4) that $b_\beta = 0$. So $Z = \sum_1^{n'} a_i H_{\gamma_i} \in \mathfrak{h}'$ and for any $Y = \Sigma a_i' H_{\gamma_i} \in \mathfrak{h}$,

$$(Z, Y)' = \sum_{i,j=1}^{n'} a_i a_j' (H_{\gamma_i}, H_{\gamma_j})' = 0.$$

Let $\mathfrak{h}_0'$ be the space of all real linear combinations of H_α ($\alpha \in \Sigma'$). If $H = \sum_1^{n'} b_i H_{\gamma_i}$ (b_i's are real), then

$$(H, H)' = \sum_{\varphi \in \Sigma'} \varphi(H)^2.$$

and

$$\varphi(H) = (H, H_\varphi) = \sum_1^{n'} b_i(\gamma_i, \varphi).$$

Since (γ_i, φ) $(i = 1, \ldots n')$ are rational $(H, H)' \geqslant 0$. If $(H, H) = 0$, then for all φ, $\varphi(H) = 0$; in particular,

$$\gamma_j(H) = \sum_1^{n'} b_i(\gamma_i, \gamma_j) = 0, \quad j = 1, \ldots, n'.$$

Now $\gamma_1, \ldots, \gamma_{n'}$ are linearly independent, so $\det((\gamma_i, \gamma_j))_{1 \leqslant i, j \leqslant n'} \neq 0$. Thus $b_i = 0$ $(1 \leqslant i \leqslant n')$ and $H = 0$. This proves that the restriction of the Killing form of $\mathfrak{g}'$ to $\mathfrak{h}_0'$ induces a positive definite matrix

$$((H_{\gamma_i}, H_{\gamma_j})')_{1 \leqslant i, j \leqslant n},$$

thus it is non-singular. Hence the restriction of the Killing form of $\mathfrak{g}'$ to $\mathfrak{h}_0'$ is non-degenerate. Thus from $(Z, Y)' = 0$ $(\forall, Y \in \mathfrak{h}')$ it follows that $Z = 0$. This proves that the Killing form of $\mathfrak{g}'$ is non-degenerate, thus $\mathfrak{g}'$ is semisimple. Now it is obvious that $\mathfrak{h}'$ is a Cartan subalgebra of $\mathfrak{g}'$ and the root system of $\mathfrak{g}'$ with respect to $\mathfrak{h}'$ is similar to Σ'.

CHAPTER 8

AUTOMORPHISMS OF SEMISIMPLE LIE ALGEBRAS†

8.1. The group of automorphisms and the derivation algebra of a Lie algebra

Let $\mathfrak{g}$ be a Lie algebra over the complex numbers. If a non-singular linear transformation A of $\mathfrak{g}$ satisfies

$$A[X, Y] = [AX, AY], \quad \text{for all } X, Y \in \mathfrak{g},$$

then A is said to be an automorphism of $\mathfrak{g}$. The set of all automorphisms of $\mathfrak{g}$ forms a group called the group of automorphisms of $\mathfrak{g}$; it is denoted by Aut $\mathfrak{g}$. Clearly, Aut $\mathfrak{g}$ is a subgroup of the complex-analytic group $GL(r, C) = GL(\mathfrak{g})$ of all non-singular linear transformations over $\mathfrak{g}$ ($r = \dim \mathfrak{g}$).

Since Aut $\mathfrak{g}$ is closed in $GL(r, C)$, it is a complex Lie group.

A linear transformation D of $\mathfrak{g}$ is said to be a *derivation* if it satisfies:

$$D[X, Y] = [DX, Y]+[X, DY] \quad \text{for all } X, Y \in \mathfrak{g}.$$

The set of all derivations on $\mathfrak{g}$ forms a subalgebra of the Lie algebra $\mathfrak{gl}(r, C)$; this Lie algebra is denoted by aut $\mathfrak{g}$.

THEOREM 1. aut $\mathfrak{g}$ *is the Lie algebra of* Aut $\mathfrak{g}$.

Proof. $\mathfrak{gl}(r, C)$ is the Lie algebra of $GL(r, C)$. Since Aut $\mathfrak{g}$ is a closed subgroup of $GL(r, C)$, the Lie algebra of it can be considered as a subalgebra $\mathfrak{a}$ of $\mathfrak{gl}(r, C)$.

Choose any $D \in \mathfrak{a}$, then for any complex number t, $\exp tD \in$ Aut $\mathfrak{g}$, i.e. $\exp t D$ is an automorphism of $\mathfrak{g}$. We have

$$\exp tD[X, Y] = [(\exp tD)X, (\exp tD)Y], \quad \text{for all } X, Y \in \mathfrak{g}.$$

Write $\exp tD$ as

$$\exp tD = I+tD+t^2D_t,$$

† Some knowledge of the theory of Lie groups is presupposed. For example, the reader should understand the material in Vol. 1 of *Theory of Lie Groups* (Princeton University Press, 1946) by C. Chevelley. The reader may omit this chapter and read the following chapters.

where D_t is a matrix and $D_t \to D^2/2!$ as $t \to 0$. Thus

$$(I+tD+t^2D_t)[X, Y] = [(I+tD+t^2D_t)X, (I+tD+t^2D_t)Y].$$

Expanding this equation, we get

$$\begin{aligned}[X, Y]+tD[X, Y]+t^2D_t[X, Y] \\ &= [X, Y]+[tDX, Y]+[X, tDY]+t^2[D_tX, Y+tDY+t^2D_tY] \\ &\quad +t^2[DX, DY+tD_tY]+t^2[X, D_tY].\end{aligned}$$

Now dividing both sides by t and letting $t \to 0$, we get

$$D[X, Y] = [DX, Y]+[X, DY].$$

Thus D is a derivation. This proves that $\mathfrak{a} \subset \operatorname{aut} \mathfrak{g}$.

Now let $D \in \operatorname{aut} \mathfrak{g}$, i.e.

$$D[X, Y] = [DX, Y]+[X, DY].$$

By induction,

$$D^p[X, Y] = \sum_{k=0}^{p} \frac{p!}{k!(p-k)!}[D^kX, D^{p-k}Y],$$

thus

$$\begin{aligned}\exp tD[X, Y] &= \sum_{p=0}^{\infty} \frac{1}{p!} \sum_{k=0}^{p} \frac{p!}{k!(p-k)!}[D^kX, D^{p-k}Y] \\ &= [\exp tDX, \exp tDY].\end{aligned}$$

and $\exp tD \in \operatorname{Aut} \mathfrak{g}$. Hence $D \in \mathfrak{a}$ and $\operatorname{aut} \mathfrak{g} \subseteq \mathfrak{a}$. The proof of Theorem 1 is now completed.

COROLLARY. *The group of automorphisms of a simply connected Lie group $\mathfrak{G}$ is isomorphic to a Lie group (the group of automorphisms of the Lie algebra), the Lie algebra of this Lie group is a subalgebra of the derivation algebra of the Lie algebra of $\mathfrak{G}$.*

If $\mathfrak{g}$ is a Lie algebra and $X \in \mathfrak{g}$, then $\operatorname{ad} X$ is an inner derivation of $\mathfrak{g}$. The set of all inner derivations form a Lie algebra which is called the algebra of inner derivations of $\mathfrak{g}$ and is denoted by $\operatorname{ad} \mathfrak{g}$. Naturally, $\operatorname{ad} \mathfrak{g}$ is a subalgebra of $\operatorname{aut} \mathfrak{g}$, thus $\operatorname{ad} \mathfrak{g}$ is the Lie algebra of an analytic subgroup of the analytic group formed by component of the identity in $\operatorname{Aut} \mathfrak{g}$. This analytic subgroup is denoted by $\operatorname{Ad} \mathfrak{g}$ and is said to be the group of inner automorphisms of $\mathfrak{g}$.

If $\mathfrak{g}$ is the Lie algebra of the analytic group $\mathfrak{G}$ and A is the adjoint representation:

$$\sigma \xrightarrow{A} d\alpha_\sigma(\alpha_\sigma : t \longrightarrow \sigma t \sigma^{-t}),$$

then A maps $\mathfrak{G}$ into $GL(\mathfrak{g})$. Since all $d\alpha_\sigma$ ($\sigma \in \mathfrak{G}$) are automorphisms of $\mathfrak{g}$, A maps $\mathfrak{G}$ into $\operatorname{Aut} \mathfrak{g}$. We know that

$$X \xrightarrow{dA} \operatorname{ad} X \quad \text{for all } X \in \mathfrak{g},$$

i.e. dA maps $\mathfrak{g}$ onto ad $\mathfrak{g}$. Therefore, ad $\mathfrak{g}$ is the Lie algebra of the analytic group $\{d\alpha_\sigma | \sigma \in \mathfrak{C}\}$. Since $\{d\alpha_\sigma | \sigma \in \mathfrak{C}\}$ and Ad $\mathfrak{g}$ are both analytic subgroups of $GL(\mathfrak{g})$ and they have the same Lie algebra, thus

$$\text{Ad}\,\mathfrak{g} = \{d\alpha_\sigma | \sigma \in \mathfrak{C}\}.$$

Now $\{d\alpha_\sigma | \sigma \in \mathfrak{C}\}$ is isomorphic to the group of inner automorphisms of $\mathfrak{C}$; this explains why Ad $\mathfrak{g}$ is said to be the group of inner automorphisms of $\mathfrak{g}$.

In general, Ad $\mathfrak{g}$ is not a closed subgroup of Aut $\mathfrak{g}$, but for semisimple Lie algebras, Ad $\mathfrak{g}$ is closed in Aut $\mathfrak{g}$. This follows from the following theorem:

THEOREM 2. *All derivations of a semisimple Lie algebra* $\mathfrak{g}$ *are inner, i.e.* aut $\mathfrak{g}$ = ad $\mathfrak{g}$.

Proof. Let D be a derivation of $\mathfrak{g}$. Define a Lie algebra $\mathfrak{g}_1$ on $\{(X, \lambda D) | X \in \mathfrak{g}, \lambda \in C\}$ by

$$(X, \lambda D) + (Y, \mu D) = (X+Y, (\lambda+\mu)D),$$
$$[(X, \lambda D), (Y, \mu D)] = ([X, Y] + \lambda D(Y) - \mu D(X), 0).$$

Clearly, $\mathfrak{g}_1$ is a linear space and commutation is linear with respect to both elements. We now show that the Jacoby identity holds. We have

$$\begin{aligned}
&[[(X, \lambda D), (Y, \mu D)], (Z, \nu D)] \\
&= [([X, Y] + \lambda D(Y) - \mu D(X), 0), (Z, \nu D)] \\
&= [[X, Y] + \lambda D(Y) - \mu D(X), Z] - \nu D([X, Y] + \lambda D(Y) - \mu D(X)) \\
&= [[X, Y], Z] + \lambda [D(Y), Z] - \mu [D(X), Z] - \nu D[X, Y] - \lambda \nu D^2(Y) + \nu \mu D^2(X).
\end{aligned}$$

Thus the Jacobi identity holds.

We claim that $\mathfrak{g}_1$ is not semisimple. Suppose $\mathfrak{g}_1$ is semisimple and let $\mathfrak{g}_0 = \{(X, 0) | X \in \mathfrak{g}\}$. Clearly, $\mathfrak{g}_0$ is isomorphic to $\mathfrak{g}$ and thus it is a semisimple ideal of $\mathfrak{g}_1$. Hence $\mathfrak{g}_1$ has the decomposition:

$$\mathfrak{g}_1 = \{(X, 0) | X \in \mathfrak{g}\} \dot{+} \mathfrak{a}.$$

It follows that $\mathfrak{a}$ is one-dimensional. This contradicts the semisimplicity of $\mathfrak{g}_1$.

Now let $\mathfrak{r}$ be a maximal ideal of $\mathfrak{g}_1$. Since $\mathfrak{g} \cap \mathfrak{r}$ is a solvable ideal of $\mathfrak{g}$, $\mathfrak{g} \cap \mathfrak{r}$ must be $\{0\}$. Thus

$$\mathfrak{g}_1 = \mathfrak{g} \dot{+} \mathfrak{r} \quad \text{(direct sum of ideals)}.$$

Naturally, $\mathfrak{r}$ is one-dimensional. Let

$$(0, D) = (A, 0) \dot{+} T, \qquad T \in \mathfrak{r}$$

then

$$[(0, D), (0, D)] = [(A, 0), (X, 0)],$$

i.e.

$$(DX, 0) = ([A, X], 0).$$

Thus

$$DX = \text{ad}\, AX.$$

This proves that

$$D = \text{ad}\, A.$$

COROLLARY 1. *The group of inner automorphisms* $Ad\,\mathfrak{g}$ *of a semisimple Lie algebra* $\mathfrak{g}$ *is the component of the identity of* $Aut\,\mathfrak{g}$, *thus it is a closed subgroup of* $Aut\,\mathfrak{g}$.

COROLLARY 2. *The group of automorphisms of a simply connected semisimple Lie group* $\mathfrak{G}$ *is a Lie group; the Lie algebra of the Lie group is isomorphic to the Lie algebra* $\mathfrak{g}$ *of* $\mathfrak{G}$. *Moreover, the component of the identity of the group of automorphisms is the group of inner automorphisms of* $\mathfrak{G}$.

Proof. The first part follows from the Corollary to Theorem 1. The second part follows from Corollary 1.

Let $\mathfrak{g}$ be a semisimple Lie algebra; $\text{Aut}\,\mathfrak{g}/\text{aut}\,\mathfrak{g}$ is said to be the group of outer automorphisms. We will study the structure of this group in the next section.

THEOREM 3. *The centers of* $Aut\,\mathfrak{g}$ *and* $Ad\,\mathfrak{g}$ *of a semisimple Lie algebra* $\mathfrak{g}$ *are* $\{e\}$.

Proof. Let A be in the center of Aut $\mathfrak{g}$ or Ad $\mathfrak{g}$. For any $X \in \mathfrak{g}$ and any complex number t $e^{t\,\text{ad}\,X}$ is an inner automorphism of $\mathfrak{g}$. Thus

$$Ae^{t\,\text{ad}\,X} = e^{t\,\text{ad}\,X}A.$$

Comparing the linear terms of both sides, we have

$$A\,\text{ad}\,X = \text{ad}\,XA,$$

i.e.

$$(A\,\text{ad}\,X)\,Y = (\text{ad}\,XA)Y \quad \text{for any } Y \in \mathfrak{g},$$

or

$$[AX, AY] = [X, AY].$$

Since A is an automorphism, $AX = X$ implies that A is the identity mapping.

8.2. The group of outer automorphisms of a semisimple Lie algebra

Let $\mathfrak{g}$ be a semisimple Lie algebra, Aut $\mathfrak{g}$ be the group of automorphisms and Ad $\mathfrak{g}$ be the group of inner automorphisms. Ad $\mathfrak{g}$ is the component of the identity of Aut $\mathfrak{g}$, thus it is a normal subgroup. In this section, we study the group $\text{Aut}\,\mathfrak{g}/\text{Ad}\,\mathfrak{g}$.

Let $\mathfrak{h}$ be a fixed Cartan subalgebra of $\mathfrak{g}$. If $\mathfrak{A}$ is the set of all automorphisms of $\mathfrak{g}$ that maps $\mathfrak{h}$ onto $\mathfrak{h}$, then $\mathfrak{A}$ is a subgroup of Aut $\mathfrak{g}$. For any $A \in \mathfrak{A}$, let

$$A \to A\,\text{ad}\,\mathfrak{g}. \tag{8.1}$$

This is a homomorphism from $\mathfrak{A}$ to $\text{Aut}\,\mathfrak{g}/\text{Ad}\,\mathfrak{g}$. We show that this homomorphism is onto. Let $B \in \text{Aut}\,\mathfrak{g}$, then $B^{-1}(\mathfrak{h})$ is also a Cartan subalgebra of $\mathfrak{g}$. By Theorem 6 of Chapter 3, there exists $U \in \text{Ad}\,\mathfrak{g}$ such that $U(\mathfrak{h}) = B^{-1}(\mathfrak{h})$. Thus $BU(\mathfrak{h}) = \mathfrak{h}$, i.e. $A = BU \in \mathfrak{A}$. Hence $A\,\text{ad}\,\mathfrak{g} = B\,\text{ad}\,\mathfrak{g}$ and the mapping (8.1) is onto. Let $\mathfrak{A}_0$ be the kernel of (8.1), then

$$\mathfrak{A}/\mathfrak{A}_0 \approx \text{Aut}\,\mathfrak{g}/\text{Ad}\,\mathfrak{g}.$$

Thus we can study $\mathfrak{A}/\mathfrak{A}_0$ instead of $\text{Aut}\,\mathfrak{g}/\text{Ad}\,\mathfrak{g}$. Let $A \in \mathfrak{A}$, then A induces a transformation on $\mathfrak{h}$. Moreover, we have the following:

LEMMA 1. *If A is an automorphism of $\mathfrak{g}$ and $A(\mathfrak{h}) = \mathfrak{h}$, then A induces an orthogonal transformation on $\mathfrak{h}_0$ and a permutation on the roots of $\mathfrak{g}$ (it is assumed that the roots of $\mathfrak{g}$ are embedded into $\mathfrak{h}$ by means of the Killing form).*

Proof. Let Σ be the root system of $\mathfrak{g}$ and $\alpha \in \Sigma$. If $E_\alpha \in \mathfrak{g}^\alpha$, then for any $H \in \mathfrak{h}$,

$$A[H, E_\alpha] = \alpha(H)AE_\alpha = [AH, AE_\alpha]. \tag{8.2}$$

Thus AE_α is also a root vector. On the other hand, the equation

$$\alpha^*(H) = (H, AE_\alpha), \qquad H \in \mathfrak{h}$$

defines a linear function on $\mathfrak{h}$ and $AH_\alpha = H_{\alpha^*}$. Hence

$$\alpha(H) = (H, H_\alpha) = (AH, AH_\alpha) = (AH, H_{\alpha^*}) = \alpha^*(AH);$$

from (8.2), it follows that $[AH, AE_\alpha] = \alpha^*(AH)AE_\alpha$. Since $A\mathfrak{h} = \mathfrak{h}$,

$$[H, AE_\alpha] = \alpha^*(H)AE_\alpha,$$

i.e. AE_α is the root vector corresponding to the root α^*. Thus $AH_\alpha = H_{\alpha^*} \in \mathfrak{h}_0$ and $A\mathfrak{h}_0 = \mathfrak{h}_0$. Now the restriction of the Killing form to $\mathfrak{h}_0$ is a Euclidean metric on $\mathfrak{h}_0$ and A preserves the Killing form, thus A induces an orthogonal transformation on $\mathfrak{h}_0$. From $AH_\alpha = H_{\alpha^*}$ or $A_\alpha = \alpha^*$, it follows that A induces a permutation on the roots of $\mathfrak{g}$. This completes the proof of Lemma 1.

By Lemma 1, any $A \in \mathfrak{A}$ induces an orthogonal transformation on $\mathfrak{h}_0$ and a permutation on the roots of $\mathfrak{g}$. Denote the orthogonal transformation induced by A by $f(A)$, then the mapping

$$A \to f(A) \tag{8.3}$$

is a homomorphism. Let $\mathfrak{T}$ be the image of the mapping (8.3). We show first that $\mathfrak{T}$ is the set of all orthogonal transformations on $\mathfrak{h}_0$ that induce permutations on the roots. In fact, if σ is such a transformation, then by Theorem 6 of Chapter 5, σ can be extended to an automorphism A_σ of $\mathfrak{g}$ such that $AH_\alpha = \sigma H_\alpha$, for all $\alpha \in \Sigma$. Obviously, $A_\sigma \in \mathfrak{A}$ and $f(A_\sigma) = \sigma$. If $\mathfrak{A}^0$ is the kernel of (8.3), then

$$\mathfrak{A}/\mathfrak{A}^0 \approx \mathfrak{T}. \tag{8.4}$$

If $\mathfrak{S}$ is the image of $\mathfrak{A}_0$ under (8.3), then

$$\mathfrak{A}_0/\mathfrak{A}_0 \cap \mathfrak{A}^0 \approx \mathfrak{S}. \tag{8.5}$$

LEMMA 2. *An automorphism A of $\mathfrak{g}$ leaves every element of $\mathfrak{h}$ fixed iff $A = e^{\operatorname{ad} \tilde{H}}$ for some $\tilde{H} \in \mathfrak{h}$.*

Proof. Sufficiency is obvious. To show necessity, let $A \in \operatorname{Aut} \mathfrak{g}$ and $AH = H\ (\forall H \in \mathfrak{h})$, then every root is fixed under A. Thus

$$AE_\alpha = \nu_\alpha E_\alpha, \qquad \nu_\alpha \neq 0.$$

Choose root vectors $E_{\pm\alpha}$ such that

$$[E_\alpha, E_{-\alpha}] = H_\alpha,$$

then

$$\nu_\alpha \nu_{-\alpha} = 1. \tag{8.6}$$

From $[E_\alpha, E_\beta] = N_{\alpha\beta} E_{\alpha+\beta}$ $(\beta \neq \pm\alpha)$, it follows that

$$\nu_\alpha \nu_\beta = \nu_{\alpha+\beta}, \quad \text{for} \quad \alpha, \beta, \alpha+\beta \in \Sigma. \tag{8.7}$$

Let $\{\alpha_1, \ldots, \alpha_n\}$ be a fundamental system of roots. For any $\alpha \in \Sigma$, let $\alpha = \sum_1^n m_i \alpha_i$ where $m_1, \ldots, m_n$ are all non-positive or all non-negative integers. By (8.6), (8.7) and induction, it can be shown that

$$\nu_\alpha = \prod_1^n \nu_{\alpha_i}^{m_i}.$$

If $\tilde{H} \in \mathfrak{h}$, then the automorphism $B = e^{\operatorname{ad} \tilde{H}} = \sum_{r=0}^\infty (1/r!)(\operatorname{ad} \tilde{H})^r$ fixes every element in $\mathfrak{h}$ and for any $\alpha \in \Sigma$,

$$\begin{aligned} BE_\alpha &= \sum_{r=0}^\infty \frac{1}{r!}[\tilde{H}, [\tilde{H} \ldots [\tilde{H}, E_\alpha]] \\ &= \sum_{r=0}^\infty \frac{1}{r!} \alpha(\tilde{H})^r E_\alpha = e^{\alpha(\tilde{H})} E_\alpha. \end{aligned}$$

Choose $\tilde{H}$ such that $\alpha_i(\tilde{H}) = \log \nu_{\alpha_i}$ $(1 \leq i \leq n)$; since $\nu_{\alpha_i} \neq 0$, so $\tilde{H}$ is uniquely determined by $\log \nu_{\alpha_i}$ $(1 \leq i \leq n)$. Thus

$$BE_{\alpha_i} = e^{\alpha_i(\tilde{H})} E_{\alpha_i} = e^{\log \nu_{\alpha_i}} E_{\alpha_i} = \nu_{\alpha_i} E_{\alpha_i}.$$

In general,

$$\begin{aligned} BE_\alpha &= e^{\alpha(\tilde{H})} E_\alpha = e^{\sum_1^n m_i \alpha_i(\tilde{H})} E_\alpha \\ &= \prod_1^n (e^{\alpha_i(\tilde{H})})^{m_i} E_\alpha = \prod_1^n \nu_{\alpha_i}^{m_i} E_\alpha = \nu_\alpha E_\alpha. \end{aligned}$$

This proves that $A = B = e^{\operatorname{ad} \tilde{H}}$.

From Lemma 2, it follows that $\mathfrak{A}^0 \subset \mathfrak{A}_0$. Thus (8.5) becomes

$$\mathfrak{A}_0/\mathfrak{A}^0 \approx \mathfrak{S}. \tag{8.8}$$

From (8.4) and (8.8),

$$\mathfrak{A}/\mathfrak{A}_0 \approx \mathfrak{T}'/\mathfrak{S}.$$

LEMMA 3. $\mathfrak{S} = W$, *i.e. the Weyl group of* $\mathfrak{g}$ *is the set of all linear transformations on* $\mathfrak{h}_0$ *induced by inner automorphisms of* $\mathfrak{g}$ *that maps* $\mathfrak{h}$ *onto* $\mathfrak{h}$.

Proof. We show first that $W \subset S$. It suffices to show that any S_α $(\alpha \in \Sigma)$ is induced (on $\mathfrak{h}_0$) by an automorphism of the form $U = e^{\operatorname{ad} X}$. Let

$$X = \frac{i\pi}{\sqrt{\{2(\alpha, \alpha)\}}}(E_\alpha + E_{-\alpha}),$$

then for any $H \in \mathfrak{h}$,

$$UH = H + \sum_{p=0}^\infty \frac{1}{(2p+1)!}(\operatorname{ad} X)^{2p+1} H + \sum_{p=0}^\infty \frac{1}{(2p+2)!}(\operatorname{ad} X)^{2p+2} H.$$

It is easy to verify that

$$(\mathrm{ad}\,X)^{2p+1}H = \frac{(i\pi)^{2p+1}}{\sqrt{\{2(\alpha,\alpha)\}}}\alpha(H)(-E_\alpha+E_{-\alpha}),$$

$$(\mathrm{ad}\,X)^{2p+2}H = \frac{(i\pi)^{2p+2}}{(\alpha,\alpha)}\alpha(H)\,H_\alpha.$$

Thus

$$\sum_{p=0}^{\infty}\frac{1}{(2p+1)!}(\mathrm{ad}\,X)^{2p+1}H = \sum_{p=0}^{\infty}\frac{1}{(2p+1)!}\frac{(i\pi)^{2p+1}}{\sqrt{\{2(\alpha,\alpha)\}}}\alpha(H)(-E_\alpha+E_{-\alpha})$$
$$= \left(\sum_{p=0}^{\infty}\frac{(-1)^p\pi^{2p+1}}{(2p+1)!}\right)\frac{i}{\sqrt{\{2(\alpha,\alpha)\}}}\alpha(H)(-E_\alpha+E_{-\alpha})$$
$$= \sin\pi\,\frac{i}{\sqrt{\{2(\alpha,\alpha)\}}}\alpha(H)(-E_\alpha+E_{-\alpha}) = 0,$$

$$\sum_{p=0}^{\infty}\frac{1}{(2p+2)!}(\mathrm{ad}\,X)^{2p+2}H = \sum_{p=0}^{\infty}\frac{1}{(2p+2)!}\frac{(i\pi)^{2p+2}}{(\alpha,\alpha)}\alpha(H)\,H_\alpha$$
$$= \frac{\cos\pi-1}{(\alpha,\alpha)}\alpha(H)\,H_\alpha.$$

Thus for any $H \in \mathfrak{h}$, $UH \in \mathfrak{h}$. By Lemma 1, U induces an orthogonal transformation on $\mathfrak{h}_0$ and permutation on the roots. According to the computations above, for all $H \in P_\alpha$ (P_α is defined by $\alpha(H) = 0$), $UH = H$. We also have

$$UH_\alpha = H_\alpha+\frac{\cos\pi-1}{(\alpha,\alpha)}\alpha(H_\alpha)\,H_\alpha = -H_\alpha.$$

Thus the transformation induced by U on $\mathfrak{h}_0$ is the reflection S_α with respect to P_α.

We now show that $\mathfrak{S} = W$. Let $\tau \in \mathfrak{S}$, then τ induces a permutation on the roots Σ; a fundamental system of roots $\{\alpha_1, \ldots, \alpha_n\}$ is transformed to another fundamental system of roots $\{\tau\alpha_1, \ldots, \tau\alpha_n\}$. Thus the Weyl chamber

$$C_0 = \{H \in \mathfrak{h}_0 \,|\, \alpha_i(H) = 0,\ i = 1, \ldots, n\}$$

is transformed to

$$C_1 = \{H \in \mathfrak{h}_0 \,|\, \tau\alpha_i(H) = 0,\ i = 1, \ldots, n\}.$$

Since the Weyl group W is transitive on the Weyl chambers, there exists $S \in W$ such that $S(C_0) = C_1$. Thus $S^{-1}\tau(C_0) = C_0$. Let $\tau_1 = S^{-1}\tau$, then $\tau_1 \in \mathfrak{S}$ and $\tau_1(C_0) = C_0$. It suffices to show that $\tau_1 \in W$.

Extend τ_1 to an automorphism A_{τ_1} of $\mathfrak{g}$ such that

$$A_{\tau_1}(H_\alpha) = \tau_1(H_\alpha), \quad \text{for all} \quad \alpha \in \Sigma.$$

then $A_{\tau_1}\,\mathfrak{h} = \mathfrak{h}$ and $A_{\tau_1} \in \mathfrak{A}_0$. If we can show the existence of $\tilde{H} \in \mathfrak{h}$ such that $A_{\tau_1} = e^{\mathrm{ad}\,\tilde{H}}$, then A_{τ_1} leaves elements of $\mathfrak{h}$ fixed, thus τ_1 is the identity mapping on $\mathfrak{h}_0$ and $\tau_1 \in W$.

τ_1 can be considered as an orthogonal transformation on $\mathfrak{h}_0^*$, thus it induces a permutation on the roots ($\tau_1\xi = \tau_1 H_\xi$, for all $\xi \in \mathfrak{h}_0^*$). Write τ_1 as product of different cycles $\tau_1 = \sigma_1 \ldots . \sigma_p$ Corresponding to this decomposition of τ_1, write

$$\mathfrak{g} = \mathfrak{h} + \sum_1^p \mathfrak{g}^{\sigma_i},$$

where $\mathfrak{g}^{\sigma_i}$ is the direct sum of all root spaces of the roots contained in σ_i. Then $\mathfrak{g}^{\alpha_i}$ is an invariant space of $A_{\tau_{i_1}}$. If σ is one of $\sigma_1, \ldots, \sigma_p$ and $\sigma = (\alpha, \sigma(\alpha), \ldots, \sigma^{q_1}(\alpha))$, then $E_\alpha, E_{\sigma(\alpha)}, \ldots, E_{\sigma^{q-1}(\alpha)}$ form a basis of $\mathfrak{g}^\alpha$. Let

$$A_{\tau_1} E_{\sigma^k(\alpha)} = \nu_{k+1} E_{\sigma^{k+1}(\alpha)}, \quad k = 0, 1, \ldots, q-1,$$

then the matrix of A_{τ_1} in $\mathfrak{g}^\sigma$ with respect to the basis $\{E_\alpha, E_{\sigma(\alpha)}, \ldots, E_{\sigma^{q-1}(\alpha)}\}$ is

$$\begin{bmatrix} 0 & 0 & 0 & \ldots & 0 & \nu_q \\ \nu_1 & 0 & 0 & \ldots & 0 & 0 \\ 0 & \nu_2 & 0 & \ldots & 0 & 0 \\ \vdots & \vdots & \vdots & & \vdots & \vdots \\ 0 & 0 & 0 & \ldots & \nu_{q-1} & 0 \end{bmatrix}$$

Thus the characteristic equation of A_{τ_1}, in $\mathfrak{g}^\sigma$ is

$$\lambda^q - \nu_1 \ldots \nu_q = 0$$

If A_{τ_1}, is replaced by $A_{\tau_1}B$ ($B = e^{\operatorname{ad} H}$, $H \in \mathfrak{h}$) in the discussion above, then $A_{\tau_1}B \in \mathfrak{A}_0$ and the permutations on the roots induced by $A_{\tau_1}B$ and A_{τ_1} are the same. Let

$$\lambda^q - \nu_q', \ldots, \nu_1' = 0$$

be the characteristic equation of $A_{\tau_1} B$ in $\mathfrak{g}^\sigma$. Now ad $HE_{\sigma^k(\alpha)} = \sigma^k(\alpha)(H)E_{\sigma^k(\alpha)}$, thus $e^{\operatorname{ad} H} E_{\sigma^k(\alpha)} = e^{\sigma^k(\alpha)(H)} E_{\sigma^k(\alpha)}$ and $\nu_k' = e^{\sigma^{k-1}(\alpha)(H)} \nu_k$. Hence the characteristic equation of $A_{\tau_1} B$ in $\mathfrak{g}^\sigma$ is

$$\lambda^q - e^{\sum_1^q \sigma^{k-1}(\alpha)(H)} \nu_1, \ldots, \nu_q = 0.$$

Let

$$C_0 = \{\xi \in \mathfrak{h}_0^* \mid (\alpha_i, \xi) = 0, i = 1, \ldots, n\}.$$

Since $\tau_1(C_0) = C_0$, we have $\tau_1(\Sigma_+) = \Sigma_+$ and $\tau_1(\Sigma_-) = \Sigma_-$. Thus $\alpha, \sigma(\alpha), \ldots, \sigma^{q-1}(\alpha)$ are all positive or all negative. Hence $\sum_{k=1}^q \sigma^{k-1}(\alpha) \neq 0$. So there exists $H \in \mathfrak{h}$ such that

$$e^{\sum_1^q \sigma^{k-1}(\alpha)(H)} \nu_1, \ldots, \nu_q \neq 1$$

holds for $\sigma = \sigma_1, \sigma_2, \ldots$ or σ_p. Therefore, one is not an eigenvalue of $A_{\tau_1}B$ in $\sum_1^p \mathfrak{g}^{\alpha_i}$ and

$$\mathfrak{g}'_{A_{\tau_1}B} \cap \left(\sum_1^p \mathfrak{g}^{\alpha_i}\right) = \{0\},$$

where

$$\mathfrak{g}'_{A_{\tau_1}B} = \{X \in \mathfrak{g} \mid (A_{\tau_1}B - I)^m X = 0 \quad \text{for some } m\},$$

thus $\mathfrak{g}'_{A_{\tau_1}B} \subset \mathfrak{h}$.

We show first the following:

LEMMA 4. *If* $A \in$ *Ad* $\mathfrak{g}$, *then dim* $\mathfrak{g}'_A \geqslant n =$ *the rank of* $\mathfrak{g} =$ *dim* $\mathfrak{h}$.

Proof. Suppose $A = e^{\mathrm{ad}\, X}$ for some $X \in \mathfrak{g}$, then $\mathfrak{g}'_A \supset \mathfrak{g}^0_X$. Since dim $\mathfrak{g}^0_X \geqslant n$, Lemma 4 is true. Thus Lemma 4 is true for all A in a sufficiently small neighborhood of the identity I in Ad $\mathfrak{g}$.

Choose any $A \in$ Ad $\mathfrak{g}$. Since Ad $\mathfrak{g}$ is a connected Lie group, A and I can be connected by a curve which is covered by a finite number of open neighborhoods $U_1, \ldots, U_m$. Moreover, $I \in U_1$, $A \in U_m$ and $U_i \cap U_{i+1} \neq \Phi$ $(i = 1, 2, \ldots, m-1)$. Now assume that Lemma 3 is true for all automorphisms $A \in U_1$. If $A \in U_2$, write

$$\det(\lambda I - A) = \sum_{j=0}^{r} (\lambda - 1)^j \varphi_j(A),$$

where $r = \dim \mathfrak{g}$ and $\varphi_0(A), \ldots, \varphi_r(A)$ are analytic functions of A. If $A \in U_1 \cap U_2$, then $\varphi_0(A) \equiv \varphi_1(A) \equiv \ldots \equiv \varphi_{n-1}(A) \equiv 0$. Thus for $A \in U_2$, we also have $\varphi_0(A) \equiv \ldots \equiv \varphi_{n-1}(A) \equiv 0$. Now by repeating the same argument for U_2 and $U_3, \ldots, U_{m-1}$ and U_m. Lemma 4 can be proved.

We now continue to prove Lemma 3. Applying Lemma 4 to $A_{\tau_1}B$ we get dim $\mathfrak{g}'_{A_{\tau_1}B} \geqslant$ dim $\mathfrak{h}$. We have shown that $\mathfrak{g}'_{A_{\tau_1}B} \subset \mathfrak{h}$, thus $\mathfrak{g}'_{A_{\tau_1}B} = \mathfrak{h}$. Hence $A_{\tau_1}B$ induces an orthogonal transformation on $\mathfrak{h}_0$ with eigenvalues one; since elementary divisors of orthogonal transformations are simple, $A_{\tau_1} B$ induces the identity transformation on $\mathfrak{h}_0$, thus it also induces the identity transformation on $\mathfrak{h}$. By Lemma 2, there exists $H' \in \mathfrak{h}$ such that $A_{\tau_1} B = e^{\mathrm{ad}\, H'}$. Hence $A_{\tau_1} = e^{\mathrm{ad}\,(H'-H)}$. The proof of Lemma 3 is now completed.

The discussions above can be summarized in the following theorem.

THEOREM 4. *Let* $\mathfrak{g}$ *be a semisimple Lie algebra, Aut* $\mathfrak{g}$ *be the group of automorphisms of* $\mathfrak{g}$ *and Ad* $\mathfrak{g}$ *be the group of inner automorphisms of* $\mathfrak{g}$. *If* $\mathfrak{h}$ *is a Cartan subalgebra of* $\mathfrak{g}$, $\mathfrak{T}$ *is the group of orthogonal transformations in* $\mathfrak{h}_0$ *that permute the roots of* $\mathfrak{g}$ *and* W *is the Weyl group of* $\mathfrak{g}$, *then* W *is a normal subgroup of* $\mathfrak{T}$ *and*

$$\text{Aut}\, \mathfrak{g}/\text{Ad}\, \mathfrak{g} \approx \mathfrak{T}/W.$$

COROLLARY. *The group of outer automorphisms of a semisimple Lie algebra* $\mathfrak{g}$ *is isomorphic to a finite group* G *which consists of all equivalent mappings of a fundamental system of roots* Π *onto itself.*

Proof. It suffices to show that $\mathfrak{T}/W \approx G$. Since Π is a basis of $\mathfrak{h}_0^*$, G is a subgroup of $\mathfrak{T}$. The mapping

$$\tau \to \tau W \tag{8.9}$$

is a homomorphism from G to $\mathfrak{T}/W$ with kernel $G \cap W$. Since W is transitive and regular on the Weyl chambers, $G \cap W = \{I\}$. We now show that (8.9) is onto. Let $\sigma \in \mathfrak{T}$ and $\sigma^{-1}(C_0) = C_1$, where C_0 is the Weyl chamber determined by Π. Now there exists $S \in W$ such that $S(C_0) = C_1$. Thus $\sigma S(C_0) = C_0$. Therefore, $\sigma_1 = \sigma S$ is an equivalent mapping of Π, i.e. $\sigma_1 \in G$. Obviously, $\sigma_1 W = \sigma S W = \sigma W$. This proves that (8.9) is onto, thus

$$G \approx \mathfrak{T}/W.$$

THEOREM 5. *Under the assumption of Theorem 4, if for every $\alpha_i \in \Pi$, vectors $E_{\pm\alpha_i}$ are chosen such that $(E_{\alpha_i}, E_{-\alpha_i}) = 1$, then every equivalent mapping σ of Π can be extended uniquely to an automorphism A_σ of $\mathfrak{g}$ satisfying*

$$A_\sigma(\pm E_{\alpha_i}) = E_{\pm\sigma(\alpha_i)}, \quad i = 1, \ldots, n. \tag{8.10}$$

The set of all such automorphisms form a group G' which is isomorphic to the group of equivalent mappings of Π onto itself and

$$\text{Aut } \mathfrak{g} = G' \text{ Ad } \mathfrak{g}, \quad G' \cap \text{Ad } \mathfrak{g} = \{I\}.$$

Proof. By Theorem 3 of Chapter 6, every equivalent mapping σ of Π onto itself can be uniquely extended to an automorphism A_σ of $\mathfrak{g}$ satisfying (8.10). If G' is the group of automorphisms obtained in this way, then it is obviously isomorphic to the group G of equivalent mappings of Π. By the Corollary to Theorem 4,

$$\text{Aut } \mathfrak{g}/\text{Ad } \mathfrak{g} \approx G'.$$

Since G' is a subgroup of Aut $\mathfrak{g}$ and the only inner automorphism in G' is the identity mapping,

$$\text{Aut } \mathfrak{g} = G' \text{ Ad } \mathfrak{g}, \quad G' \cap \text{Ad } \mathfrak{g} = \{I\}.$$

This proves Theorem 5.

By Theorem 5 or the Corollary to Theorem 4, the groups of outer automorphisms of simple Lie algebras can be determined.

COROLLARY. *Groups of outer automorphisms of simple Lie algebras are as listed below:*

Simple Lie algebra	A_1	A_n $(n > 1)$	B_n $(n \geqslant 1)$	C_n $(n \geqslant 1)$	D_n $(n \geqslant 3, n \neq 4)$
Group of outer automorphisms	$\{I\}$	Cyclic group of order 2	$\{I\}$	$\{I\}$	Cyclic group of order 2

D_4	E_6	E_7	E_8	F_4	G_2
Symmetric group on three letters	Cyclic group of order 2	$\{I\}$	$\{I\}$	$\{I\}$	$\{I\}$

Finally, we explicitly determine the groups of automorphisms of the classical Lie algebras.

We know that A_n is the Lie algebra of the group $SL(n+1, C)$ of all $(n+1)\times(n+1)$ complex matrices of determinant one. Inner automorphisms of $SL(n+1, C)$ have the form

$$Z \to PZP^{-1}, \quad \text{for all } Z \in SL(n+1, C),$$

where $P \in SL(n+1, C)$. The differential of this automorphism is

$$X \to PXP^{-1}, \quad \text{for all } X \in A_n.$$

Hence the group of inner automorphisms of A_n is the quotient group of $SL(n+1, C)$ with respect to the center. Denote this group by $PSL(n+1, C)$. It is easy to show that the center of $SL(n+1, C)$ is the cyclic subgroup of order $n+1$ generated by $e^{2\pi i/(n+1)}\, I_{n+1}$. When $n > 1$, the mapping

$$X \to -X'$$

is an outer automorphism of A_n.

B_n is the Lie algebra of the group $SO(2n+1, C)$ of $(2n+1)\times(2n+1)$ matrices of determinant one and satisfying

$$Z\begin{bmatrix} 1 & 0 & 0 \\ 0 & 0 & I_n \\ 0 & I_n & 0 \end{bmatrix} Z' = \begin{bmatrix} 1 & 0 & 0 \\ 0 & 0 & I_n \\ 0 & I_n & 0 \end{bmatrix},$$

It can be proved that the center of $SO(2n+1, C)$ is the identity. Thus the groups of inner automorphisms and outer automorphisms are both $SO(2n+1, C)$.

C_n is the Lie algebra of the group $Sp(2n, C)$ of $2n\times 2n$ matrices satisfying

$$Z\begin{bmatrix} 0 & I_n \\ -I_n & 0 \end{bmatrix} Z' = \begin{bmatrix} 0 & I_n \\ -I_n & 0 \end{bmatrix}.$$

It can be proved that the center of $Sp(2n, C)$ is $\{I_{2n}, I_{-2n}\}$. Denote the quotient group of $Sp(2n, C)$ with respect to $\{I_{2n}, I_{-2n}\}$ by $PSp(2n, C)$. Then the groups of inner automorphisms and outer automorphisms are both equal to $PSp\,(2n, C)$.

D_n is the Lie algebra of the group $SO(2n, C)$ of $2n\times 2n$ matrices of determinant one satisfying

$$Z\begin{bmatrix} 0 & I_n \\ I_n & 0 \end{bmatrix} Z' = \begin{bmatrix} 0 & I_n \\ I_n & 0 \end{bmatrix}. \tag{8.11}$$

It can be proved that the center of $SO(2n, C)$ is $\{I_{2n}, I_{-2n}\}$. Denote the quotient group of $SO(2n, C)$ with respect to $\{I_{2n}, I_{-2n}\}$ by $PSO(2n, C)$. Then $PSO(2n, C)$ is the group of inner automorphisms of D_n. The set of all $2n\times 2n$ matrices satisfying (8.11) forms a group $O(2n, C)$. If $P \in O(2n, C)$ and $\det P = -1$, then

$$X \to PXP^{-1} \quad (X \in D_n)$$

is an outer automorphism of D_n. When $n \geqslant 3$ and $n \neq 4$, the group of outer automorphisms of D_n is the quotient group $PO(2n, C)$ of $O(2n, C)$ with respect to $\{I_{2n}, I_{-2n}\}$.

With the discussion above, we can obtain some isomorphic relations among classical groups of small dimensions. For example, from $A_1 \approx B_1 \approx C_1$, it follows that $PSL(2, C) \approx SO(3, C) \approx PSp(2, C)$. From $B_2 \approx C_2$, it follows that $SO(5, C) \approx PSp\,(4, C)$ From $A_3 \approx D_3$ it follows that $PSL(4, C) \approx PSO(6, C)$.

CHAPTER 9

REPRESENTATIONS OF LIE ALGEBRAS

9.1. Fundamental concepts

Let $\mathfrak{g}$ be a Lie algebra and V be a finite-dimensional vector space over C. A homomorphism ϱ from $\mathfrak{g}$ into $\mathfrak{gl}(V)$

$$X \to \varrho(X) \tag{9.1}$$

is said to be a *linear representation* or simply a representation of $\mathfrak{g}$. V is said to be the representation space and dim V is the dimension of the representation. If $\{e_1, \ldots, e_N\}$ is a basis of V and

$$\varrho(X)e_j = \sum_{i=1}^{N} a_{ij}(X)e_i, \quad j = 1, \ldots, N,$$

then

$$X \to (a_{ij}(X))_{1 \leq i,\, j \leq N} \tag{9.2}$$

is a homomorphism from $\mathfrak{g}$ to $\mathfrak{gl}(N, C)$; this homomorphism is said to be a *matrix representation* of $\mathfrak{g}$. If $\{e_1', \ldots, e_N'\}$ is another basis of V, then

$$X \to (b_{ij}(X))_{1 \leq i,\, j \leq N} \tag{9.3}$$

is also a matrix representation. Let (p_{ij}) be the matrix relating the two bases, i.e.

$$e_j' = \sum_{i=1}^{N} p_{ij}e_i, \quad j = 1, \ldots, N,$$

then

$$(b_{ij}(X)) = (p_{ij})^{-1}(a_{ij}(X))(p_{ij}), \quad \forall X \in \mathfrak{g}. \tag{9.4}$$

In general, two matrix representations (9.2) and (9.3) are said to be *equivalent* if for some non-singular $N \times N$ matrix (p_{ij}), (9.4) is satisfied.

Let ϱ_1 and ϱ_2 be two linear representations of $\mathfrak{g}$ with representation spaces V_1 and V_2, respectively. If there exists a one–one linear mapping P from V_1 onto V_2 such that

$$P\varrho_1(X) = \varrho_2(X)P, \quad \forall X \in \mathfrak{g}$$

then the representations ϱ_1 and ϱ_2 are said to be equivalent. Representation spaces of equivalent representations have the same dimensions. If $\{e_{11}, \ldots, e_{1N}\}$ and $\{e_{21}, \ldots, e_{2N}\}$ are bases of the spaces V_1 and V_2, $\tilde{\varrho}_i(X)$ is the matrix of $\varrho_i(X)$ with respect to the basis $\{e_{i1}, \ldots, e_{iN}\}$ $(i = 1, 2)$ and (p_{ij}) is the matrix of P with respect to these two bases, i.e.

$$Pe_{1j} = \sum_{i=1}^{N} p_{ij}e_{2i}, \quad j = 1, \ldots, N,$$

then

$$\tilde{\varrho}_1(X) = (p_{ij})^{-1}\, \varrho_2(X)\, (p_{ij}).$$

In particular, if $e_{21} = Pe_{11}, \ldots, e_{2N} = Pe_{1N}$, then

$$\tilde{\varrho}_1(X) = \tilde{\varrho}_2(X).$$

In what follows, representations are understood to be linear representations.

A basic problem of the representation theory of Lie algebras is to find all the representations (up to equivalence) of a Lie algebra.

A representation ϱ of $\mathfrak{g}$ is said to be *irreducible* if the representation space V is irreducible under $\varrho(\mathfrak{g})$. Otherwise, ϱ is said to be reducible.

If a representation ϱ of $\mathfrak{g}$ is reducible, then the representation space contains a proper invariant subspace V_1. For any $v+V_1 \in V/V_1$ and $X \in \mathfrak{g}$, define

$$\tilde{\varrho}(X)(v+V_1) = \varrho(X)\, v+V_1.$$

It can be proved that this definition is independent of the choice of v in $v+V_1$. Moreover, the mapping

$$X \to \tilde{\varrho}(X)$$

is a representation of $\mathfrak{g}$ with representation space V/V_1.

If the representation space V of ϱ is the direct sum of the invariant spaces $V_1, \ldots, V_s$, then ϱ induces a representation ϱ_i in V_i by the definition

$$\varrho_i(X)\, v_i = \varrho(X)\, v_i, \quad X \in \mathfrak{g}, \quad v_i \in V_i.$$

ϱ is then said to be the direct sum of $\varrho_1, \ldots, \varrho_s$ and this is denoted by $\varrho = \varrho_1 + \ldots + \varrho_s$. If ϱ is the direct sum of some irreducible representations, then it is said to be completely reducible.

Let $\mathfrak{g}$ be a Lie algebra, then the mapping

$$X \to \operatorname{ad} X, \quad X \in \mathfrak{g}$$

is a representation of $\mathfrak{g}$ with representation space $\mathfrak{g}$. This representation is said to be the *regular representation* of $\mathfrak{g}$. Invariant subspaces of the regular representation are ideals of $\mathfrak{g}$. If $\mathfrak{g}$ is semisimple, then the regular representation is completely reducible.

THEOREM 1. *If a representation ϱ of a Lie algebra $\mathfrak{g}$ is completely reducible and $\varrho = \varrho_1 + \ldots + \varrho_r = \varrho_1' + \ldots + \varrho_s'$ are two decompositions of ϱ into irreducible representations, then $r = s$ and $\varrho_1', \ldots, \varrho_s'$ can be rearranged so that ϱ_i and ϱ_i' are equivalent* $(i = 1, \ldots, r)$.

Proof. Corresponding to the decomposition $\varrho = \varrho_1 + \ldots + \varrho_r$, the representation space V can be decomposed into $V = V_1 \dot{+} \ldots \dot{+} V_r$, where the V_i's are irreducible. Corresponding to $\varrho = \varrho_1' + \ldots + \varrho_s'$, there is a similar decomposition $V = V_1' \dot{+} \ldots \dot{+} V_s'$.

Consider $V = V_1 + V_1' + \ldots + V_s'$. For any $k\,(1 \leqslant k \leqslant s)$, since V_k' is irreducible, $(V_1 + V_1' + \ldots + V_{k-1}') \cap V_k' = \{0\}$ or $(V_1 + V_1' + \ldots + V_{k-1}') \cap V_k' = V_k'$. In the first case, $V_1 + V_1' + \ldots + V_{k-1}' + V_k' = (V_1 + V_1' + \ldots + V_{k-1}') \dot{+} V_k'$; in the second case, $V_1 + V_1' + \ldots + V_{k-1}' + V_k' = V_1 + V_1' + \ldots + V_{k-1}'$. Therefore, by omitting some of $V_1', \ldots, V_s'$ and rearranging $V_1', \ldots, V_s'$, we can assume that

$$V = V_1 \dot{+} V_i' \dot{+} \ldots \dot{+} V_s'.$$

Thus

$$V_1 \approx V/V_i' \dot{+} \ldots \dot{+} V_s \approx V_1' \dot{+} \ldots \dot{+} V_{i-1}'.$$

Since V_1 is irreducible, $i = 2$. Hence

$$V = V_1 \dot{+} V_2' \dot{+} V_3' \dot{+} \ldots \dot{+} V_s'$$

and ϱ_1 is equivalent to ϱ_1'.

Now consider $V = V_1 + V_2 + V_2' + \ldots V_s'$. By the same argument, we can omit V_2' of $V_2', \ldots, V_s'$ so that

$$V = V_1 \dot{+} V_2 \dot{+} V_3' \dot{+} \ldots \dot{+} V_s'$$

and ϱ_2 is equivalent to ϱ_2'.

By repeating the same argument, it can be proved that $r = s$ and ϱ_i is equivalent to ϱ_i' $(i = 1, \ldots, r)$. The proof of Theorem 1 is now completed.

By Theorem 1, if any representation of a Lie algebra is completely reducible, then the problem of finding all inequivalent representations is reduced to finding all inequivalent irreducible representations.

We now define some operations on representations:

(1) Let ϱ_1 and ϱ_2 be representations of $\mathfrak{g}$ with representation spaces V_1 and V_2 respectively. Let $V = V_1 \dot{+} V_2$. For any $v_1 \in V_1$, $v_2 \in V_2$ and $X \in \mathfrak{g}$, define

$$\varrho(X)(v_1 + v_2) = \varrho_1(X)\,v_1 + \varrho_2(X)\,v_2.$$

Then

$$X \to \varrho(X)$$

is a representation with representation space $V = V_1 \dot{+} V_2$. Clearly, $\varrho = \varrho_1 + \varrho_2$. We say that ϱ is the *sum* of ϱ_1 and ϱ_2.

(2) Let ϱ_1 and ϱ_2 be representations of $\mathfrak{g}$ with representation spaces V_1 and V_2. Let $V = V_1 \otimes V_2$ (Kronecker product). For any $v_1 \in V_1$, $v_2 \in V_2$, and $X \in \mathfrak{g}$, define

$$\varrho(X)(v_1 \otimes v_2) = \varrho_1(X)\,v_1 \otimes v_2 + v_1 \otimes \varrho(X)\,v_2.$$

Then

$$X \to \varrho(X)$$

is also a representation of $\mathfrak{g}$; this representation is said to be the Kronecker product of ϱ_1 and ϱ_2.

(3) Let ϱ be a representation of $\mathfrak{g}$ with representation space V. Let V^* denote the dual space of V. For any $X \in \mathfrak{g}$, define $\varrho^*(X)$ by

$$\big(v, \varrho(X)\, v^*\big) = -\big(\varrho(X)\, v, v^*\big), \quad v \in V, \quad v^* \in V^*.$$

Then ϱ^* is a representation of $\mathfrak{g}$; this representation is said to be the *star representation* or the *contragredient representation* of ϱ.

9.2. Schur's lemma

LEMMA 1. *If ϱ is an irreducible representation of a Lie algebra $\mathfrak{g}$, then a linear transformation A that commutes with every $\varrho(X)$ $(X \in \mathfrak{g})$ is a scalar multiple of the identity I. In other words, if*

$$\varrho(X)\, A = A\varrho(X), \quad \text{for all } X \in \mathfrak{g}, \tag{9.5}$$

then

$$A = \lambda I, \quad \text{for some } \lambda \in C.$$

Proof. Let V be the representation space of ϱ and λ be an eigenvalue of A. If $V_1 = \{v \in V \mid Av = \lambda v\}$, then $V_1 \neq \{0\}$. If $v \in V_1$, then from (9.5),

$$A\big(\varrho(X)\, v\big) = \varrho(X)\, Av = \varrho(X)\, \lambda v = \lambda\varrho(X)\, v,$$

i.e. $p(X)\, v \in V_1$. Thus V_1 is invariant under $\varrho(\mathfrak{g})$. Since ϱ is irreducible, $V_1 = V$. Hence $A = \lambda I$.

LEMMA 2. *Let ϱ_1 and ϱ_2 be two irreducible representations of a Lie algebra $\mathfrak{g}$ with representation spaces V_1 and V_2. If A is a linear transformation from V_2 to V_1 such that*

$$\varrho_1(X)\, A = A\varrho_2(X) \tag{9.6}$$

then $A = 0$ or A is invertible. If A is invertible, then ϱ_1 and ϱ_2 are equivalent. Hence if ϱ_1 and ϱ_2 are not equivalent, then $A = 0$.

Proof. Let V_2' be the kernel of A, then V_2' is a subspace of V_2. For any $v_2 \in V_2'$,

$$A\big(\varrho_2(X)\, v_2\big) = \varrho_1(X)\, Av_2 = 0,$$

thus $\varrho_2(X)v_2 \in V_2'$. Therefore, V_2' is invariant under $\varrho_2(\mathfrak{g})$. Now ϱ_2 is irreducible, so either $V_2' = V_2$ or $V_2' = \{0\}$. If $V_2' = V_2$, then $A = 0$. In what follows, we assume $V_2' = \{0\}$, then A is one–one.

Consider the image $AV_2 = V_1'$ of A, V_1' is a subspace of V_1. Let $v_1 = Av_2 \in V_1'$, then by (9.6)

$$\varrho_1(X)\, v_1 = \varrho_1(X)\, Av_2 = A\big(\varrho_2(X)\, v_2\big) \in AV_2 = V_1'.$$

Thus V_1' is invariant under $\varrho_1(\mathfrak{g})$. Since ϱ_1 is irreducible, $V_1' = \{0\}$ or $V_1' = V_1$. If $V_1' = \{0\}$, then $A = 0$. If $V_1' = V_1$, then A is one–one and onto, thus ϱ_1 and ϱ_2 are equivalent.

9.3. Representations of the three-dimensional simple Lie algebra

The set of all skew-symmetric 3×3 complex matrices form a three-dimensional simple Lie algebra $\mathfrak{g}_3$. Let

$$M_1 = \begin{bmatrix} 0 & 0 & 0 \\ 0 & 0 & -1 \\ 0 & 1 & 0 \end{bmatrix}, \quad M_2 = \begin{bmatrix} 0 & 0 & 1 \\ 0 & 0 & 0 \\ -1 & 0 & 0 \end{bmatrix},$$

$$M_3 = \begin{bmatrix} 0 & -1 & 0 \\ 1 & 0 & 0 \\ 0 & 0 & 0 \end{bmatrix},$$

then M_1, M_2 and M_3 form a basis of $\mathfrak{g}_3$. The structure formulas are

$$[M_1, M_2] = M_3, \quad [M_2, M_3] = M_1, \quad [M_3, M_1] = M_2.$$

Let

$$H = iM_3, \quad E_1 = i(M_1+iM_2), \quad E_{-1} = i(M_1-iM_2),$$

then

$$[H, E_1] = E_1, \quad [H, E_{-1}] = -E_{-1}, \quad [E_1, E_{-1}] = 2H.$$

Thus $\mathfrak{h} = \{\lambda H | \lambda \in C\}$ is a Cartan subalgebra of $\mathfrak{g}_3$ and the two roots of $\mathfrak{h}$ are λ and $-\lambda$ with root vectors E_1 and E_{-1}, i.e.

$$[\lambda H, E_1] = \lambda E_1, \quad [\lambda H, E_{-1}] = -\lambda E_1.$$

Let ϱ be a representation of $\mathfrak{g}_3$ with representation space V. We call the eigenvalues of $\varrho(H)$ the *weights* of ϱ and non-zero eigenvectors of $p(H)$ the *weight vectors* of ϱ.

LEMMA 3. *Let v be a weight vector corresponding to the weight m, i.e. $\varrho(H)\,v = mv$. If $\varrho(E_1)v \neq 0$, then $\varrho(E_1)\,v$ is a weight vector corresponding to the weight $m+1$; if $\varrho(E_{-1})v \neq 0$, then $\varrho(E_{-1})v$ is a weight vector corresponding to the weight $m-1$.*

Proof. We have

$$\begin{aligned} \varrho(H)\,\varrho(E_1)\,v &= \varrho([H, E_1])\,v + \varrho(E_1)\,\varrho(H)\,v \\ &= \varrho(E_1)\,v + \varrho(E_1)\,mv = (m+1)\,\varrho(E_1)\,v. \end{aligned}$$

Similarly,

$$\varrho(H)\,\varrho(E_{-1})v = (m-1)\,\varrho(E_{-1})\,v.$$

Lemma 3 follows immediately from these two equations.

Since V is finite-dimensional, by Lemma 3, there exists a weight vector v such that $\varrho(E_1)v = 0$. Let j be the weight of v and denote v by v_j, i.e.

$$\varrho(H)\,v_j = jv_j.$$

Let

$$v_{j-1} = \varrho(E_{-1})\,v_j, \quad v_{j-2} = \varrho(E_{-1})\,v_{j-1}, \;\ldots$$

We have

$$\begin{aligned} \varrho(E_1)\, v_j &= 0, \\ \varrho(E_1)\, v_{j-1} &= \varrho(E_1)\, \varrho(E_{-1})\, v_j = \varrho([E_1, E_{-1}])\, v_j + \varrho(E_{-1})\, \varrho(E_1) v_j \\ &= \varrho(2H)\, v_j = 2j v_j. \end{aligned}$$

In general,

$$\varrho(E_1)\, v_m = (j-m)(j+m+1)\, v_{m+1}.$$

In fact, if this equation holds for m, then for $m-1$, we have

$$\begin{aligned} \varrho(E_1)\, v_{m-1} &= \varrho(E_1)\, \varrho(E_{-1})\, v_m \\ &= \varrho([E_1, E_{-1}])\, v_m + \varrho(E_{-1})\, \varrho(E_1)\, v_m \\ &= \varrho(2H)\, v_m + \varrho(E_{-1})(j-m)(j+m+1)\, v_{m+1} \\ &= 2m v_m + (j-m)(j+m+1)\, v_m \\ &= (j-m+1)(j+m)\, v_m. \end{aligned}$$

If j' is the first integer such that

$$v_{j'} \neq 0 \quad \text{and} \quad \varrho(E_{-1})\, v_{j'} = v_{j'-1} = 0,$$

then from $\varrho(E_1)\, v_{j'-1} = 0$, we have

$$(j-j'+1)(j+j') = 0.$$

Since $j-j' \geqslant 0$, we must have $j+j' = 0$ or $j = -j'$. Thus j is a non-negative integer or half integer. Moreover, it can be proved that

$$v_j, \ldots, v_{-j}$$

generate an invariant subspace of ϱ. For if V_j is the subspace generated by these vectors, then

$$\left.\begin{aligned} \varrho(H) v_m &= m v_m, \\ \varrho(E_{-1}) v_m &= v_{m-1}, \\ \varrho(E_1) v_m &= (j-m)(j+m+1) v_{m+1} \end{aligned} \qquad (m = j, j-1, \ldots, -j) \right\} \tag{9.7}$$

where $v_{j+1} = v_{-j-1} = 0$. Thus V_j is invariant under $\varrho(\mathfrak{g}_3)$. If $V' \neq \{0\}$ is an invariant subspace contained in V_j, then $\varrho(H)$ has an eigenvalue k in V' and k is one of $j, j-1, \ldots, -j$. Since all eigenvalues of $\varrho(H)$ in V_j are of multiplicity one, $v_k \in V'$. By (9.7), $v_j, v_{j-1}, \ldots, v_j$ all belong to V'. Thus $V' = V_j$. This proves that V_j is irreducible. In particular, if ϱ is irreducible, then $V_j = V$; in this case, j is said to be the highest weight of ϱ. We have proved the following:

THEOREM 2. *If ϱ is an irreducible representation of $\mathfrak{g}_3$ with representation space V, then dim $V = 2j+1$, where j is a non-negative integer or half integer. Moreover, there exists a basis $\{v_j, \ldots, v_{-j}\}$ of V such that*

$$\left.\begin{aligned} \varrho(H) v_m &= m v_m, \\ \varrho(E_{-1}) v_m &= v_{m-1}, \\ \varrho(E_1) v_m &= (j-m)(j+m+1) v_{m-1} \end{aligned} \qquad (m = j, j-1, \ldots, -j) \right\}$$

where $v_{j+1} = v_{-j-1} = 0$. Thus two irreducible representations of $\mathfrak{g}_3$ are equivalent iff their highest weights are equal. Conversely, we have the following:

THEOREM 3. *For any non-negative integer or half integer, the equations* (9.7) *define an irreducible representation* ϱ *of* $\mathfrak{g}_3$ *with highest weight* j.

Proof. It can be verified that

$$\begin{aligned}
[\varrho(H), \varrho(E_1)] &= \varrho(E_1), \\
[\varrho(H), \varrho(E_{-1})] &= -\varrho(E_{-1}), \\
[\varrho(E_1), \varrho(E_{-1})] &= 2\varrho(H).
\end{aligned}$$

Hence ϱ is a representation of $\mathfrak{g}_3$. Irreducibility of ϱ follows from the proof of irreducibility of V_j in Theorem 2.

Theorem 2 and Theorem 3 solve the problem of finding all irreducible representations of $\mathfrak{g}_3$.

The irreducible representation of $\mathfrak{g}_3$ with highest weight j will be denoted by ϱ_j, the representation space of ϱ_j will be denoted by V_j. The dimension of ϱ_j is $2j+1$.

Theorem 2 has the following Corollary:

COROLLARY. *Let* v *be a vector of the weight* r *in* V_j. *If* p, q *are the largest non-negative integers such that* $\varrho_j(E_{-1})^p v \neq 0$ and $\varrho_j(E_1)^q v \neq 0$, *then* $2r = -(q-p)$ *and* $p+q = 2j$. *Moreover*, $\varrho_j(E_{-1})^i \varrho_j(E_1)^q v \neq 0$ $(0 \leq i \leq p+q)$ and $\varrho_j(E_{-1})^i \varrho_j(E_1)^q v$ *is a vector of weight* $r+q-i$.

Proof. Since the eigenvalues of $\varrho_j(H)$ are of multiplicity one, v and v_r are linearly dependent, thus we may let $v = v_r$. By (9.7), if $\varrho_j(E_1)^i v_r \neq 0$ then $\varrho_j(E_1)^i v_r$ is of weight $r+i$, thus v_{r+i} and $\varrho_j(E_1)^i v_r$ are dependent. Hence from $\varrho_j(E_1)\,\varrho_j(E_1)^q v_r = 0$, it follows that $r+q = j$. Similarly, $r-p = -j$. By adding and subtracting these two equations, we get $2r = -(q-p)$ and $p+q = 2j$.

The following Theorem assures the complete reducibility of any representation of $\mathfrak{g}_3$. Thus the problem of finding all representations of $\mathfrak{g}_3$ is solved by this Theorem together with Theorems 2 and 3.

THEOREM 4. *Any representation of* $\mathfrak{g}_3$ *is completely reducible*.

Proof. We define the Casimir operator of a representation ϱ of $\mathfrak{g}_3$ by

$$\begin{aligned}
\varrho(G) &= -\tfrac{1}{2}(\varrho(M_1)^2 + \varrho(M_2)^2 + \varrho(M_3)^2) \\
&= \tfrac{1}{4}(\varrho(E_1)\,\varrho(E_{-1}) + \varrho(E_{-1})\,\varrho(E_1)) + \tfrac{1}{2}\varrho(H)^2.
\end{aligned}$$

It is easy to show that $\varrho(G)$ commutes with every linear transformation in $\varrho(\mathfrak{g}_3)$, e.g.

$$\begin{aligned}
2[\varrho(G), \varrho(M_1)] &= -[\varrho(M_2)^2), \varrho(M_1)] - [\varrho(M_3)^2, \varrho(M_1)] \\
&= -\varrho(M_2)\,[\varrho(M_2), \varrho(M_1)] - [\varrho(M_2), \varrho(M_1)]\,\varrho(M_2) - \varrho(M_3)\,[\varrho(M_3), \varrho(M_1)] \\
&\quad - [\varrho(M_3), \varrho(M_1)]\,\varrho(M_3) \\
&= \varrho(M_2)\,\varrho(M_3) + \varrho(M_3)\,\varrho(M_2) - \varrho(M_3)\,\varrho(M_2) - \varrho(M_2)\,\varrho(M_3) = 0.
\end{aligned}$$

Therefore, if ϱ is irreducible, then by Schur's lemma, $\varrho(G)$ is a scalar matrix. In particular, if j is the highest weight of ϱ_j, then

$$\varrho_j(G) = \tfrac{1}{2}j(j+1)I.$$

This is because

$$\begin{aligned}\varrho_j(G)v_m &= \{\tfrac{1}{4}(\varrho_j(E_1)\,\varrho_j(E_{-1})+\varrho_j(E_{-1})\,\varrho_j(E_1))+\tfrac{1}{2}\varrho_j(H)^2\}v_m\\ &= \tfrac{1}{4}\varrho_j(E_1)v_{m-1}+\tfrac{1}{4}\varrho_j(E_{-1})\,(j-m)\,(j+m+1)v_{m+1}+\tfrac{1}{2}m^2v_m\\ &= \tfrac{1}{4}(j-m+1)\,(j+m)v_m+\tfrac{1}{4}(j-m)\,(j+m+1)v_m+\tfrac{1}{2}m^2v_m\\ &= \tfrac{1}{2}j(j+1)v_m, \quad (m = j, j-1, \ldots, -j).\end{aligned}$$

We now prove a special case of Theorem 4.

LEMMA 4. *Suppose the representation space V of ϱ contains two irreducible representations, i.e. in the representation space V of ϱ there is an irreducible invariant subspace V_0 such that V/V_0 is irreducible, then there exists an invariant subspace V' of V such that $V = V_0 \dot{+} V'$.*

Proof. Let ϱ_j and $\varrho_{j'}$ be the two representations contained in ϱ. Choose a basis in V so that the matrix of any transformation in $\varrho(\mathfrak{g}_3)$ has the form

$$\varrho(X) = \begin{bmatrix}\varrho_j(X) & B(X)\\ 0 & \varrho_{j'}(X)\end{bmatrix}, \quad X \in \mathfrak{g}_3.$$

Consider the two cases $j \neq j'$ and $j = j'$:

(1) $j \neq j'$. In this case, $\frac{1}{2}j(j+1) \neq \frac{1}{2}j'(j'+1)$ and

$$\varrho(G) = \begin{bmatrix}\frac{1}{2}j(j+1)I & K\\ 0 & \frac{1}{2}j'(j'+1)I\end{bmatrix}.$$

Use the matrix

$$P = \begin{bmatrix}I & (\frac{1}{2}j(j+1)-\frac{1}{2}j'(j'+1))^{-1}K\\ 0 & I\end{bmatrix}$$

to transform $\varrho(G)$ and all matrices in $\varrho(\mathfrak{g}_3)$. We get

$$P\varrho(G)P^{-1} = \begin{bmatrix}\frac{1}{2}j(j+1)I & 0\\ 0 & \frac{1}{2}j'(j'+1)I\end{bmatrix}$$

and that all matrices $P\varrho(X)P^{-1}$ $(X \in \mathfrak{g}_3)$ must commute with $P\varrho(G)P^{-1}$. Thus

$$P\varrho(X)P^{-1} = \begin{bmatrix}\varrho_j(X) & 0\\ 0 & \varrho_{j'}(X)\end{bmatrix}, \quad X \in \mathfrak{g}_3.$$

This proves that Lemma 4 is true if $j \neq j'$.

(2) $j = j'$. In this case, ϱ_j and $\varrho_{j'}$ are equivalent. V contains an irreducible invariant subspace V_j that has a basis $\{v_j, \ldots, v_{-j}\}$ such that the equations (9.7) hold. V also contains a set of vectors $v_j', \ldots v_{-j}'$, such that $v_j'+V_j, \ldots, v_{-j}'+V_j$ form a basis of V/V_j and the equations (9.7) hold for the representation of $\mathfrak{g}_3$ in V/V_j. Choose v_j' so that

$$\varrho(H)v_j' = jv_j'+\mu v_j,$$

then define $v'_{j-1}, \ldots, v'_{-j}$ by

$$\varrho(E_{-1})v'_m = v'_{m-1}, \quad (m = j, \ldots, -j+1).$$

Thus we have

$$\begin{aligned}\varrho(H)v'_{j-1} &= \varrho(H)\,\varrho(E_{-1})v'_j\\ &= \varrho([H, E_{-1}])v'_j + \varrho(E_{-1})\,\varrho(H)v'_j\\ &= -\varrho(E_{-1})v'_j + \varrho(E_{-1})\,(jv'_j + \mu v_j)\\ &= (j-1)v'_{j-1} + \mu v_{j-1},\end{aligned}$$

$$\varrho(E_1)v'_j = 0$$

(since $\varrho(E_{-1})v'_j$ is an eigenvector of $\varrho(H)$ of eigenvalue $j+1$) and

$$\begin{aligned}\varrho(E_1)v'_{j-1} &= \varrho(E_1)\,\varrho(E_{-1})v'_j\\ &= \varrho(2H)v'_j + \varrho(E_{-1})\,\varrho(E_1)v'_j\\ &= 2jv'_j + 2\mu v_j.\end{aligned}$$

In general, we have

$$\begin{aligned}\varrho(H)v'_m &= mv'_m + \mu v_m,\\ \varrho(E_1)v'_m &= (j-m)(j+m+1)v'_{m+1} + 2\mu(j-m)v_{m+1}.\end{aligned}$$

In fact, if this equation holds for m, then

$$\begin{aligned}\varrho(H)v'_{m-1} &= \varrho(H)\,\varrho(E_{-1})v'_m\\ &= \varrho([H, E_{-1}])v'_m + \varrho(E_{-1})\,\varrho(H)v'_m\\ &= -\varrho(E_{-1})v'_m + \varrho(E_{-1})\,(mv'_m + \mu v_m)\\ &= (m-1)v'_{m-1} + \mu v_{m-1},\end{aligned}$$

$$\begin{aligned}\varrho(E_1)v'_{m-1} &= \varrho(E_1)\,\varrho(E_{-1})v'_m\\ &= \varrho(2H)v'_m + \varrho(E_{-1})\,\varrho(E_1)v'_m\\ &= 2mv'_m + 2\mu v_m + \varrho(E_{-1})\,\{(j-m)(j+m+1)v'_{m+1} + 2\mu(j-m)v_{m+1}\}\\ &= 2mv'_m + 2\mu v_m + (j-m)(j+m+1)v'_m + 2\mu(j-m)v_m\\ &= (j-m+1)(j+m)v'_m + 2\mu(j-m+1)v_m.\end{aligned}$$

Since $-j-1$ is not an eigenvalue of $\varrho(H)$, when $m = -j-1$, $v'_{-j-1} = 0$. Thus from

$$\varrho(E_1)v'_{-j-1} = 0,$$

we get

$$2\mu(j-(-j+1)) = 0.$$

Hence

$$\mu = 0.$$

Therefore,

$$\begin{aligned}\varrho(H)v'_m &= mv'_m,\\ \varrho(E_{-1})v'_m &= v'_{m-1}, \qquad m = j, j-1, \ldots, -j\\ \varrho(E_1)v'_m &= (j-m)(j+m+1)v'_{m+1}.\end{aligned}$$

This proves that $v'_{-j}, v'_{-j+1}, \ldots, v'_j$ generate an invariant subspace V'_j of V and $V = V_j \dot{+} V'_j$.

We now continue to prove Theorem 4. Let ϱ be a representation of $\mathfrak{g}_3$ with representation space V and $V_0\,(\subset V)$ be an irreducible invariant subspace of V. Assume that Theorem 4 is true for representation spaces of dimensions $\leqslant \dim V$, then

$$V/V_0 = \bar{V}_1 \dot{+} \dots \dot{+} \bar{V}_m,$$

where $\bar{V}_1, \dots, \bar{V}_m$ are irreducible invariant subspaces of V/V_0. Let U_i $(i = 1, \dots, m)$ be the subspace of V mapped onto V/V_0 under the canonical mapping, then

$$U_i/V_0 = \bar{V}_i, \quad i = 1, \dots, m.$$

By Lemma 4, U_i contains an irreducible invariant subspace V_i such that

$$U_i = V_0 \dot{+} V_i.$$

Thus

$$V = V_0 \dot{+} V_1 \dot{+} \dots \dot{+} V_m.$$

This proves that ϱ is completely reducible.

By Theorem 4, the Corollary to Theorem 2 can be generalized to the following:

LEMMA 5. *Let ϱ be a representation of $\mathfrak{g}_3$, then the weights of ϱ are integers or half integers. Let v be in the representation space V of ϱ of weight r. If p, q are largest non-negative integers such that $\varrho(E_{-1})^p v \neq 0$ and $\varrho(E_1)^q v \neq 0$, then $2r = -(q-p)$. Moreover, $\varrho(E^{-1})^i\, \varrho(E_1)^q v \neq 0$ $(0 \leqslant i \leqslant p+q)$ and it is of weight $r+q-i$.*

Proof. By Theorem 4, ϱ can be decomposed into the sum of irreducible representations. By Theorem 2, every component of ϱ is equivalent to some ϱ_j. The representation space V is correspondingly decomposed into the direct sum of irreducible invariant subspaces. With respect to this decomposition of V, v has the decomposition

$$v = \sum_i v_i$$

where v_i belongs to an irreducible invariant subspace of highest weight j (j depends on i). Obviously, every v_i has the same weight r as v, thus r is an integer or a half integer.

Now for all i, $\varrho(E_1)^{q+1} v_i = 0$, thus $\varrho(E_1)^q v_i \neq 0$ for at least one i. If $\varrho(E_1)^q v_i \neq 0$ and p' is the largest non-negative integer such that $\varrho(E_{-1})^{p'} v_i \neq 0$, then by the Corollary to Theorem 2, $2r = -(q-p')$. Hence $p' = 2r+q$. Let $\varrho(E_1)^q v_k = 0$. If p'', q'' are largest non-negative integers such that $\varrho(E_{-1})^{p''} v_k \neq 0$ and $\varrho(E_1)^{q''} v_k \neq 0$, then by the Corollary to Theorem 2, $2r = -(q''-p'')$, thus $p'' = 2r+q''$; since $q'' < q$, we have $p'' < p'$. Therefore, for those i such that $\varrho(E_1)^q v_i \neq 0$, $\varrho(E_{-1})^p v_i$ is not the zero vector. By the Corollary to Theorem 2, $2r = -(q-p)$.

CHAPTER 10

REPRESENTATIONS OF SEMISIMPLE LIE ALGEBRAS

10.1. Irreducible representations of semisimple Lie algebras

Let $\mathfrak{g}$ be a semisimple Lie algebra and $\mathfrak{h}$ be a fixed Cartan subalgebra. Let σ be a representation of $\mathfrak{g}$ with representation space V. A linear function $\omega = \omega(H)$ defined on $\mathfrak{h}$ is said to be a *weight* of ϱ if there exists a non-zero $x \in V$ such that

$$\varrho(H)x = \omega(H)x, \quad \text{for all } H \in \mathfrak{h};$$

the vector x is said to be a weight vector corresponding to the weight ω. Two weights ω_1 and ω_2 of ϱ are said to be equivalent if there exists $S \in W$ (Weyl group of $\mathfrak{g}$) such that $S(\omega_1) = \omega_2$.

Obviously, the roots of $\mathfrak{g}$ are the weights of the regular representation.

For any linear function ω defined on $\mathfrak{h}$, let

$$V_\omega = \{x \in V \mid \varrho(H)x = \omega(H)\,x, \quad \text{for all} \quad H \in \mathfrak{h}\},$$

then V_ω is a subspace of V, V_ω contains all root vectors of the root ω and

$$V_\omega \subseteq V^\omega = V^\omega_{\varrho(\mathfrak{h})}.$$

Moreover, ω is a weight of ϱ iff $V_\omega \neq \{0\}$. The multiplicity of ω is defined to be $\dim V^\omega$.

E. Cartan† studied representations of simple Lie algebras separately, but most of his work can be generalized to semisimple Lie algebras.

THEOREM 1. *Let ϱ be an irreducible representation of a semisimple Lie algebra $\mathfrak{g}$ with representation space V, $\mathfrak{h}$ be a Cartan subalgebra of $\mathfrak{g}$ and $\Pi = \{\alpha_1, \ldots, \alpha_n\}$ be a fundamental system of roots of $\mathfrak{g}$. Then*

1° *V is the direct sum of weight subspaces, i.e. $V = \dot{\sum}_\omega V_\omega$. Thus $V_\omega = V^\omega$.*

2° *If ω is a weight of ϱ and α is a root, then $2(\omega, \alpha)/(\alpha, \alpha)$ is an integer,*

$$\omega - \frac{2(\omega, \alpha)}{(\alpha, \alpha)}\alpha$$

† E. Cartan, Les groupes projectifs qui ne laissent invariante aucune multiplicité plane, *Bull. Soc. Math. France*, **41** (1913), 53–96.

is also a weight of ϱ *and* $\dim V_\omega = \dim V_\alpha$. *Thus equivalent weights have the same multiplicity.*

3° *There exists a unique weight* ω_0 *(depending on the choice of* Π*) called the highest weight such that all weights of* ϱ *are of the form*

$$\omega_0 - \alpha_{i_1} - \ldots - \alpha_{i_k}, \quad \alpha_i, \ldots \alpha_{i_k} \in \Pi$$

and $\dim V_{\omega_0} = 1$.

Proof. (1) We show first that ϱ has at least one weight.

Since $\mathfrak{h}$ is abelian, $\varrho(\mathfrak{h})$ is an abelian linear Lie algebra. Thus $\varrho(\mathfrak{h})$ is solvable. By Lie's theorem, ϱ has a weight.

(2) Secondly, we prove the following:

LEMMA 1. *Let* ω *be a weight of* ϱ *and* $x \in V_\omega$. *If* $\alpha \in \Sigma$ *and* $\varrho(E_\alpha)x \neq 0$, *then* $\omega+\alpha$ *is also a weight of* ϱ *and* $\varrho(E_\alpha)x \in V_{\omega+\alpha}$. *If* $\omega+\alpha$ *is not a weight of* ϱ, *then* $\varrho(E_\alpha)x = 0$.

Proof. We have

$$\begin{aligned}\varrho(H)\varrho(E_\alpha)x &= \varrho([H, E_\alpha])x + \varrho(E_\alpha)\varrho(H)x \\ &= \varrho\big(\alpha(H)E_\alpha\big)x + \varrho(E_\alpha)\omega(H)x \\ &= (\omega+\alpha)(H)\varrho(E_\alpha)x.\end{aligned}$$

Thus Lemma 1 follows.

By Theorem 3 of Chapter 2, Lemma 1 can be generalized to the following:

LEMMA 1'. *Let* V' *be subspace of* V *which is invariant under* $\varrho(\mathfrak{h})$ *and* $\varrho(E_\alpha)$ *for some* $\alpha \in \Sigma$. *Suppose* ω *is a weight of* ϱ *and there exists* $x \in V - V_1$ *such that*

$$\varrho(H)x \equiv \omega(H)x \pmod{V_1}, \quad \text{for all} \quad H \in \mathfrak{h}.$$

If $\varrho(E_\alpha) \notin V_1$, *then* $\omega+\alpha$ *is also a weight of* ϱ *and*

$$\varrho(H)\varrho(E_\alpha)x \equiv (\omega+\alpha)(H)\varrho(E_\alpha)x \pmod{V_1}, \text{ for all } H \in \mathfrak{h}.$$

If $\omega+\alpha$ *is not a weight, then* $\varrho(E_\alpha)x \in V_1$.

(3) V is generated by weight vectors, i.e. there exists a basis of V such that every $\varrho(H)$ ($H \in \mathfrak{h}$) is diagonal.

Let ω be a weight of ϱ and x be a weight vector corresponding to ω. Consider the vectors of the form

$$\varrho(E_{\varphi_1}) \ldots \varrho(E_{\varphi_s})x, \quad \varphi_1, \ldots, \varphi_s \in \Sigma, \quad s \geq 0.$$

If $\varrho(E_{\varphi_1}) \ldots \varrho(E_{\varphi_s})x \neq 0$, then by Lemma 1, it is a weight vector corresponding to the weight $\omega+\varphi_1+\ldots+\varphi_s$. Let V_1 be the subspace spanned by these vectors, then V_1 is an invariant subspace of $\varrho(\mathfrak{g})$. Now $V_1 \neq \{0\}$ and ϱ is irreducible, thus $V = V_1$. This proves the assertion.

Now the proof of 1° is completed.

(4) For every weight ω of ϱ, embed ω into $\mathfrak{h}$ by means of the Killing form, i.e. let H_ω be such that

$$(H, H_\omega) = \omega(H), \quad \forall H \in \mathfrak{h}.$$

(5) We now prove the following:

LEMMA 2. *If ω is a weight of ϱ and $\alpha \in \Sigma$, then*

(i) $2(\omega, \alpha)/(\alpha, \alpha)$ *is an integer and* $\omega - \big(2(\omega, \alpha)/(\alpha, \alpha)\big)\alpha$ *is also a weight of* ϱ.

(ii) ω *and* $\omega - \big(2(\omega, \alpha)/(\alpha, \alpha)\big)\alpha$ *have the same multiplicity in* ϱ. *Therefore, equivalent weights have the same multiplicity.*

(iii) *If* $\dim V_\omega = 1$ *and* p, q *are largest non-negative integers such that* $\omega + k\alpha$ $(-p \leq k \leq q)$ *are weights of* ϱ, *then*

$$\frac{2(\omega, \alpha)}{(\alpha, \alpha)} = -(q-p)$$

and $\omega + k\alpha$ *is not a weight of* ϱ *if* $k > q$ or $k < -p$.

Proof. Choose root vectors E_α and $E_{-\alpha}$ such that $(E_\alpha, E_{-\alpha}) = 1$, then $E_\alpha, E_{-\alpha}$ and H_α form a three-dimensional subalgebra with structure formulas

$$[H_\alpha E_\alpha] = (\alpha, \alpha) E_\alpha, \quad [H, E_{-\alpha}] = -(\alpha, \alpha) E_{-\alpha}, \quad [E_\alpha, E_{-\alpha}] = H_\alpha.$$

Let

$$H_1 = \frac{H_\alpha}{(\alpha, \alpha)}, \quad E_1 = \sqrt{\left(\frac{2}{(\alpha, \alpha)}\right)} E_\alpha, \quad E_{-1} = \sqrt{\left(\frac{2}{(\alpha, \alpha)}\right)} E_{-\alpha},$$

then H, E_1, E_{-1} also form a basis of this subalgebra and

$$[H_1, E_1] = E_1, \quad [H_1, E_{-1}] = -E_{-1}, \quad [E_1, E_{-1}] = 2H_1.$$

Thus this subalgebra is the Lie algebra $\mathfrak{g}_3$ discussed in Section 9.3.

Clearly, ϱ induces a representation of $\mathfrak{g}_3$. Let v be a vector of weight ω; by Lemma 5 of Chapter 9, if p, q are largest positive integers such that $\varrho(E_{-1})^p v \neq 0$ and $\varrho(E_1)^q v \neq 0$, then $2\omega(H_1) = -(q-p)$. This proves that

$$\frac{2(\omega, \alpha)}{(\alpha, \alpha)} = \frac{2\omega(H_\alpha)}{(\alpha, \alpha)} = 2\omega(H_1) = -(q-p)$$

is an integer. By the same Lemma, we have $\varrho(E_{-1})^i\, \varrho(E_1)^q v \neq 0$ $(0 \leq i \leq p+q)$. By Lemma 1, $\varrho(E_{-1})^i\, \varrho(E_1)^q v$ is of weight $\omega + (q-i)\alpha$; let $i = p$, then

$$\omega - \frac{2(\omega, \alpha)}{(\alpha, \alpha)}\alpha = \omega + (q-p)\alpha$$

is also a weight of ϱ. This proves (i).

We now prove (ii). Suppose ω is not a simple weight and let V_1 be the subspace spanned by $\varrho(E_1)^q v, \ldots, \varrho(E_1)v, v, \varrho(E_{-1})v, \ldots, \varrho(E_{-1})^p v$. Clearly, V_1 is invariant under $\varrho(\mathfrak{h})$, $\varrho(E_1)$

and $\varrho(E_{-1})$. Since ω is not simple, there exists $v_1 \in V - V_1$ such that $\varrho(H)v_1 \equiv \omega(H)v_1$ (mod V_1) for all $H \in \mathfrak{h}$. Notice that ϱ induces a representation of $\mathfrak{g}_3$ in V/V_1. If p', q' are largest integers such that $\varrho(E_{-1})^{p'}v_1 \notin V_1$ and $\varrho(E_{-1})^{q'}v_1 \notin V_1$, then by Lemma 5 of Chapter 9, $2\omega(H) = -(q'-p')$. Similarly,

$$\frac{2(\omega, \alpha)}{(\alpha, \alpha)} = -(q'-p').$$

By the same Lemma, $\varrho(E_{-1})^i \varrho(E_1)^q v_1 \notin V_1$ $(0 \leqslant i \leqslant p'+q')$. By Lemma 1′,

$$\varrho(H)\,\varrho(E_{-1})^i (E_1)^{q'} v_1 \equiv (\omega+(q'-i)\alpha)(H)\,\varrho(E_{-1})^i\,\varrho(E_1)^{q'} v_1 \pmod{V_1},$$

for all $H \in \mathfrak{h}$. Set $i = p'$, then

$$\omega+(q'-p')\alpha = \omega - \frac{2(\omega, \alpha)}{(\alpha, \alpha)}\alpha$$

is not a simple root of ϱ.

By repeating the argument, it can be proved that the multiplicity of

$$\omega' = \omega - \frac{2(\omega, \alpha)}{(\alpha, \alpha)}\alpha$$

is not smaller than dim V_ω. Now

$$\omega' - \frac{2(\omega', \alpha)}{(\alpha, \alpha)}\alpha = \omega,$$

so we can also show that dim $V_\omega \geqslant$ dim $V_{\omega'}$. Thus dim $V_\omega =$ dim $V_{\omega'}$. Since

$$S_\alpha(\omega) = \omega - \frac{2(\omega, \alpha}{(\alpha, \alpha)}\alpha$$

and W is generated by all S_α $(\alpha \in \Sigma)$, equivalent weights must have the same multiplicity. This proves (ii).

Finally, we prove (iii). Suppose ω is simple and $v\ (\neq 0) \in V_\omega$. Let p, q be largest non-negative integers such that $\varrho(E_{-1})^p\, v \neq 0$ and $\varrho(E_1)^q\, v \neq 0$. If p_0, q_0 are largest integers such that $\omega - p_0\alpha$ and $\omega + q_0\alpha$ are weights of ϱ, then $p_0 \geqslant p$, $q_0 \geqslant q$. It suffices to show that $p = p_0$ and $q = q_0$. If u is of weight $\omega + q_0\alpha$, then $\varrho(E_1)u = 0$. Let s be the largest non-negative integer such that $\varrho(E_{-1})^s u \neq 0$, then by Lemma 5 of Chapter 9, $2(\omega+q_0\alpha)(H_1) = s$; substituting $2\omega(H_1)$ by $-(q-p)$, we get $-q+p+2q_0 = s$. Since $q_0 \geqslant q$, we get $s \geqslant p+q_0 \geqslant q_0$ and $\varrho(E_{-1})^{q_0} u \neq 0$. Now ω is simple, so $\varrho(E_{-1})^{q_0} u$ and v are linearly dependent. If p, s are largest non-negative integers such that $\varrho(E_{-1})^p v \neq 0$ and $\varrho(E_{-1})^s\, u \neq 0$, then $s - q_0 = p$. Substituting $s - q_0 = p$ into $-q+p+2q_0 = s$, we get $q = q_0$. Similarly, $p = p_0$.

The proofs of Lemma 2 and 2° of Theorem 1 are now completed.

(6) By a proof similar to the one of Theorem 3 of Chapter 5, we can prove the following:

LEMMA 3. *If ω is a weight of some representation ϱ of $\mathfrak{g}$*, then $\omega \in \mathfrak{h}_0^*$.

(7) Finally, we prove 3°.

It is known that $\Pi = \{\alpha_1, \ldots, \alpha_n\}$ is a basis of $\mathfrak{h}_0^*$. Define a partial ordering on $\mathfrak{h}_0^*$ by the definition:

$$\lambda \geqslant \mu \quad \text{iff} \quad \lambda - \mu \quad \text{is linear combination of } \alpha_1, \ldots, \alpha_n$$

with non-negative coefficients.

If $\lambda \geqslant \mu$, then we say that λ is higher than or equal to μ; if $\lambda \geqslant \mu$ and $\lambda \neq \mu$, then λ is said to be higher than μ. λ is said to be positive if $\lambda > 0$; if $\lambda < 0$, then it is said to be negative. With respect to this partial ordering, the root system Σ is again divided into positive roots Σ_+ and negative roots Σ_-. In what follows, the ordering on $\mathfrak{h}_0^*$ is understood to be this partial ordering.

Let ω_0 be a weight of ϱ such that no other weight of ϱ is higher than ω_0. Let x be a non-zero vector of weight ω_0 and consider the vectors

$$\varrho(E_{-\alpha_{i_1}}) \ldots \varrho(E_{-\alpha_{i_k}})x, \quad \alpha_{\alpha_{i_1}}, \ldots, \alpha_{i_k} \in \Pi, \quad k \geqslant 0.$$

If $\varrho(E_{-\alpha i_1}) \ldots \varrho(E_{-\alpha_{i_k}}) x \neq 0$, then it corresponds to the weight

$$\omega_0 - \alpha_{\alpha_{i_1}} - \ldots - \alpha_{i_k}.$$

The set of such vectors forms a subspace V_1. We will show that V_1 is invariant under $\varrho(\mathfrak{g})$. Clearly, V_1 is invariant under $\varrho(H)\,(\forall H \in \mathfrak{h})$ and $\varrho(E_{-\alpha})\,(\forall \alpha \in \Pi)$. From

$$[E_\alpha, E_\beta] = \delta_{\alpha\beta} H_\alpha, \quad \alpha, \beta \in \Pi$$

we have

$$\varrho(E_\alpha)\,\varrho(E_{-\beta}) = \delta_{\alpha\beta}\varrho(H_\alpha) + \varrho(E_{-\beta})\,\varrho(E_\alpha),$$

thus V_1 is also invariant under $\varrho(E_\alpha)\,(\forall \alpha \in \Pi)$. Suppose $\gamma \in \Sigma_-$ and $\gamma = -(m_1\alpha_1 + \ldots + m_n\alpha_n)$, where $m_1, \ldots, m_n \geqslant 0$. We use induction on $\sum_1^n m_i$ to show that V_1 is invariant under $\varrho(E_\gamma)$. When $\sum_1^n m_i = 1$, we have $-\gamma \in \Pi$, thus the result is obvious. If $\sum_1^n m_i > 1$ and $\gamma = \beta + (-\alpha)$ where $\alpha \in \Pi$, then by the induction assumption, V_1 is invariant under $\varrho(E_\beta)$. From

$$E_\gamma = \frac{1}{N_{\beta,-\alpha}}[E_\beta, E_{-\alpha}],$$

it follows that

$$\varrho(E_\gamma) = \frac{1}{N_{\beta,-\alpha}}\{\varrho(E_\beta)\,\varrho(E_{-\alpha}) - \varrho(E_{-\alpha})\,\varrho(E_\beta)\}.$$

Thus V_1 is invariant under $\varrho(E_\gamma)\,(\gamma < 0)$. Similarly, V_1 is invariant under $\varrho(E_\alpha)\,(\alpha > 0)$. Hence V_1 is invariant under $\varrho(\mathfrak{g})$, so $V_1 = V$. It is now obvious that ω_0 is simple and higher than any other weights of ϱ. The proof of Theorem 1 is now completed.

THEOREM 2. *Let $\mathfrak{g}$ be a semisimple Lie algebra, $\mathfrak{h}$ be a Cartan subalgebra of $\mathfrak{g}$ and $\Pi = \{\alpha_1, \ldots, \alpha_n\}$ be a fundamental system of roots of $\mathfrak{g}$. If ϱ and ϱ' are two irreducible representations with equal highest weight with respect to Π, then ϱ and ϱ' are equivalent.*

Proof. Let x_0 and x_0' be of highest weight ω_0 in V of ϱ and V' of ϱ' respectively. Consider the vectors of the form

$$\varrho(E_{-\alpha_{i_1}}) \cdots \varrho(E_{-\alpha_{i_k}})x_0 \tag{10.1}$$

and

$$\varrho'(E_{-\alpha_{i_1}}) \ldots \varrho'(E_{-\alpha_{i_k}})x'_0 \tag{10.2}$$

in V and V' respectively. From the proof of Theorem 1, these vectors span V and V' respectively; the weights of these vectors are also known. We now prove that the mapping defined by

$$\Sigma c_{i_1, \ldots, i_k}\varrho(E_{-i_1}) \ldots \varrho(E_{-i_k})x_0 \xrightarrow{f} \Sigma c_{i_1, \ldots, i_k}\varrho'(E_{-i_1}) \ldots \varrho'(E_{-i_k})x'_0$$

is an isomorphism between V and V'. Suppose

$$c_1v_1 + c_2v_2 + \ldots = 0 \tag{10.3}$$

where $v_1, v_2, \ldots$ are of the form (10.1). Consider the vector

$$u' = c_1v'_1 + c_2v'_2 + \ldots, \tag{10.4}$$

where $v'_1 = f(v_1)$, $v'_2 = f(v_2)$, ... The set of all vectors (10.4) forms a subspace V'_1 of V'. It is easy to see that V'_1 is invariant under $\varrho'(\mathfrak{g})$. Thus either $V'_1 = V'$ or $V'_1 = 0$. Now notice that $x_0 \notin V'_1$. Otherwise, all vectors $v_1, v_2, \ldots$ in (10.3) are of weight ω_0 and $v_1 = v_2 = \ldots = x_0$, so

$$c_1 + c_2 + \ldots = 0;$$

correspondingly, in (10.4), we have $v'_1 = v'_2 = \ldots = x'_0$ and

$$x'_0 = (c_1 + c_2 + \ldots)x'_0 = 0.$$

This is a contradiction. Hence $V'_1 = 0$. It is now clear that ϱ and ϱ' are equivalent.

From Theorem 2, we know that an irreducible representation of a semisimple Lie algebra is uniquely determined by the highest weight of it. Therefore, to find all irreducible representations of the Lie algebra, it is necessary to determine the elements in $\mathfrak{h}_0^*$ that can be highest weights of irreducible representations.

THEOREM 3. *Let $\mathfrak{g}$ be a semisimple Lie algebra, $\mathfrak{h}$ be a Cartan subalgebra and $\Pi = \{\alpha_1, \ldots, \alpha_n\}$ be a fundamental system of roots of $\mathfrak{g}$. If ω is the highest weight of an irreducible representation of $\mathfrak{g}$, then*

$$\omega_{\alpha_i} = \frac{2(\omega, \alpha_i)}{(\alpha_i, \alpha_i)}, \qquad i = 1, \ldots, n,$$

are non-negative integers. Conversely, if $\omega \in \mathfrak{h}_0^*$ and ω_{α_i} $(i = 1, \ldots, n)$ are non-negative integers, then ω is the highest weight of some irreducible representation of $\mathfrak{g}$.

Proof. For now, we only prove the first part. For the classical Lie algebras, the proof of the second part will be given in the following two chapters; for the general case, the proof will be given in Chapter 13.

Let ω be the highest weight of an irreducible representation ϱ of $\mathfrak{g}$. If for some $\alpha \in \Pi$,

$$\omega_\alpha = \frac{2(\omega, \alpha)}{(\alpha, \alpha)} < 0,$$

then by Theorem 1, $\omega-\omega_\alpha\alpha$ is also a weight of ϱ. Clearly, $\omega < \omega-\omega_\alpha\alpha$; this contradicts the assumption that ω is the highest weight of ϱ. Therefore, $\omega_\alpha \geqslant 0\ (\forall \alpha \in \Pi)$.

In principle, Theorem 2 and Theorem 3 solve the problem of obtaining all irreducible representations for semisimple Lie algebras.

Since the highest weight ω of an irreducible representation ϱ is determined by the numbers $\omega_\alpha\ (\forall\alpha \in \Pi)$, ϱ can be denoted by assigning the numbers ω_α to the corresponding dots of the Dynkin diagram of $\mathfrak{g}$. Thus two irreducible representations of a semisimple Lie algebra are equivalent iff they have the same diagram.

Definition. An element $\mu \in \mathfrak{h}^*$ is said to be an integral linear function if

$$\frac{2(\mu, \alpha_i)}{(\alpha_i, \alpha_i)}, \qquad i = 1, \ldots, n$$

are integers; μ is said to be a dominant linear function if

$$\frac{2(\mu, \alpha_i)}{(\alpha_i, \alpha_i)}, \qquad i = 1, \ldots, n$$

are non-negative real numbers.

Clearly, integral linear functions and dominant linear functions are in $\mathfrak{h}_0^*$. Theorem 3 can be stated as:

THEOREM 3′. *Under the assumptions of Theorem* 3, $\omega \in \mathfrak{h}^*$ *is the dominant weight (relative to* Π*) of an irreducible representation of* $\mathfrak{g}$ *iff* ω *is a dominant integral linear function (relative to* Π*).*

Finally, we give a characterization of dominant linear functions.

LEMMA 4. *An element* $\mu \in \mathfrak{h}_0^*$ *is a dominant linear function iff* $\mu \geqslant S(\mu)\ (\forall S \in W)$.

Proof. Suppose $\mu \geqslant S(\mu)\ (\forall S \in W)$, then $\mu \geqslant S_i(\mu)\ (i = 1, \ldots, n)$. Now

$$S_i(\mu) = \mu - \frac{2(\mu, \alpha_i)}{(\alpha_i, \alpha_i)}\alpha_i, \qquad i = 1, \ldots, n$$

thus

$$\frac{2(\mu, \alpha_i)}{(\alpha_i, \alpha_i)} \geqslant 0, \qquad i = 1, \ldots, n.$$

Conversely, suppose

$$\frac{2(\mu, \alpha_i)}{(\alpha_i, \alpha_i)}, \qquad i = 1, \ldots, n$$

are non-negative real numbers, then $\mu \geqslant S_i(\mu)$. Let $S \in W$, and $S = S_{i_1} \ldots S_{i_k}$ where $1 \leqslant i_1, \ldots, i_k \leqslant n$. We use induction on k to show that $\mu \geqslant S(\mu)$. Write $S = S'S_{i_k}$, where $S' = S_{i_1} \ldots S_{i_{k-1}}$. From

$$S_{i_k}(\mu) = \mu - \frac{2(\mu, \alpha_{i_k})}{(\alpha_{i_k}, \alpha_{i_k})}\alpha_{i_k},$$

it follows that

$$S(\mu) = S'(\mu) - \frac{2(\mu, \alpha_{i_k})}{(\alpha_{i_k}, \alpha_{i_k})} \alpha_{i_k}.$$

If $S'(\alpha_k) > 0$, then $S'(\mu) \geqslant S(\mu)$ since

$$\frac{2(\mu, \alpha_{i_k})}{(\alpha_{i_k}, \alpha_{i_k})} \geqslant 0.$$

By the induction assumption, $\mu \geqslant S'(\mu)$, thus $\mu \geqslant S(\mu)$.

Now suppose $S'(\alpha_{i_k}) < 0$. Let $\beta_k = \alpha_{i_k}$ and $\beta_t = S_{i_t} S_{i_{t+1}} \ldots . S_{i_{k-1}}(\alpha_{i_k})$ $(t = 1, 2, \ldots, k-1)$, then $\beta_k = \alpha_{i_k} > 0$ and $\beta_1 = S'(\alpha_{i_k}) < 0$. Thus for some j $(1 < j \leqslant k)$, $\beta_t > 0$ (for $t \geqslant j$) and $\beta_{j-1} < 0$. Therefore, $\beta_j > 0$ and $\beta_{j-1} = S_{i_{j-1}}(\beta_j) < 0$. By Lemma 6 of Chapter 6, $\beta_j = \alpha_{i_{j-1}}$. Let $T = S_{i_1}, \ldots, S_{i_{j-2}}$ and $T' = S_{i_j}, \ldots, S_{i_{k-1}}$, then $S' = TS_{i_{j-1}}T'$ and $\beta_j = T'(\alpha_{i_k}) = \alpha_{i_{j-1}}$. Therefore, $T'S_{i_k}T'^{-1} = S_{i_{j-1}}$, i.e. $T'S_{i_k} = S_{i_{j-1}}T'$. Hence $S = S'S_{i_k} = TS_{i_{j-1}}T'S_{i_k} = TT'S_{i_k}^2 = TT'$. Now S is the product of $k-1$ (possibly with repetition) of S_i $(i = 1, \ldots, n)$, thus $\mu \geqslant S(\mu)$.

COROLLARY 1. *$\mu \in \mathfrak{h}_0^*$ is a dominant integral function iff for all $\alpha \in \Sigma_+$, $2(\mu, \alpha)/(\alpha, \alpha)$ are non-negative integers.*

Proof. If $2(\mu, \alpha)/(\alpha, \alpha)$ are non-negative integers for all $\alpha \in \Sigma_+$, then μ is naturally a dominant integral function.

Conversely, if μ is a dominant integral function, then by definition, $2(\mu, \alpha_i)/(\alpha_i, \alpha_i)$ $(i = 1, \ldots, n)$ are non-negative integers. By Lemma 4,

$$\mu \geqslant S(\mu) = \mu - \frac{2(\mu, \alpha)}{(\alpha, \alpha)} \alpha, \quad \text{for all } \alpha \in \Sigma_+.$$

Hence $2(\mu, \alpha)/(\alpha, \alpha)$ are non-negative. Now it remains to show that $2(\mu, \alpha)/(\alpha, \alpha)$ $(\forall \alpha \in \Sigma_+)$ are integers. We show first that if μ is an integral linear function, then $S(\mu)$ $(\forall S \in W)$ are also integral linear functions. When $S = S_i$,

$$S(\mu) = S_i(\mu) = \mu - \frac{2(\mu, \alpha_i)}{(\alpha_i, \alpha_i)} \alpha_i,$$

thus

$$\frac{2(S(\mu), \alpha_j)}{(\alpha_j, \alpha_j)} = \frac{2(\mu, \alpha_j)}{(\alpha_i, \alpha_i)} - \frac{2(\mu, \alpha_i)}{(\alpha_i, \alpha_i)} \cdot \frac{2(\alpha_i, \alpha_j)}{(\alpha_j, \alpha_j)}, \quad j = 1, 2, \ldots n,$$

are integers and $S(\mu)$ is an integral linear function. Now W is generated by $S_1, \ldots, S_n$, thus for any $S \in W$, $S(\mu)$ is an integral linear function. For any $\alpha \in \Sigma_+$, by Lemma 5 of Chapter 6, there exists $S \in W$ such that $S(\alpha) = \alpha_i$ for some i. Therefore,

$$\frac{2(\mu, \alpha)}{(\alpha, \alpha)} = \frac{2(S(\mu), S(\alpha))}{(S(\alpha), S(\alpha))} = \frac{2(S(\mu), \alpha))}{(\alpha_i, \alpha_i)}$$

and $2(\mu, \alpha)/(\alpha, \alpha)$ is an integer.

COROLLARY 2. *If ϱ is an irreducible representation of a semisimple Lie algebra $\mathfrak{g}$, then in any equivalent class of weights of ϱ, there exists a dominant weight ω, i.e. ω satisfies $\omega \geqslant S(\omega)$ $(\forall S \in W)$.*

Proof. If ω_1 is a weight of ϱ and

$$\frac{2(\omega_1, \alpha_i)}{(\alpha_i, \alpha_i)} \geqslant 0 \qquad (i = 1, \ldots, n).$$

then ω_1 is a dominant weight. If

$$\frac{2(\omega_1, \alpha_j)}{(\alpha_j, \alpha_j)} \leqslant 0$$

for some j $(1 \leqslant j \leqslant n)$, then

$$\omega_1 \leqslant S_j(\omega_1) = \omega_1 - \frac{2(\omega_1, \alpha_j)}{(\alpha_j, \alpha_j)} \alpha_j.$$

Now consider $S_j(\omega_1)$, if

$$\frac{2(S_j(\omega_1), \alpha_i)}{(\alpha_i, \alpha_i)} \geqslant 0, \qquad i = 1, \ldots, n,$$

then $S_j(\omega_1)$ is dominant weight. Otherwise, there exists a weight which is equivalent to and higher than $S_j(\omega_1)$. By repeating the same argument and noticing that there are only a finite number of weights in ϱ that are equivalent to ω, a dominant weight in the class of ω is obtained.

10.2. Theorem of complete reducibility

THEOREM 4 *(H. Weyl†). Any representation of a semisimple Lie algebra $\mathfrak{g}$ is completely reducible.*

This is a famous theorem of H. Weyl; the original proof of Weyl made use of the theory of Lie groups. The following algebraic proof is due to Casimir and van der Waerden.‡

Notice that if Lemma 4 of Chapter 9 is true for semisimple Lie algebras, then this theorem will follow from a proof similar to the proof of Theorem 4 of Section 9.3. Thus the proof of this theorem is reduced to proving Lemma 4 of Chapter 9 for semisimple Lie algebras.

We first define an operator for $\mathfrak{g}$ which is a generalization of the Casimir operator for $\mathfrak{g}_3$. Let $X_\mu, \ldots$ be a basis of $\mathfrak{g}$ and

$$g_{\lambda\mu} = \operatorname{Tr} \operatorname{ad} X_\lambda \operatorname{ad} X_\mu .$$

By Cartan's criterion for semisimplicity, the matrix

$$(g_{\lambda\mu})$$

is non-degenerate. Let

$$(g_{\lambda\mu})^{-1} = (g^{\lambda\mu}).$$

If ϱ is a representation of $\mathfrak{g}$, then let

$$\varrho(G) = \sum_{\lambda, \mu} g^{\lambda\mu} \varrho(X_\lambda)\, \varrho(X_\mu)$$

† See footnote of Theorem 6 in Chapter 5.

‡ H. Casimir and B. L. van der Waerden, *Math. Ann.* **111** (1935), 1–12.

and call $\varrho(G)$ the Casimir operator of ϱ. It can be shown that $\varrho(G)$ commutes with every element in $\varrho(\mathfrak{g})$. In fact,

$$[\varrho(G), \varrho(X_\nu)] = \sum_{\lambda,\mu} g^{\lambda\mu}[\varrho(X_\lambda)\,\varrho(X_\mu), \varrho(X_\nu)]$$
$$= \sum_{\lambda,\mu} g^{\lambda\mu}\{\varrho(X_\lambda)\,[\varrho(X_\mu), \varrho(X_\nu)] + [\varrho(X_\lambda), \varrho(X_\nu)]\,\varrho(X_\mu)\}.$$

If

$$[X_\lambda, X_\mu] = \sum_\nu c^\nu_{\lambda\mu} X_\nu,$$

then

$$[\varrho(G), \varrho(X_\nu)] = \sum_{\lambda,\mu} g^{\lambda\mu}\left\{\varrho(X_\lambda)\sum_\tau c^\tau_{\mu\nu}\varrho(X_\tau) + \sum_\tau c^\tau_{\lambda\nu}\varrho(X_\tau)\,\varrho(X_\mu)\right\}$$
$$= \sum_{\lambda,\mu,\beta} \varrho(X_\lambda)\,\varrho(X_\tau)\,\{g^{\lambda\mu}c^\tau_{\mu\nu} + g^{\mu\tau}c^\lambda_{\mu\nu}\}.$$

If it can be shown that $(\sum_\mu g^{\lambda\mu}c^\tau_{\mu,\nu})_{\lambda,\tau}$ is a skew-symmetric matrix, then $[\varrho(G), \varrho(X_\nu)] = 0$. Notice that $g_{\tau\sigma} = \sum_{\alpha,\beta} c^\beta_{\tau\alpha}c^\alpha_{\sigma\beta}$, thus

$$\sum_{\lambda,\tau}\left(g_{\varrho\lambda}\sum_\mu g^{\lambda\mu}c^\tau_{\mu\nu}g_{\tau\sigma}\right) = \sum_{\mu,\tau}\delta^\mu_\varrho c^\tau_{\mu\nu}g_{\tau\sigma} = \sum_\tau c^\tau_{\varrho\nu}g_{\tau\sigma} = \sum_{\tau,\alpha,\beta} c^\tau_{\varrho\nu}c^\beta_{\tau\alpha}c^\alpha_{\sigma\beta}$$
$$= -\sum_{\tau,\alpha,\beta} c^\tau_{\nu\alpha}c^\beta_{\tau\varrho}c^\alpha_{\sigma\beta} - \sum_{\tau,\alpha,\beta} c^\tau_{\alpha\varrho}c^\beta_{\tau\nu}c^\alpha_{\sigma\beta} = \sum_{\tau,\alpha,\beta} c^\tau_{\nu\alpha}c^\beta_{\varrho\tau}c^\alpha_{\sigma\beta} + \sum_{\tau,\alpha,\beta} c^\tau_{\alpha\varrho}c^\beta_{\tau\nu}c^\alpha_{\beta\sigma}.$$

This proves that $\sum_\tau c^\tau_{\varrho\nu}g_{\tau\nu}$ is invariant under the permutation (ν, ϱ, σ). Now $\sum_\tau c^\tau_{\varrho\nu}g_{\tau\nu}$ is skew-symmetric with respect to ϱ and ν, thus it is also skew-symmetric with respect to ϱ and ν. This proves that the matrix

$$(g_{\varrho\lambda})\left(\sum_\mu g^{\lambda\mu}c^\tau_{\mu\nu}\right)(g_{\tau\sigma})$$

is skew-symmetric, thus $(\sum_\mu g^{\lambda\mu}c^\tau_{\mu\nu})_{\lambda,\tau}$, is also skew-symmetric.

Now $\varrho(G)$ commutes with every transformation in $\varrho(\mathfrak{g})$, thus by Schur's lemma, if ϱ is an irreducible representation of $\mathfrak{g}$, then $\varrho(G)$ is a scalar matrix. Notice that the eigenvalues of $\varrho_1(G)$ and $\varrho_2(G)$ may be equal even if ϱ_1 and ϱ_2 are not equivalent; this differs from the case of $\mathfrak{g}_3$. For example, consider A_n $(n \geqslant 2)$. Let

$$\varrho_1 : X \to X, \qquad \text{for all } X \in A_n$$

and

$$\varrho_2 : X \to -X', \qquad \text{for all } X \in A_n,$$

then ϱ_1 and ϱ_2 are irreducible representations of A_n with highest weights λ_1 and $-\lambda_{n+1}$ respectively. Thus ϱ_1 and ϱ_2 are not equivalent. Now

$$\varrho_1(G) = \sum_{\lambda,\mu} g^{\lambda\mu}\varrho_1(X_\lambda)\,\varrho_1(X_\mu) = \sum_{\lambda,\mu} g^{\lambda\mu}X_\lambda X_\mu,$$
$$\varrho_2(G) = \sum_{\lambda,\mu} g^{\lambda\mu}\varrho_2(X_\lambda)\,\varrho_2(X_\mu) = \sum_{\lambda,\mu} g^{\lambda\mu}(-X'_\lambda)(-X'_\mu)$$
$$= \sum_{\lambda,\mu} g^{\lambda\mu}(X_\mu X_\lambda)' = \sum_{\lambda,\mu} g^{\lambda\mu}(X_\lambda X_\mu)',$$

because $(g^{\lambda\mu})$ is symmetric. Since $\varrho_1(G)$ and $\varrho_2(G)$ are scalar matrices and

$$\varrho_1(G)' = \varrho_2(G),$$

we have

$$\varrho_1(G) = \varrho_2(G).$$

LEMMA 5. *Suppose* $\mathfrak{g}$ *is a semisimple Lie algebra,* ϱ *is a representation of* $\mathfrak{g}$ *with representation space* V *and* $V_1 \subset V$ *is an invariant subspace under* $\varrho(\mathfrak{h})$, $\varrho(E_\alpha)$, $\varrho(E_{-\alpha})$, *where* $E_{\pm\alpha}$ *are root vectors of the roots* $\pm\alpha$ *and* $[E_\alpha, E_{-\alpha}] = H_\alpha$. *Let* ω *be a weight of* ϱ *in the quotient space* V/V_1 *such that* $\omega+\alpha$ *is not a weight of* ϱ *in* V/V_1, *i.e. there exists* $e_0 \in V-V_1$ *such that*

$$\varrho(H)\,e_0 \equiv \omega(H)\,e_0 \pmod{V_1}, \quad \textit{for all} \quad H \in \mathfrak{h}$$

and there exists no $v \in V-V_1$ *such that*

$$\varrho(H)\,v \equiv (\omega+\alpha)(H)\,v \pmod{V_1}, \quad \textit{for all} \quad H \in \mathfrak{h}.$$

If

$$e_{-1} = \varrho(E_{-\alpha})\,e_0,\ \ldots,\ e_{-i} = \varrho(E_{-\alpha})^i\,e_0,\ \ldots$$

and p *is the first non-negative integer such that*

$$e_{-p} \not\equiv 0 \pmod{V_1} \quad \textit{and} \quad e_{-p-1} = \varrho(E_{-\alpha})\,e_{-p} \equiv 0 \pmod{V_1},$$

then $e_0, e_{-1}, \ldots, e_{-p}$ *are of weight* $\omega, \omega-\alpha, \ldots, \omega-p\alpha$, *respectively, and*

$$p = \frac{2(\omega, \alpha)}{(\alpha, \alpha)}.$$

Moreover,

$$\varrho(H)\,e_{-i} \equiv (\omega-i\alpha)(H)\,e_{-i} \pmod{V_1} \tag{10.5}$$

and

$$\varrho(E_\alpha)\,e_{-i} \equiv \frac{i(p-i+1)}{2}(\alpha, \alpha)\,e_{-i+1} \pmod{V_1}, \tag{10.6}$$

where $e_1 = 0$.

Proof. It is only necessary to prove (10.6). This will be proved by induction. Since $\omega+\alpha$ is not a weight in V/V_1, (10.6) is true for $i = 0$. Suppose (10.6) is true for $k \geqslant 0$, i.e.

$$\varrho(E_\alpha)\,e_{-k} \equiv \frac{k(p-k+1)}{2}(\alpha, \alpha)\,e_{-k+1} \pmod{V_1},$$

then

$$\begin{aligned}\varrho(E_\alpha)\,e_{-(k+1)} &\equiv \varrho(E_\alpha)\,\varrho(E_{-\alpha})\,e_{-k} \equiv \varrho([E_\alpha, E_{-\alpha}])\,e_{-k} + \varrho(E_{-\alpha})\,\varrho(E_\alpha)\,e_{-k} \\ &\equiv \varrho(H_\alpha)\,e_{-k} + \varrho(E_{-\alpha})\frac{k(p-k+1)}{2}(\alpha, \alpha)\,e_{-k+1} \\ &\equiv (\omega-k\alpha, \alpha)\,e_{-k} + \frac{k(p-k+1)}{2}(\alpha, \alpha)\,e_{-k} \\ &\equiv \frac{(k+1)(p-k)}{2}(\alpha, \alpha)\,e_{-k} \pmod{V_1},\end{aligned}$$

i.e. (10.6) is also true for $k+1$. The proof is thus completed.

Now define a partial ordering on $\mathfrak{h}_0^*$ by means of $\Pi = \{\alpha_1, \ldots, \alpha_n\}$ and suppose that no other weight of ϱ (with representation space V) is higher than ω. If $\alpha \in \Sigma_+$, then $\omega+\alpha$ is not a weight of ϱ. By Lemma 5, a series of vectors $e_0, e_{-1}, \ldots, e_{-p}$ can be obtained such that

$$e_{-i} = \varrho(E_{-\alpha})^i e_0, \quad e_{-p} \neq 0, \quad e_{-p-1} = 0;$$

the vectors $e_0, e_{-1}, \ldots, e_{-p}$ are of weights $\omega, \omega-\alpha, \ldots, \omega-p\alpha$ respectively. Moreover,

$$p = \frac{2(\omega, \alpha)}{(\alpha, \alpha)}$$

and

$$\varrho(E_\alpha) e_{-i} = \frac{i(p-i+1)}{2}(\alpha, \alpha) e_{-i+1}, \quad i = 0, 1, \ldots, p,$$

where $e_1 = 0$. Then there exists a weight ω' in $V/\langle e_0, e_{-1}, \ldots, e_{-p}\rangle$ which is not lower than any other weight. Let e_0' be a vector of weight ω', then

$$\varrho(H) e_0' \equiv \omega'(H) e_0' \ (\text{mod } \langle e_0, e_{-1}, \ldots, e_{-p}\rangle).$$

Starting with e_0', we obtain a series of vectors $e_0', e_{-1}', \ldots, e_{-p}$, where

$$e_{-i}' = \varrho(E_{-\alpha})^i e_0', \quad e_{-p'}' \not\equiv 0,$$
$$e_{-p'-1}' \equiv 0 \ (\text{mod } \langle e_0, e_{-1}, \ldots, e_{-p}\rangle);$$

the vectors $e_0', e_{-1}', \ldots, e_{-p'}'$, are of weights $\omega', \omega'-\alpha, \ldots, \omega'-p'\alpha$ respectively. Similarly

$$p' = \frac{2(\omega', \alpha)}{(\alpha, \alpha)}$$

and

$$\varrho(E_\alpha) e_{-i}' \equiv \frac{i(p'-i+1)}{2}(\alpha, \alpha) e_{-i+1}' \ (\text{mod } \langle e_0, \ldots, e_{-p}\rangle)$$
$$(i = 0, 1, 2, 3, \ldots, p'),$$

where $e_1' = 0$. By repeating the same argument, we obtain vectors $e_0, e_{-1}, \ldots, e_{-p}, e_0', e_{-1}', e_{-p'}', \ldots,$ that span the representation space V.

We can now prove Lemma 4 of Chapter 9 for any semisimple Lie algebra $\mathfrak{g}$. Let a representation ϱ of $\mathfrak{g}$ contain two irreducible representations ϱ_{ω_1} and ϱ_{ω_2} with highest weights ω_2 and ω_2 respectively. We consider the following three cases:

(1) ϱ_{ω_1} and ϱ_{ω_2} are not equivalent, but ω_1 is a weight of ϱ_{ω_2}.
Let

$$\varrho_{\omega_1}(G) = g_{\omega_1} I \quad \text{and} \quad \varrho_{\omega_2}(G) = g_{\omega_2} I,$$

where g_{ω_1} and g_{ω_2} are complex numbers. We will show that $g_{\omega_1} < g_{\omega_2}$.

Choose a basis of $\mathfrak{g}$ which is formed by a basis $H_1, \ldots, H_n$ of $\mathfrak{h}$ and root vectors E_α corresponding to the roots $\alpha \in \Sigma$ such that

$$(E_\alpha, E_{-\alpha}) = 1, \quad \alpha \in \Sigma.$$

Let

$$q_{ik} = \text{Tr ad } H_i \text{ ad } H_k,$$

and

$$(q^{ik}) = (q_{ik})^{-1},$$

then the matrix of the Casimir operator in a representation ϱ is

$$\varrho(G) = \sum_{i,k} q^{ik}\varrho(H_i)\,\varrho(H_k) + \sum_{\alpha} \varrho(E_\alpha)\,\varrho(E_{-\alpha}).$$

Let V and U be the representation spaces of ϱ_{ω_1} and ϱ_{ω_2} respectively. Let the weight space of ω_1 in ϱ_{ω_2} be U_{ω_1}. It is known that U_{ω_1} is invariant under $\varrho_{\omega_2}(H_i)\,\varrho_{\omega_2}(H_k)$ and $\varrho_{\omega_2}(E_\alpha)\,\varrho_{\omega_2}(E_{-\alpha})$. Let $r = \dim V_{\omega_1}$; then

$$g_{\omega_2} = \frac{1}{r}\operatorname{Tr}_{U_{\omega_1}}\varrho_{\omega_2}(G)$$

$$= \frac{1}{r} q^{ik} \operatorname{Tr}_{U_{\omega_1}} \varrho_{\omega_2}(H_i)\,\varrho_{\omega_2}(H_k) + \frac{1}{r}\sum_{\alpha} \operatorname{Tr}_{U_{\omega_1}} \varrho_{\omega_2}(E_\alpha)\,\varrho_{\omega_2}(E_{-\alpha}).$$

Let V_{ω_1} be the weight space of ω_1 in ϱ_{ω_1}. Since ω_1 is the highest weight of ϱ_1, V_{ω_1} is one-dimensional. Therefore,

$$g_{\omega_1} = \operatorname{Tr}_{V_{\omega_1}} \varrho_{\omega_1}(G)$$

$$= \sum_{i,k} q^{ik} \operatorname{Tr}_{V_{\omega_1}}\varrho_{\omega_1}(H_i)\,\varrho_{\omega_1}(H_k) + \sum_{\alpha} \operatorname{Tr}_{V_{\omega_1}} \varrho_{\omega_1}(E_\alpha)\,\varrho_{\omega_1}(E_{-\alpha}).$$

Obviously,

$$\operatorname{Tr}_{U_{\omega_1}} \varrho_{\omega_2}(H_i)\,\varrho_{\omega_2}(H_k) = r\omega_1(H_i)\,\omega_1(H_k).$$

Hence the first summand of the expressions g_{ω_1} and g_{ω_2} are equal. To show $g_{\omega_1} < g_{\omega_2}$, it suffices to show that for any $\alpha \in \Sigma$

$$\frac{1}{r}\operatorname{Tr}_{U_{\omega_1}} \varrho_{\omega_2}(E_\alpha)\,\varrho_{\omega_2}(E_{-\alpha}) \geqslant \operatorname{Tr}_{V_{\omega_1}} \varrho_{\omega_1}(E_\alpha)\,\varrho_{\omega_1}(E_{-\alpha}),$$

and that for at least one $\alpha \in \Sigma$, the inequality is strict, i.e. "$>$" is true. Since $\operatorname{Tr} \varrho(E_{-\alpha})\varrho(E_\alpha) = \operatorname{Tr} \varrho(E_\alpha)\,\varrho(E_{-\alpha})$, it suffices to consider the positive roots only.

Let α be a positive root; by our previous consideration, we can obtain a series of vectors

$$e_0,\ e_{-1},\ \ldots,\ e_{-p},\ e_0',\ e_{-1}',\ \ldots,\ e_{-p'}',\ \ldots, \tag{10.7}$$

that span the space U. Since ϱ_{ω_2} is irreducible, U can be generated by weight vectors, thus the vectors (10.7) can be assumed to be weight vectors. Let $\dot{e}_1, \ldots, \dot{e}_r$ be vectors in (10.7) that span U_{ω_1}. Suppose $\dot{e}_\nu$ belongs to a series of vectors of length $p_\nu + 1$ and is indexed by $-i_\nu$ in this series, then

$$\varrho_{\omega_2}(E_\alpha)\,\varrho_{\omega_2}(E_{-\alpha})\dot{e}_\nu = \frac{(i_\nu+1)(p_\nu-i_\nu)}{2}(\alpha, \alpha)\dot{e}_\nu.$$

Therefore,

$$\operatorname{Tr}_{U_{\omega_1}} \varrho_{\omega_2}(E_\alpha)\,\varrho_{\omega_2}(E_{-\alpha}) = \sum_{\nu=1}^{r} \frac{(i_\nu+1)(p_\nu-i_\nu)}{2}(\alpha, \alpha).$$

Similarly,

$$\mathrm{Tr}_{V_{\omega_1}} \varrho_{\omega_1}(E_\alpha)\, \varrho_{\omega_1}(E_{-\alpha}) = \frac{(0+1)(p-0)}{2} (\alpha, \alpha) = \frac{p}{2}(\alpha, \alpha),$$

where

$$p = \frac{2(\omega_1, \alpha)}{(\alpha, \alpha)}.$$

If the vector $\dot{e}_\nu$ indexed by $-i_\nu$ in a series of vectors of U is of weight ω_1, then

$$p_\nu = \frac{2(\omega_1 + i_\nu\alpha, \alpha)}{(\alpha, \alpha)} = p + 2i_\nu.$$

Therefore,

$$(i_\nu+1)(p_\nu - i_\nu) = (i_\nu+1)(p+i_\nu) \geqslant p$$

and

$$\mathrm{Tr}_{U_{\omega_1}} \varrho_{\omega_2}(E_\alpha)\, \varrho_{\omega_2}(E_{-\alpha}) \geqslant \sum_{\nu=1}^{r} \frac{p}{2}(\alpha, \alpha) = r\frac{p}{2}(\alpha, \alpha).$$

We have proved that

$$\frac{1}{r}\mathrm{Tr}_{U_{\omega_1}} \varrho_{\omega_2}(E_\alpha)\, \varrho_{\omega_2}(E_{-\alpha}) \geqslant \mathrm{Tr}_{V_{\omega_1}} \varrho_{\omega_1}(E_\alpha)\, \varrho_{\omega_1}(E_{-\alpha}).$$

Moreover, if equality holds for all α, then $i_\nu = 0$ ($\nu = 1, \ldots, r$). Then all the α-string of weights containing ω_1 in ϱ_{ω_2} start with ω_1. This cannot be true for all α. For if e_0 is of highest weight ω_2 in U,

$$e_1 = \varrho(E_{-\alpha_{i_1}}) \ldots \varrho(E_{-\alpha_{i_s}})e_0, \quad \alpha_{i_1}, \ldots, \alpha_{i_s} \in \Pi$$

is of weight ω_1 in U and $v = \varrho(E_{-\alpha_{i_2}}) \ldots \varrho(E_{-\alpha_{i_s}})e_0$, then $e_1 = \varrho(E_{-\alpha_{i_1}})v$ and $v \neq 0$. Notice that the weight of v is $\omega_1 + \alpha_{i_1}$, which is higher than ω_1. Thus e_1 cannot be the first of a series of weight vectors corresponding to an α_{i_1}-string of weights containing ω_1. Therefore, for at least one α,

$$\frac{1}{r}\mathrm{Tr}_{U_{\omega_1}} \varrho_{\omega_2}(E_\alpha)\, \varrho_{\omega_2}(E_{-\alpha}) > \mathrm{Tr}_{V_{\omega_1}} \varrho_{\omega_1}(E_\alpha)\, \varrho_{\omega_1}(E_{-\alpha}).$$

This proves that $g_{\omega_1} < g_{\omega_2}$.

Since $g_{\omega_1} \neq g_{\omega_2}$, it can be proved as in the case of $\mathfrak{g}_3$ that ϱ decomposes into the direct sum of ϱ_{ω_1} and ϱ_{ω_2}.

(2) ϱ_{ω_1} and ϱ_{ω_2} are not equivalent and none of the two highest weights is a weight of the other representation.

Let V be the representation space of ϱ and V_{ω_1} be the invariant subspace that is transformed according to ϱ_{ω_1}. Then V/V_{ω_1} is transformed according to ϱ_{ω_2}. Suppose e_{ω_i} ($i = 1, 2$) is a weight vector in V of weight ω_i ($i = 1, 2$). By assumption, ω_2 is not a weight of ϱ_{ω_1}, thus $e_{\omega_2} \notin V_{\omega_1}$.

Consider all vectors of the form

$$\varrho(E_{-\alpha_{i_1}}) \ldots \varrho(E_{-\alpha_{i_s}})\, \varrho(E_{\beta_{j_1}}) \ldots \varrho(E_{\beta_{j_t}})e_{\omega_2},$$

where $\alpha_{i_1}, \ldots, \alpha_{i_s}, \beta_{j_1}, \ldots, \beta_{j_t} \in \Pi$. If a vector of this form is non-zero, then it is of weight

$$\omega_2 - \alpha_{i_1} - \ldots - \alpha_{i_s} + \beta_{j_1} + \ldots + \beta_{j_t}.$$

These vectors form an invariant subspace V' of V. If $V' \cap V_{\omega_1} = \{0\}$, then $V = V_{\omega_1} + V'$ and Lemma 4 of Chapter 9 is true. We now show that $V' \cap V_{\omega_1} \neq \{0\}$ is impossible.

Suppose $V' \cap V_{\omega_1} \neq \{0\}$. Since V_{ω_1} is irreducible, $V_{\omega_1} \subset V'$. Thus

$$e_{\omega_1} = \Sigma c_\nu A_\nu e_{\omega_2},$$

where each c_ν is a complex number and each A_ν is an operator of the form $\varrho(E_{-\alpha_{i_1}}) \ldots \varrho(E_{-\alpha_{i_s}})\, \varrho(E_{\beta_{j_1}}) \ldots \varrho(E_{\beta_{j_t}})$. Since vectors of different weights are independent, we may assume that $A_1 e_{\omega_2}$ ($\neq 0$) is of weight ω_1. Since ω_1 is simple (for ω_1 is the highest weight of ϱ_1 and it is not a weight of ϱ_2),

$$A_1 e_{\omega_2} = a_1 e_{\omega_1}, \qquad a_1 \neq 0.$$

Let $A_1 = \varrho(E_{-\alpha_{i_1}}) \ldots \varrho(E_{-\alpha_{i_s}})\, \varrho(E_{\beta_{j_1}}) \ldots \varrho(E_{\beta_{j_t}})$, $\alpha_{i_1}, \ldots, \alpha_{i_s}, \beta_{j_1}, \ldots, \beta_{j_t} \in \Pi$. Then

$$\omega_1 = \omega_2 - \alpha_{i_1} - \ldots - \alpha_{i_s} + \beta_{j_1} + \ldots + \beta_{j_t}.$$

Now we distinguish the cases $\omega_1 > \omega_2$ and $\omega_1 < \omega_2$.

If $\omega_1 > \omega_2$, then $t > 0$. Thus e_{ω_2} is of weight ω_2 which is not a weight of ϱ_{ω_1} and $\omega_2 + \beta_j$ is a weight of ϱ_{ω_1} which is not a weight of ϱ_{ω_2} because $\varrho(E_{\beta_{j_t}}) e_{\omega_2}$ is of weight $\omega_2 + \beta_{j_t} > \omega_2$.

If $\omega_1 < \omega_2$, then $s > 0$. Thus ω_1 is a weight of ϱ_{ω_1} but not ϱ_{ω_2} and $\omega_1 + \alpha_{i_1}$ is weight of ϱ_{ω_2} but not ϱ_{ω_1}, because $\varrho(E_{-\alpha_{i_2}}) \ldots \varrho(E_{-\alpha_{i_s}})\, \varrho(E_{\beta_{j_1}}) \ldots \varrho(E_{\beta_{j_t}}) e_{\omega_2}$ is of weight $\omega_1 + \alpha_{i_1}$.

The proof is now reduced to proving the following:

LEMMA 6. *Let ϱ_1 and ϱ_2 be two representations of a semisimple Lie algebra $\mathfrak{g}$ and $\alpha \in \Sigma$. If ω is a linear function defined on $\mathfrak{h}$, then it is impossible that ω is a weight of ϱ_1 but not of ϱ_2 and $\omega + \alpha$ is a weight of ϱ_2 but not of ϱ_1.*

Proof. Suppose ω is a weight of ϱ_1 but not of ϱ_2. Construct an α-string of weights for ϱ_1 starting with ω,

$$\omega, \omega - \alpha, \ldots, \omega - p\alpha,$$

where

$$p = \frac{2(\omega, \alpha)}{(\alpha, \alpha)}.$$

Suppose $\omega + \alpha$ is a weight of ϱ_2 but not of ϱ_1. Construct an α-string of weights for ϱ_2 starting with $\omega + \alpha$,

$$\omega + \alpha, \omega + 2\alpha, \ldots, \omega + (q+1)\alpha,$$

where

$$-q = \frac{2(\omega + \alpha, \alpha)}{(\alpha, \alpha)} = 2\frac{(\omega, \alpha)}{(\alpha, \alpha)} + 2.$$

Then $p + q = -2$ and this is impossible.

(3) ϱ_{ω_1} and ϱ_{ω_2} are equivalent. Let $\omega = \omega_1 = \omega_2$.

Let V be the representation space of ϱ, then V contains an irreducible invariant subspace V_1 which contains a vector e_0 of weight ω_1 and is generated by the vectors of the form

$$\varrho(E_{-\alpha_{i_1}}) \dots \varrho(E_{-\alpha_{i_s}})e_0, \quad \alpha_{i_1}, \dots, \alpha_{i_s} \in \Pi.$$

V/V_1 also contains a vector $\bar{e}_0'$ of weight ω. Choose e_0' such that

$$\varrho(H)e_0' = \omega(H)e_0' + \mu(H)e_0.$$

We will show that $\mu = 0$.

Let $\alpha \in \Sigma_+$ and construct the series of vectors

$$e_0', e_{-1}' = \varrho(E_{-\alpha})e_0', \dots, e_{-i}' = \varrho(E_{-\alpha})^i e_0', \dots.$$

If p' is the first non-negative integer such that

$$e_{-p'}' \not\equiv 0 \,(\text{mod } V_1), \quad e_{-p'-1}' = \varrho(E_{-\alpha})e_{-p}' \equiv 0 \,(\text{mod } V_1),$$

then

$$p' = \frac{2(\omega, \alpha)}{(\alpha, \alpha)} = p.$$

By induction, it can be shown that

$$\varrho(H)e_{-i}' = (\omega - i\alpha)(H)e_{-i}' + \mu(H)e_{-i}, \tag{10.8}$$

$$\varrho(E_\alpha)e_{-i}' = \frac{i(p-i+1)}{2}(\alpha, \alpha)e_{-i+1}' + i(\mu, \alpha)e_{-i+1} \tag{10.9}$$

$$(i = 0, 1, 2, \dots, p+1),$$

where $e_1 = e_1' = 0$.

Setting $i = p+1$ in (10.8), we get

$$\varrho(H)e_{-(p+1)}' = \big(\omega - (p+1)\alpha\big)(H)e_{-(p+1)}'.$$

Hence $e_{-(p+1)}'$ is of weight $\omega - (p+1)\alpha$. Now $\omega - (p+1)\alpha$ is not a weight of ϱ, thus $e_{-(p+1)}' = 0$ Setting $i = p+1$ in (10.9), we get

$$0 = (p+1)(\mu, \alpha)e_p.$$

Since $e_p \neq 0$, $(\mu, \alpha) = 0$. Now α is arbitrary, so $(\mu, \alpha) = 0$ for all $\alpha \in \Sigma$ and $\mu = 0$. This proves that

$$\varrho(H)e_0' = \omega(H)e_0',$$

i.e. e_0' is also a weight vector of weight ω.

Consider the vectors of the form

$$\varrho(E_{-\alpha_{i_1}}) \dots \varrho(E_{-\alpha_{i_s}})e_0', \quad \alpha_{i_1}, \dots, \alpha_{i_s} \in \Pi.$$

If such a vector is non-zero, then it is of weight $\omega - \alpha_{i_1} - \dots - \alpha_{i_s}$. These vectors generate an invariant subspace V_2. If $V_1 \cap V_2 \neq \{0\}$, then $V_1 \subset V_2$. Hence

$$e_0 = \Sigma c_\nu A_\nu e_0',$$

where each A_ν is an operator of the form $\varrho(E_{-\alpha_{i_1}}) \ldots \varrho(E_{-\alpha_{i_s}})$. Since vectors of different weights are independent,

$$e_0 = ce_0', \qquad c \neq 0.$$

contradiction. Thus $V_1 \cap V_2 = \{0\}$ and $V = V_1 \dotplus V_2$.

Now we have proved that Lemma 4 of Chapter 9 is true for any semisimple Lie algebra $\mathfrak{g}$. As mentioned before, complete reducibility of representations of $\mathfrak{g}$ can be proved as in the case of $\mathfrak{g}_3$.

10.3. Fundamental representations of semisimple Lie algebras

THEOREM 5. *Let $\mathfrak{g}$ be a semisimple Lie algebra, $\mathfrak{h}$ be a Cartan subalgebra and $\Pi = \{\alpha_1, \ldots, \alpha_n\}$ be a fundamental system of roots. If ϱ_i $(i = 1, 2)$ is an irreducible representation of $\mathfrak{g}$ with highest weight ω_i $(i = 1, 2)$ and representation space V_i $(i = 1, 2)$, then $\varrho = \varrho_1 \otimes \varrho_2$ contains a unique irreducible component with highest weight $\omega_1 + \omega_2$ (this irreducible component is denoted by $\overline{\varrho_1 \otimes \varrho_2}$).*

Proof. Let $e_i^{(1)}, \ldots, e_i^{(s_i)}$ be a basis of V_i such that each of these vectors is a weight vector and $e_i^{(1)}$ is of weight ω_i $(i = 1, 2)$. Then $e_1^{(j)} \otimes e_2^{(k)}$ are weight vectors of $\varrho = \varrho_1 \otimes \varrho_2$. For if $e_1^{(j)}$ is of weight $\omega_1^{(j)}$ and $e_2^{(k)}$ is of weight $\omega_2^{(k)}$, then

$$\begin{aligned} \varrho(H)(e_1^{(j)} \otimes e_2^{(k)}) &= \varrho_1(H)e_1^{(j)} \otimes e_2^{(k)} + e_1^{(j)} \otimes \varrho_2(H)e_2^{(k)} \\ &= (\omega_1^{(j)}(H) + \omega_2^{(k)}(H)(e_1^{(j)} \otimes e_2^{(k)}), \end{aligned}$$

i.e. $e_1^{(j)} \otimes e_2^{(k)}$ is of weight $\omega_1^{(j)} + \omega_2^{(k)}$. In particular, $e_1^{(1)} \otimes e_2^{(1)}$ is of weight $\omega_1 + \omega_2$. Since ω_i is the highest weight of ϱ_i $(i = 1, 2)$, ω_i is simple. Hence $\omega_1 + \omega_2$ is also simple and higher than any other weight of ϱ.

Let V' be the smallest invariant subspace containing $e_1^{(1)} \otimes e_2^{(1)}$. We now show that V' is irreducible. Suppose V' is reducible; by Theorem 4, we may let

$$V' = U_1 \dotplus U_2,$$

where both U_1 and U_2 are non-zero invariant subspaces. Let

$$e_1^{(1)} \otimes e_2^{(1)} = v_1 + v_2 \qquad v_1 \in U_1, \quad v_2 \in U_2;$$

substituting $v_1 + v_2 = e_1^{(1)} \otimes e_2^{(1)}$ into

$$\varrho(H)(e_1^{(1)} \otimes e_2^{(1)}) = (\omega_1 + \omega_2)(H)(e_1^{(1)} \otimes e_2^{(1)}),$$

we get

$$\varrho(H)v_1 + \varrho(H)v_2 = (\omega_1 + \omega_2)(H)v_1 + (\omega_1 + \omega_2)(H)v_2.$$

Therefore,

$$\varrho(H)v_1 = (\omega_1 + \omega_2)(H)v_1$$

and

$$\varrho(H)v_2 = (\omega_1 + \omega_2)(H)v_2$$

Since $\omega_1 + \omega_2$ is simple, $v_1 = 0$ or $v_2 = 0$; v_1 and v_2 cannot be zero simultaneously. Let $v_2 = 0$, then $e_1^{(1)} \otimes e_2^{(1)} = v_1$. Since V' is the smallest invariant subspace containing $e_1^{(1)} \otimes e_2^{(1)}$,

$V' \subset U_1$, so $V' = U_1$. Thus $U_2 = 0$—— contradiction. This proves that V' is an irreducible subspace of ϱ. Since $e_1^{(1)} \otimes e_2^{(1)} \in V'$ and $e_1^{(1)} \otimes e_2^{(1)}$ is of weight $\omega_1+\omega_2$, ϱ induces an irreducible representation $\overline{\varrho_1 \otimes \varrho_2}$ on V' with highest weight $\omega_1+\omega_2$. The proof of Theorem 5 is now completed.

Under the assumptions of Theorem 5, we define the following n linear functions $\omega_1, \ldots, \omega_n$:

$$\frac{2(\omega, \alpha_i)}{(\alpha_i, \alpha_i)} = \delta_{ij}, \qquad i, j = 1, \ldots, n.$$

Then $\omega_1, \ldots, \omega_n$ are dominant integral linear functions. Suppose $\varrho_1, \ldots, \varrho_n$ are irreducible representations of $\mathfrak{g}$ with highest weights $\omega_1, \ldots, \omega_n$ respectively. If ω is a dominant integral linear function, i.e.

$$\frac{2(\omega, \alpha_i)}{(\alpha_i, \alpha_i)} = r_i, \qquad i = 1, \ldots, n,$$

are non-negative integers, then by Lemma 5,

$$\underbrace{\varrho_1 \otimes \ldots \otimes \varrho_1}_{r_1} \otimes \underbrace{\varrho_2 \otimes \ldots \otimes \varrho_2}_{r_2} \ldots \otimes \underbrace{\varrho_n \otimes \ldots \otimes \varrho_n}_{r_n}$$

contains a unique component with highest weight ω. Therefore, in order to prove the second part of Theorem 3, it suffices to show the existence of $\varrho_1, \ldots, \varrho_n$. These representations are said to be the fundamental representations of $\mathfrak{g}$.

THEOREM 6. *Let a semisimple Lie algebra $\mathfrak{g}$ be the direct sum of two semisimple subalgebras $\mathfrak{g}_1$ and $\mathfrak{g}_2$, i.e. $\mathfrak{g} = \mathfrak{g}_1 \dotplus \mathfrak{g}_2$. Let the Cartan subalgebra $\mathfrak{h}$ of $\mathfrak{g}$ be correspondingly decomposed as $\mathfrak{h} = \mathfrak{h}_1 \dotplus \mathfrak{h}_2$, where $\mathfrak{h}_i$ is a Cartan subalgebra of $\mathfrak{g}_i$ $(i = 1, 2)$. Suppose the fundamental system of root Π of $\mathfrak{g}$ with respect to $\mathfrak{h}$ is decomposed into $\Pi = \Pi_1 \cup \Pi_2$, where $\Pi_1 = \{\alpha_1, \ldots \alpha_n\}$ and $\Pi_2 = \{\beta_1, \ldots, \beta_m\}$ and Π_i is the fundamental system of roots of $\mathfrak{g}_i$ with respect to $\mathfrak{h}_i$ $(i = 1, 2)$. Then*

(i) *Suppose ϱ_i is an irreducible representation of $\mathfrak{g}_i$ with highest weight ω_i and representation space V_i $(i = 1, 2)$. If*

$$\varrho(X)(v_1 \otimes v_2) = \varrho_1(X_1)v_1 \otimes v_2 + v_1 \otimes \varrho_2(X_2)v_2,$$

$$X = X_1 + X_2, \quad X_i \in \mathfrak{g}_i, \qquad i = 1, 2,$$

then $\varrho(X)$ is an irreducible representation of $\mathfrak{g}$ with highest weight $\omega_1+\omega_2$, where

$$(\omega_1+\omega_2)(H) = \omega_1(H_1)+\omega_2(H_2), \quad H = H_1+H_2, \quad H_i \in \mathfrak{h}_i, \qquad i = 1, 2.$$

(ii) *Conversely, every irreducible representation of $\mathfrak{g}$ is obtained as in* (i).

Proof. Let $X = X_1+X_2 \in \mathfrak{g}$, $X_i \in \mathfrak{g}_i$ $(i = 1, 2)$.
Let

$$\varrho_1'(X)v_1 = \varrho_1(X_1)v_1, \quad v_1 \in V_1,$$
$$\varrho_2'(X)v_2 = \varrho_2(X_2)v_2, \quad v_2 \in V_2,$$

then ϱ_i' $(i = 1, 2)$ is an irreducible representation of $\mathfrak{g}$ with representation space V_i and highest weight ω_i $(i = 1, 2)$. If for any $H = H_1+H_2 \in \mathfrak{h}$, $H_i \in \mathfrak{h}_i$ $(i = 1, 2)$, we let

$$\omega_1(H) = \omega_1(H_1) \quad \text{and} \quad \omega_2(H) = \omega_2(H_2),$$

then

$$\varrho = \varrho_1' \otimes \varrho_2'.$$

Thus in order to prove (i), it suffices to show the irreducibility of ϱ.

Let $\{e_i^{(1)}, \ldots, e_i^{(s_i)}\}$ be a weight basis of V_i $(i = 1, 2)$ where $e_i^{(k)}$ is of weight $\omega_i^{(k)}$ $(i = 1, 2;$ $k = 1, \ldots, s_i)$ and $\omega_1^{(1)} = \omega_1$, $\omega_2^{(1)} = \omega_2$. By the proof of Theorem 2, every $e_1^{(k)}$ can be assumed to have the form

$$\varrho_1(E_{-\alpha_{j_1}}) \ldots \varrho_1(E_{-\alpha_{j_s}})e_1^{(1)}, \qquad s \geqslant 0$$

and every $e_2^{(k)}$ can be assumed to have the form

$$\varrho_2(E_{-\beta_{k_1}}) \ldots \varrho_2(E_{-\beta_{k_t}})e_2^{(1)}, \qquad t \geqslant 0,$$

where $\alpha_{j_1}, \ldots, \alpha_{j_s} \in \Pi_1$ and $\beta_{k_1}, \ldots, \beta_{k_t} \in \Pi_2$. If V' is the smallest invariant subspace containing $e_1^{(1)} \otimes e_2^{(1)}$, then V' contains all $e_1^{(j)} \otimes e_2^{(k)}$ $(j = 1, 2, \ldots, s_1;\ k = 1, 2, \ldots, s_2)$, so $V' = V$. By the proof of Theorem 5, V' is irreducible, so V is irreducible. This proves (i).

Now suppose ϱ is an irreducible representation of $\mathfrak{g}$ with representation space V and highest weight ω. The restriction of ϱ to $\mathfrak{g}_i$ is a representation ϱ_i' of $\mathfrak{g}_i$ $(i = 1, 2)$. Since $[\mathfrak{g}_1, \mathfrak{g}_2] = 0$,

$$\varrho(X_1)\varrho(X_2) = \varrho(X_2)\varrho(X_1), \quad \forall X_1 \in \mathfrak{g}_1, \quad X_2 \in \mathfrak{g}_2; \tag{10.10}$$

so

$$\varrho_1'(X_1)\,\varrho_2'(X_2) = \varrho_2'(X_2)\,\varrho_1'(X_1). \tag{10.10$'$}$$

Suppose ϱ_2' is reducible. Since ϱ is irreducible, by (10.10′) and Schur's lemma, ϱ_2' can be decomposed into the sum of equivalent irreducible representations, i.e.

$$\varrho_2' = \underbrace{\varrho_2 + \varrho_2 + \ldots + \varrho_2}_{s_1},$$

where ϱ_2 is an irreducible representation of $\mathfrak{g}_2$. If the dimension of ϱ_2 is s_2, then $\dim V = s_1 s_2$. A basis of V can be chosen so that the matrices in ϱ_2' have the form

$$\varrho_2'(X_2) = \begin{bmatrix} \varrho_2(X_2) & & \\ & \ddots & \\ & & \varrho_2(X_2) \end{bmatrix}_{s_1}, \qquad X_2 \in \mathfrak{g}_2.$$

By Schur's lemma, matrices in ϱ_1' have the form

$$\varrho_1'(X_1) = \begin{bmatrix} a_{11}(X_1)I_{s_2} & \ldots & a_{1s_1}(X_1)I_{s_2} \\ a_{21}(X_1)I_{s_2} & \ldots & a_{2s_1}(X_1)I_{s_2} \\ \vdots & & \vdots \\ a_{s_1 1}(X_1)I_{s_2} & \ldots & a_{s_1 s_1}(X_1)I_{s_2} \end{bmatrix}, \qquad X_1 \in \mathfrak{g}_1.$$

Let

$$\varrho_1(X_1) = \begin{bmatrix} a_{11}(X_1) & \dots & a_{1s_1}(X_1) \\ a_{21}(X_1) & \dots & a_{2s_1}(X_1) \\ \cdot & \cdot\ \cdot\ \cdot & \cdot \\ \cdot & \cdot\ \cdot\ \cdot & \cdot \\ \cdot & \cdot\ \cdot\ \cdot & \cdot \\ a_{s_1 1}(X_1) & \dots & a_{s_1 s_1}(X_1) \end{bmatrix},$$

then

$$X_1 \to \varrho_1(X_1)$$

is a representation of $\mathfrak{g}_1$ with dimension s_1 and

$$\varrho_1'(X_1) = \varrho_1(X_1) \otimes I_{s_2}.$$

Therefore, if $X = X_1 + X_2$, $X_1 \in \mathfrak{g}_1$ and $X_2 \in \mathfrak{g}_2$, then

$$\varrho(X) = \varrho_1'(X_1) + \varrho_2'(X_2) = \varrho_1(X_1) \otimes I_{s_2} + I_{s_1} \otimes \varrho_2(X_2).$$

Since ϱ is irreducible, so is ϱ_1. This proves (ii).

From the proof, it can be seen that if the highest weight ω of ϱ is decomposed into

$$\omega = \omega_1 + \omega_2, \quad \omega_1 \in \mathfrak{h}_1^*, \quad \omega_2 \in \mathfrak{h}_2^*,$$

then ϱ_1 is an irreducible representation with highest weight ω_1 and ϱ_2 is an irreducible representation with highest weight ω_2.

COROLLARY. *A fundamental representation of a semisimple Lie algebra $\mathfrak{g}$ is one–one on a simple ideal of $\mathfrak{g}$ and maps all other simple ideals of $\mathfrak{g}$ to the zero matrix.*

Proof. This follows from the fact that an irreducible representation with highest weight zero is the zero representation.

Therefore, to prove the second part of Theorem 3, it suffices to prove the existence of fundamental representations for simple Lie algebras.

10.4. Tensor representations

Let V be an N-dimensional vector space over the complex numbers and r be a positive integer. The Kronecker product

$$V^r = \underbrace{V \otimes \dots \otimes V}_{r}$$

V^r is said to be the tensor space of rank r of V. If $e_1, \dots, e_N$ form a basis of V, then

$$e_{i_1} \otimes e_{i_2} \otimes \dots \otimes e_{i_r}, \quad 1 \leqslant i_1, \dots, i_r \leqslant N,$$

form a basis of V^r.

Now suppose ϱ is a representation of a Lie algebra $\mathfrak{g}$ with representation space V. Let $x \in \mathfrak{g}$ and define

$$\{\varrho^r(x)\}(e_{i_1}\otimes \dots \otimes e_{i_r}) = \sum_{j=1}^{r} e_{i_1}\otimes \dots \otimes e_{i_{j-1}}\otimes \varrho(x)\, e_{i_j}\otimes e_{i_{j+1}} \dots \otimes e_{i_r},$$

then ϱ^r is a representation of $\mathfrak{g}$ with representation space V^r. In fact,

$$\varrho^r = \underbrace{\varrho\otimes \dots \otimes \varrho}_{r}$$

If $\mathfrak{g}$ is a semisimple Lie algebra and $e_1, \dots, e_N$ are weight vectors of weights $\omega_1, \dots, \omega_N$ respectively, then $e_{i_1}\otimes \dots \otimes e_{i_N}$ is a weight vector of ϱ^r of weight $\omega_{i_1}+\dots+\omega_{i_r}$. In particular, if ϱ is an irreducible representation with highest weight ω_1 (relative to some Π), then $r\omega$, is the highest weight of ϱ^r and

$$\underbrace{e_1\otimes \dots \otimes e_1}_{r}$$

is of weight $r\omega_1$. Therefore, by Theorem 5, the smallest invariant subspace containing $e_1\otimes \dots \otimes e_1$ is irreducible and it is the representation space of the irreducible representation with highest weight $r\omega_1$. Denote this representation by $\overline{\varrho^r}$.

Let S_r be the symmetric group on r letters $1, \dots, r$. For $p \in S_r$ and $v_1, \dots, v_r \in V$, define

$$\varphi(p)\,(v_1\otimes \dots \otimes v_r) = v_{p(1)}\otimes \dots \otimes v_{p(r)}.$$

Then

$$p \to \varphi(p)$$

is a representation of S_r with representation space V^r. This representation can be extended to a representation of the group ring R_r of S_r if for any $a = \sum_{p\in S_r} a_p\cdot p \in R_r$, $\varphi(a)$ is defined by

$$\varphi(a) = \sum_{p\in S_r} a_p\cdot \varphi(p).$$

If V is the representation space of the representation ϱ of a Lie algebra, then for $a \in R_r$ and $x \in \mathfrak{g}$,

$$\varphi(a)\,\{\varrho^r(x)\} = \{\varrho^r(x)\}\,\varphi(a).$$

Thus for any subset $\{a_1, \dots, a_m\}$ of R_r, if

$$V^r_{\{a_1,\dots,a_m\}} = \{u \in V^r \,|\, \varphi(a_1)\, u = \dots = \varphi(a_m)\, u = 0\}$$

then $V^r_{\{a_1,\dots,a_m\}}$ is an invariant subspace of ϱ^r. The representation induced by ϱ^r in this subspace is denoted by $\varrho^r_{\{a_1,\dots,a_m\}}$.

In particular, consider the subset $\{e-p|p \in S_r\}$, where e is the identity of S_r. We obtain $\Lambda^r_{\{e-p\,|\,p\in S_r\}}$ and $\varrho^r_{\{e-p\,|\,p\in S_r\}}$. A vector

$$u = \sum_{i_1,\dots,i_r=1}^{N} a_{i_1,\dots,i_r} e_{i_1}\otimes \dots \otimes e_{i_r} \in V^r$$

is in $V^r_{\{e-p \mid p\in S_r\}}$ iff for all $p \in S_r$

$$a_{i_{p(1)} i_{p(2)}, \ldots, p(r)} = a_{i_1, \ldots, i_r}.$$

Thus the tensors in $V^r_{\{e-p \mid p\in S_r\}}$ are called symmetric tensors and $V^r_{\{e-p \mid p\in S_r\}}$ is called the space of symmetric tensors of rank r (of V). $V^r_{\{e-p \mid p\in S_r\}}$ is also denoted by $V^{(r)}$ and the representation induced by ϱ^r on $V^{(r)}$ is denoted by $\varrho^{(r)}$. Let $v_1, \ldots, v_r$ be vectors in V, then the vector

$$\sum_{p\in S_r} v_{p(1)} \otimes \ldots \otimes v_{p(r)}$$

is said to be the symmetrization of $v_1 \otimes \ldots \otimes v_r$. Therefore, the symmetrizations of the vectors

$$e_{i_1} \otimes \ldots \otimes e_{i_r}, \qquad 1 \leqslant i_1 \leqslant \ldots \leqslant i_r \leqslant N,$$

form a basis of $V^{(r)}$, so

$$\dim V^{(r)} = \binom{n+r-1}{r}.$$

Moreover, if ϱ is an irreducible representation of a semisimple Lie algebra $\mathfrak{g}$ and $e_1, \ldots, e_N$ are of weights $\omega_1, \ldots, \omega_N$ respectively, then the symmetrization of

$$e_{i_1} \otimes \ldots \otimes e_{i_r}$$

is a weight vector of weight $\omega_{i_1} + \ldots + \omega_{i_r}$ of $\varrho^{(r)}$. Thus the highest weight of $\varrho^{(r)}$ is $r\omega_1$. Denote the irreducible component of $\varrho^{(r)}$ with highest weight $r\omega_1$ by $\overline{\varrho^{(r)}}$, then $\overline{\varrho^r} = \overline{\varrho^{(r)}}$.

For any $p \in S_r$, define $\operatorname{sgn}(p) = 1$ if p is even and $\operatorname{sgn}(p) = -1$ if p is odd. Consider the subset $\{e - \operatorname{sgn}(p)\, p \mid p \in S_r\}$ of R_r. As before, we obtain $V^r_{\{e-\operatorname{sgn}(p) \mid p \in S_r\}}$ and $\varrho^r_{\{e-\operatorname{sgn}(p)p \mid p \in S_r\}}$. A vector

$$u = \sum_{i_1, \ldots, i_r = 1}^{N} a_{i_1, \ldots, i_r} e_{i_1} \otimes \ldots \otimes e_{i_r} \in V^r$$

is in $V^r_{\{e-\operatorname{sgn}(p)p \mid p \in S_r\}}$ iff for all $p \in S_r$,

$$a_{i_{p(1)} i_{p(2)} \ldots i_{p(r)}} = (\operatorname{sgn}(p)) a_{i_1, \ldots, i_r}.$$

Therefore, the tensors in this space are said to be anti-symmetric tensors of rank r and the space is said to be the space of anti-symmetric tensors of rank r. Denote this space by $V^{[r]}$ and the representation induced by ϱ^r in this space by $\varrho^{[r]}$. For $v_1, \ldots, v_r \in V$, the vector

$$\sum_{p\in S_r} \operatorname{sgn}(p) v_{p(1)} \otimes \ldots \otimes v_{p(r)}$$

is said to be the anti-symmetrization of $v_1 \otimes \ldots \otimes v_r$; denote it by

$$[v_1, \ldots, v_r].$$

Therefore, the anti-symmetrizations of the vectors

$$e_{i_1} \otimes \ldots \otimes e_{i_r}, \qquad i_1 < i_2 < \ldots < i_r,$$

form a basis of $V^{[r]}$, so dim $V^{[r]} = \binom{n}{r}$. Moreover, if ϱ is an irreducible representation of a semisimple Lie algebra $\mathfrak{g}$, and $e_1, \ldots, e_N$ are vectors of weights $\omega_1, \ldots, \omega_N$ respectively, then

$$[e_{i_1}, \ldots, e_{i_r}], \qquad i_1 < i_2 \ldots < i_r,$$

is a weight vector of $\varrho^{[r]}$ of weight $\omega_{i_1} + \ldots + \omega_{i_r}$. If

$$\omega_1 \geqslant \ldots \geqslant \omega_N,$$

then the highest weight of $\varrho^{[r]}$ is

$$\omega_1 + \ldots + \omega_r.$$

Denote the irreducible component of $\varrho^{[r]}$ with highest weight $\omega_1 + \ldots + \omega_r$ by $\overline{\varrho^{[r]}}$.

10.5. Elementary representations of simple Lie algebras

Let $\mathfrak{g}$ be a semisimple Lie algebra, $\mathfrak{h}$ be a Cartan subalgebra and $\Pi = \{\alpha_1, \ldots, \alpha_n\}$ be a fundamental system of roots of $\mathfrak{g}$ with respect to $\mathfrak{h}$. Let ϱ be an irreducible representation of $\mathfrak{g}$ with highest weight ω. Recall that ϱ is a fundamental representation iff in the associated diagram there is a one at one dot and zeros at the other dots. Therefore, corresponding to every fundamental root α, there is a fundamental representation ϱ_α.

We now number the Dynkin diagrams of the simple Lie algebras as follows:

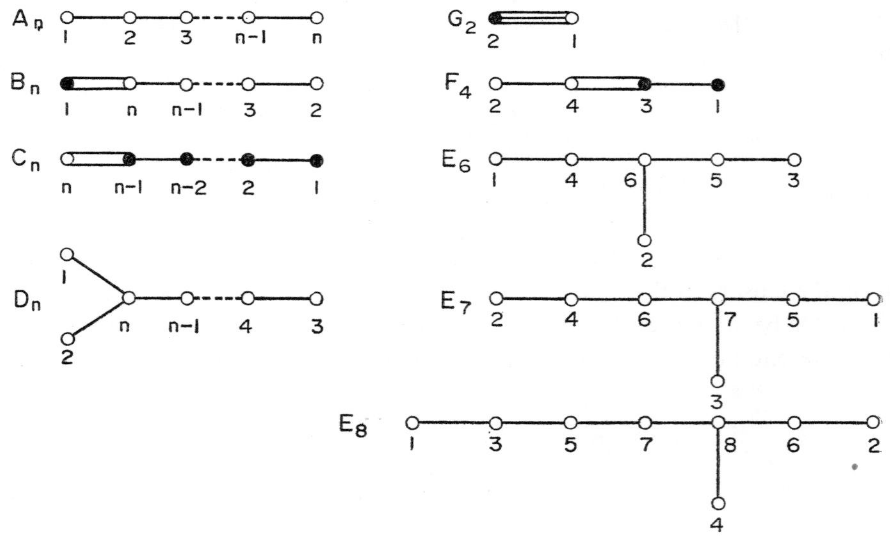

(In these diagrams, the dot ● represents vectors of shorter length.)

Now consider the Dynkin diagram of a simple Lie algebra. The dots that are connected only to one other dot are called end-points of the diagram; the corresponding fundamental representations are called elementary representations. Elementary representations of A_n are denoted by ϱ_1, ϱ_n and elementary representations of E_6 are denoted by ϱ_1, ϱ_2, ϱ_3, etc.

Let β be an end-point of the Dynkin diagram of a simple algebra $\mathfrak{g}$. A branch of β is a series of dots

$$\beta_1 = \beta, \beta_2, \ldots, \beta_k$$

that satisfy:

1° Every β_i $(i = 2, 3, \ldots, k-1)$ is only connected to β_{i-1} and β_{i+1} in the diagram.

2° The connection between β_i and β_{i+1} is of one of the following three forms:

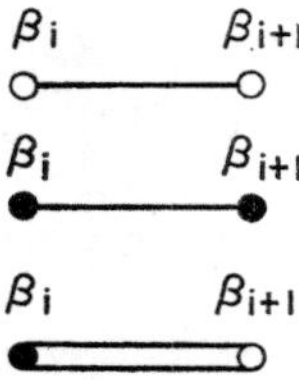

If the connection is of the last form, then $i+1 = k$.

3° The series $\beta_1, \ldots, \beta_k$ is maximal with respect to properties 1° and 2°.

For example, the simple Lie algebra A_n, B_n, C_n, D_n, E_6, E_7, E_8, F_4 and G_2 have the following branches:

$$A_n \begin{cases} 1, 2, \ldots, n, \\ n, n-1, \ldots, 1, \end{cases} \qquad G_2 \begin{cases} 1, \\ 2, \end{cases}$$

$$B_n \begin{cases} 1, n, \\ 2, 3, \ldots, n, \end{cases} \qquad F_4 \begin{cases} 1, 3, 4, \\ 2, 4, \end{cases}$$

$$C_n \begin{cases} 1, 2, \ldots, n, \\ n, \end{cases} \qquad E_6 \begin{cases} 1, 4, 6, \\ 2, 6, \\ 3, 5, 6 \end{cases}$$

$$D_n \begin{cases} 1, n, \\ 2, n, \\ 3, 4, \ldots, n, \end{cases} \qquad E_7 \begin{cases} 1, 5, 7 \\ 2, 4, 6, 7, \\ 3, 7, \end{cases}$$

$$E_8 \begin{cases} 1, 3, 5, 7, 8 \\ 2, 6, 8, \\ 4, 8. \end{cases}$$

THEOREM 7. *Let β be an end-point of a simple Lie algebra $\mathfrak{g}$ and*

$$\beta = \beta_1, \ldots, \beta_k$$

be a branch of β, then for $r = 1, 2, \ldots, k$,

$$\varrho_{\beta_r} = \overline{\varrho^{[r]}_{\beta_1}}.$$

Proof. Let ω be the highest weight of ϱ_{β_1}, i.e.

$$\frac{2(\omega, \beta_1)}{(\beta_1, \beta_1)} = 1 \quad \text{and} \quad \frac{2(\omega, \alpha)}{(\alpha, \alpha)} = 0 \quad \text{for } \alpha \in \Pi - \{\beta_1\}.$$

By Lemma 2, any $\omega-\alpha$ ($\alpha \in \Pi-\{\beta_1\}$) is not a weight of ϱ_{β_1}, $\omega-\beta_1$ is a weight and $\omega-2\beta_1$ is not a weight of ϱ_{β_1}. Hence $\omega-\beta_1$ is the second highest weight of ϱ_{β_1}. By the same lemma, ω and $\omega-\beta_1$ have the same multiplicity, thus $\omega-\beta_1$ is also simple.

Now

$$\frac{2(\omega-\beta_1, \beta_1)}{(\beta_1, \beta_1)} = -1, \quad \frac{2(\omega-\beta_1, \beta_2)}{(\beta_2, \beta_2)} = 1,$$

$$\frac{2(\omega-\beta_1, \alpha)}{(\alpha, \alpha)} = 0 \quad \text{for} \quad \alpha \in \Pi-\{\beta_1, \beta_2\}.$$

By the same argument, any $\omega-\beta_1-\alpha$ ($\alpha \in \Pi-\{\beta_2\}$) is not a weight of ϱ_{β_1}, $\omega-\beta_1-\beta_2$ is a simple weight and $\omega-\beta_1-2\beta_2$ is not a weight of ϱ_{β_1}. Thus $\omega-\beta_1-\beta_2$ is the third highest weight of ϱ_{β_1}.

By repeating the same argument, we obtain a series of weights of ϱ_{β_1}:

$$\omega, \omega-\beta_1, \omega-\beta_1-\beta_2, \ldots, \omega-\beta_1 - \ldots -\beta_k.$$

These are simple weights and are ordered decreasingly. Hence the highest weight ω_r of $\overline{\varrho_{\beta_1}^{[r]}}$ is

$$\begin{aligned}\omega_r &= \omega+(\omega-\beta_1)+ \ldots +(\omega-\beta_1- \ldots -\beta_{r-1}) \\ &= r\omega-(r-1)\beta_1-(r-2)\beta_2- \ldots -\beta_{r-1}, \\ &\qquad (1 \leqslant r \leqslant k).\end{aligned}$$

Notice that

$$\frac{2(\omega_r, \beta_i)}{(\beta_i, \beta_i)} = (r-i+1)-2(r-i)+(r-i-1) = 0,$$

$$(i = 1, 2, \ldots, r-2)$$

$$\frac{2(\omega_r, \beta_{r-1})}{(\beta_{r-1}, \beta_{r-1})} = 2-2 = 0,$$

$$\frac{2(\omega_r, \beta_r)}{(\beta_r, \beta_r)} = 1,$$

$$\frac{2(\omega_r, \alpha)}{(\alpha, \alpha)} = 0, \quad \text{if} \quad \alpha \in \Pi-\{\beta_1, \ldots, \beta_r\}.$$

This proves that ω_r is also the highest weight of ϱ_{β_r}. Thus

$$\varrho_{\beta_r} = \overline{\varrho_{\beta_1}^{[r]}}, \qquad r = 1, 2, \ldots, k.$$

It follows from Theorem 7 that in order to prove the second part of Theorem 3, it suffices to show the existence of the elementary representations for simple Lie algebras. In the following two chapters, this will be proved for the classical Lie algebras.

CHAPTER 11

REPRESENTATIONS OF THE CLASSICAL LIE ALGEBRAS

11.1. Representations of A_n

As before, let $m = n+1$. We know that

$$\mathfrak{h} = \left\{ H_{\lambda_1, \dots, \lambda_n} \middle| \sum_1^m \lambda_i = 0 \right\}$$

is a Cartan subalgebra of A_n. The roots of A_n with respect to $\mathfrak{h}$ are

$$\lambda_i - \lambda_k, \quad i \neq k, \qquad 1 \leqslant i, k \leqslant m.$$

The roots

$$\lambda_1 - \lambda_2, \ \dots, \ \lambda_n - \lambda_{n+1}$$

form a fundamental system of roots. $\mathfrak{h}_0^*$ can be embedded in an $(n+1)$-dimensional Euclidean space

$$E^{n+1} = \{a_1\lambda_1 + \ \dots \ + a_{n+1}\lambda_{n+1} \,|\, a_i\text{'s are real}\}$$

and

$$\mathfrak{h}_0^* = \left\{ a_1\lambda_1 + \ \dots \ + a_{n+1}\lambda_{n+1} \,|\, a_i\text{'s are real and } \sum_1^{n+1} a_i = 0 \right\}.$$

The metric in E^{n+1} is defined by

$$(\lambda_1, \lambda_1) = (\lambda_2 \lambda_2,) = \ \dots \ = (\lambda_{n+1}, \lambda_{n+1}) = \frac{1}{2(n+1)},$$

$$(\lambda_p, \lambda_q) = 0, \quad p \neq q.$$

Recall that a vector in E^{n+1} is said to be positive if the first non-zero coefficient of it is positive, thus an ordering is defined on E^{n+1}. The prime roots of $\mathfrak{h}_0^*$ relative to this ordering are

$$\lambda_1 - \lambda_2, \ \dots, \ \lambda_n - \lambda_{n+1}.$$

Consider an arbitrary element

$$\omega = a_1\lambda_1 + \dots + a_m\lambda_m, \quad \sum_1^m a_i = 0$$

of $\mathfrak{h}_0^*$. If ω is a weight of some representation of A_n, then

$$\frac{2(\omega, \lambda_i - \lambda_k)}{(\lambda_i - \lambda_k, \lambda_i - \lambda_k)} = \frac{\dfrac{2(a_i - a_k)}{2(n+1)}}{\dfrac{1}{(n+1)}} = a_i - a_k$$

is an integer. Thus a_ν $(1 \leqslant \nu \leqslant m)$ are fractions with denominator $n+1$ and the difference between any two of a_ν $(1 \leqslant \nu \leqslant m)$ is an integer. If ω is the highest weight of some irreducible representation, then

$$\frac{2(\omega, \lambda_i - \lambda_{i+1})}{(\lambda_i - \lambda_{i+1}, \lambda_i - \lambda_{i+1})} = a_i - a_{i+1} \geqslant 0, \qquad i = 1, 2, \dots, n,$$

i.e.

$$a_1 \geqslant a_2 \geqslant \dots \geqslant a_n \geqslant a_{n+1}.$$

If ω is the highest weight of the fundamental representation $\varrho_{\lambda_1-\lambda_2}$, then

$$a_1 - a_2 = 1, \quad a_2 - a_3 = \dots = a_n - a_{n+1} = 0.$$

From these equations and $a_1 + \dots + a_{n+1} = 0$, it follows that

$$a_1 = \frac{n}{n+1}, \quad a_2 = \frac{-1}{n+1}, \dots, a_{n+1} = \frac{-1}{n+1}.$$

Denote ω by $\omega_{\lambda_1-\lambda_2}$, then

$$\omega_{\lambda_1-\lambda_2} = \frac{n}{n+1}\lambda_1 + \frac{-1}{n+1}\lambda_2 + \dots + \frac{-1}{n+1}\lambda_{n+1}.$$

In general, if ω is the highest weight of the fundamental representation $\varrho_{\lambda_i-\lambda_{i+1}}$, then

$$a_1 - a_2 = 0, \dots, \quad a_{i-1} - a_i = 0, \quad a_i - a_{i+1} = 1, \quad a_{i+1} - a_{i+2} = 0, \dots, a_n - a_{n+1} = 0.$$

From these equations and $a_1 + \dots + a_n = 0$, it follows that

$$a_1 = a_2 = \dots = a_i = \frac{n-i+1}{n+1},$$

$$a_{i+1} = \dots = a_{n+1} = \frac{-i}{n+1}.$$

Denote ω by $\omega_{\lambda_i-\lambda_{i+1}}$, then

$$\omega_{\lambda_i-\lambda_{i+1}} = \frac{n-i+1}{n+1}\lambda_1 + \dots + \frac{n-i+1}{n+1}\lambda_i + \frac{-i}{n+1}\lambda_{i+1} + \dots + \frac{-i}{n+1}\lambda_{n+1}.$$

In particular, the highest weights of the elementary representations are

$$\omega_{\lambda_1-\lambda_2} = \frac{n}{n+1}\lambda_1 + \frac{-1}{n+1}\lambda_2 + \dots + \frac{-1}{n+1}\lambda_{n+1},$$

and

$$\omega_{\lambda_n-\lambda_{n+1}} = \frac{1}{n+1}\lambda_1 + \frac{1}{n+1}\lambda_2 + \dots + \frac{1}{n+1}\lambda_n + \frac{-n}{n+1}\lambda_{n+1}$$

respectively.

We now determine the elementary representations of A_n. For any $X \in A_n$, let

$$X \to X,$$

then we obtain an irreducible representation of A_n with highest weight $\omega_{\lambda_1-\lambda_2}$; denote this representation $\varrho_{\lambda_1-\lambda_2}$ by π_1. Secondly, for any $X \in A_n$, let

$$X \to -X',$$

then we obtain the other elementary representation $\varrho_{\lambda_n-\lambda_{n+1}}$ of A_n; denote this representation by π^1. We know that π^1 is the contragredient representation (or star representation) of π_1, i.e. $\pi^1 = (\pi_1)^*$.

We now show that for $k = 1, 2, \dots, n, \pi_1^{[k]}$ and $\pi^{1[k]}$ are both irreducible, hence $\overline{\pi_1^{[k]}} = \pi_1^{[k]}$ and $\overline{\pi^{1[k]}} = \pi^{1[k]}$. We will only prove that $\pi_1^{[k]}$ is irreducible; the irreducibility of $\pi^{1[k]}$ can be proved similarly. Let V be the representation space of π_1, then the representation space of $\pi_1^{[k]}$ is $V^{[k]}$, i.e. the space of anti-symmetric tensors of order k. We know that π_1 has $n+1$ weights.

$$\omega_1 = \frac{n}{n+1}\lambda_1 + \frac{-1}{n+1}\lambda_2 + \dots + \frac{-1}{n+1}\lambda_{n+1},$$

$$\omega_2 = \omega_1 - (\lambda_1 - \lambda_2) = \frac{-1}{n+1}\lambda_1 + \frac{n}{n+1}\lambda_2 + \frac{-1}{n+1}\lambda_3 + \dots + \frac{-1}{n+1}\lambda_{n+1},$$

$$\omega_3 = \omega_2 - (\lambda_2 - \lambda_3) = \frac{-1}{n+1}\lambda_1 + \frac{-1}{n+1}\lambda_2 + \frac{n}{n+1}\lambda_3 + \dots + \frac{-1}{n+1}\lambda_{n+1},$$

$$\cdots\cdots\cdots\cdots\cdots\cdots$$

$$\omega_{n+1} = \omega_n - (\lambda_n - \lambda_{n+1}) = \frac{-1}{n+1}\lambda_1 + \frac{-1}{n+1}\lambda_2 + \dots + \frac{-1}{n+1}\lambda_n + \frac{n}{n+1}\lambda_{n+1}.$$

Since $\dim V = n+1$, these weights are simple. Therefore, $\pi_1^{[k]}$ has $\binom{n}{k}$ simple weights

$$\omega_{i_1} + \dots + \omega_{i_k}, \qquad i_1 < \dots < i_k.$$

Let $e_1, \dots, e_{n+1}$ be weight vectors of the weights $\omega_1, \dots, \omega_{n+1}$ respectively, then

$$[e_1, \dots, e_n]$$

is of the highest weight

$$\omega_1 + \ \dots \ + \omega_k$$

of $\pi_1^{[k]}$ in $V^{[k]}$. Notice that

$$\omega_1 + \ \dots \ + \omega_k = \frac{n-k+1}{n+1}\lambda_1 + \ \dots \ + \frac{n-k+1}{n+1}\lambda_k + \frac{-k}{n+1}\lambda_{k+1} + \ \dots \ + \frac{-k}{n+1}\lambda_{n+1}$$

and

$$\omega_{i_1} + \ \dots \ + \omega_{i_k} = \sum_{j=1}^{k} \frac{n-k+1}{n+1}\lambda_{i_j} + \sum_{l \neq 1, \dots, k} \frac{-k}{n+1}\lambda_{i_l}.$$

It is easy to see that any weight $\omega_{i_1} + \ \dots \ + \omega_{i_k}$ of $\pi_1^{[k]}$ can be obtained from $\omega_1 + \ \dots \ + \omega_r$, thus $\pi_1^{[k]}$ is irreducible.

Set

$$\pi_k = \pi_1^{[k]}, \qquad k = 1, \dots, n,$$
$$\pi_k = \pi^{1[k]}, \qquad k = 1, \dots, n,$$

then the set of all fundamental representations of A_n is $\{\pi_1, \dots, \pi_n\}$ $(= \{\pi^1, \dots, \pi^n\})$ and

$$\pi_k \sim \pi^{n+1-k}, \qquad k = 1, \dots, n.$$

Finally, we consider the construction of an arbitrary irreducible representation of A_n by means of tensor spaces. Let $\pi_{k_1, \dots, k_n}$ be an irreducible representation with diagram

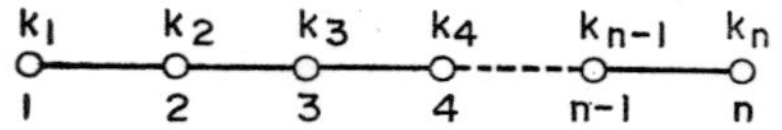

This representation can be obtained by means of the fundamental representations, i.e.

$$\Pi_{k_1, \dots, k_n} = \overline{\pi_1^{k_1} \otimes \ \dots \ \otimes \pi_n^{k_n}}.$$

Let $V_1, \dots, V_n$ be the representation spaces of $\pi_1, \dots, \pi_n$ respectively, then in

$$V_1^{k_1} \otimes \ \dots \ \otimes V_n^{k_n},$$

the vector

$$\underbrace{e_1 \otimes \ \dots \ \otimes e_1}_{k_1} \otimes \underbrace{[e_1, e_2] \otimes \ \dots \ \otimes [e_1, e_2]}_{k_2} \otimes \ \dots \ \otimes \underbrace{[e_1, \dots, e_n] \otimes \ \dots \ \otimes [e_1, \dots, e_n]}_{k_n}$$

is of highest weight

$$k_1\omega_1 + \ \dots \ + k_n\omega_n$$

($\omega_1, \dots, \omega_n$ are highest weights of $\pi_1, \dots, \pi_n$ respectively). Let $\nu = k_1 + 2k_2 + \ \dots \ + nk_n$, then this vector is in V^ν; the smallest invariant subspace of V^ν containing this vector is transformed according to $\Pi_{k_1, \dots, k_n}$.

11.2. Representations of C_n

The roots of C_n are

$$\pm\lambda_i\pm\lambda_k, \quad \pm 2\lambda_i, \qquad i, k = 1, 2, \ldots, n; i < k$$

and

$$\lambda_1-\lambda_2, \ldots, \lambda_{n-1}-\lambda_n, \quad 2\lambda_n$$

is a fundamental system of roots. The metric in

$$\mathfrak{h}_0^* = \{a_1\lambda_1 + \ldots + a_n\lambda_n \mid a\text{'s are real}\}$$

is defined by

$$(\lambda_1, \lambda_2) = (\lambda_2, \lambda_3) = \ldots = (\lambda_n, \lambda_n) = \frac{1}{4n+4},$$

$$(\lambda_p, \lambda_q) = 0, \quad \text{if } p \neq q.$$

If $\omega \in \mathfrak{h}_0^*$ is the weight of some representation of C_n, then

$$\frac{2(\omega, \pm\lambda_i\pm\lambda_k)}{(\pm\lambda_i\pm\lambda_k, \pm\lambda_i\pm\lambda_k)} = \pm a_i \pm a_k \quad \text{and} \quad \frac{2(\omega, \pm 2\lambda_i)}{(\pm 2\lambda_i, \pm 2\lambda_i)} = \pm a_i$$

are integers. Therefore $a_1, \ldots, a_n$ must be integers. If ω is the highest weight of some irreducible representation, then

$$\frac{2(\omega, \lambda_i-\lambda_{i+1})}{(\lambda_i-\lambda_{i+1}, \lambda_i-\lambda_{i+1})} = a_i - a_{i+1} \geqslant 0 \quad \text{and} \quad \frac{2(\omega, 2\lambda_n)}{(2\lambda_n, 2\lambda_n)} = a_n \geqslant 0,$$

thus

$$a_1 \geqslant \ldots \geqslant a_n \geqslant 0.$$

If $n \geqslant 2$ and $\omega_{\lambda_i-\lambda_{i+1}}$ is the highest weight of the fundamental representation $\varrho_{\lambda_i-\lambda_{i+1}}$, then

$$a_i - a_{i+1} = 1,$$

$$a_1-a_2 = a_2-a_3 = \ldots = a_{i-1}-a_i = a_{i+1}-a_{i+2} = \ldots = a_{n-1}-a_n = a_n = 0.$$

Therefore,

$$\omega_{\lambda_i-\lambda_{i+1}} = \lambda_1 + \ldots + \lambda_i, \qquad i = 1, \ldots, n-1.$$

If $\omega_{2\lambda_n}$ is the highest weight of the fundamental representation $\varrho_{2\lambda_n}$, then

$$a_1-a_2 = a_2-a_3 = \ldots = a_{n-1}-a_n = 0, \quad a_n = 1,$$

so

$$\omega_{2\lambda_n} = \lambda_1 + \ldots + \lambda_n.$$

When $n \geqslant 2$, C_n has two elementary representations $\varrho_{\lambda_1-\lambda_2}$ and $\varrho_{2\lambda_n}$. For any $X \in C$, let

$$X \to X,$$

then we obtain an irreducible representation of C_n with weights

$$\pm\lambda_1, \pm\lambda_2, \ldots, \pm\lambda_n.$$

These weights are simple and λ_1 is the highest. Thus this representation is $\varrho_{\lambda_1-\lambda_2}$; denote it by ϱ_1. Set

$$\varrho_k = \overline{\varrho_1^{[k]}}, \qquad k = 1, \ldots, n,$$

then $\varrho_1, \ldots, \varrho_n$ are fundamental representations of C_n and ϱ_n is the other elementary representation $\varrho_{2\lambda_n}$.

Let $P_{k_1, \ldots, k_n}$ be the irreducible representation with diagram

then

$$P_{k_1, \ldots, k_n} = \overline{\varrho_1^{k_1} \otimes \ldots \otimes \varrho_n^{k_n}}.$$

11.3. Representations of B_n

The roots of B_n are

$$\pm\lambda_i \pm\lambda_k, \quad i < k, \qquad \pm\lambda_i, \quad i, k = 1, \ldots, n,$$

and

$$\lambda_1-\lambda_2, \lambda_2-\lambda_3, \ldots, \lambda_{n-1}-\lambda_n, \lambda_n$$

form a fundamental system of roots of B_n. We know that the inner product defined by

$$(\lambda_1, \lambda_1) = \ldots = (\lambda_n, \lambda_n) = \frac{1}{4n-2}$$

$$(\lambda_p, \lambda_q) = 0, \qquad p \neq q$$

in

$$\mathfrak{h}_0^* = \{a_1\lambda_1 + \ldots + a_n\lambda_n \mid a_i\text{'s are real}\}$$

is positive definite. If $\omega = a_1\lambda_1 + \ldots + a_n\lambda_n$ is a weight of some representation of C_n, then

$$\frac{2(\omega, \pm\lambda_i\pm\lambda_k)}{(\pm\lambda_i\pm\lambda_k, \pm\lambda_i\pm\lambda_k)} = \pm a_i \pm a_k \quad \text{and} \quad \frac{2(\omega, \pm\lambda_i)}{(\pm\lambda_i, \pm\lambda_i)} = \pm 2a_i$$

are integers. Thus a_i $(1 \leqslant i \leqslant n)$ are all integers or half integers. If ω is the highest weight of an irreducible representation, then

$$a_i - a_{i+1} \geqslant 0 \quad \text{and} \quad 2a_n \geqslant 0.$$

Therefore,

$$a_1 \geqslant a_2 \geqslant \ldots \geqslant 2a_n \geqslant 0.$$

When $n \geqslant 2$, if ω is the highest weight of the fundamental representation $\varrho_{\lambda_i-\lambda_{i+1}}$, then

$$a_i - a_{i+1} = 1,$$
$$a_1 - a_2 = \ldots = a_{i-1} - a_i = a_{i+1} - a_{i+2} = \ldots = a_{n-1} - a_n = 2a_n = 0.$$

Hence the highest weight of $\varrho_{\lambda_i-\lambda_{i+1}}$ is

$$\omega_{\lambda_i-\lambda_{i+1}} = \lambda_1 + \ldots + \lambda_n, \qquad i = 1, \ldots, n-1.$$

If ω is the highest weight of the fundamental representation ϱ_{λ_n}, then

$$a_1 - a_2 = a_2 - a_3 = \ldots = a_{n-1} - a_n = 0, \qquad 2a_n = 1.$$

Thus the highest weight of ϱ_{λ_n} is

$$\omega_{\lambda_n} = \tfrac{1}{2}\lambda_1 + \ldots + \tfrac{1}{2}\lambda_n.$$

When $n \geqslant 2$, B_n has two elementary representations $\varrho_{\lambda_1-\lambda_2}$ and ϱ_{λ_n}. If for any $X \in B_n$,

$$X \to X,$$

then we obtain an irreducible representation of B_n with weights

$$\pm\lambda_1, \pm\lambda_2, \ldots, \pm\lambda_n.$$

These weights are simple and λ_1 is the highest, thus this representation is $\varrho_{\lambda_1-\lambda_2}$; denote it by τ_1. Let

$$\tau_k = \overline{\tau_1^{[k]}}, \qquad k = 1, 2, \ldots, n-1, n,$$

then τ_k $(k = 1, \ldots, n-1)$ are $n-1$ fundamental representations of B_n. The fundamental representation ϱ_{λ_n} is elementary and denoted by σ; σ is also called the spin representation. The highest weight of σ is $\frac{1}{2}(\lambda_1 + \ldots + \lambda_n)$ and the weights of σ are $\frac{1}{2}(\pm\lambda_1 \ldots \pm\lambda_n)$.

Let $T_{k_1, \ldots, k_{n-1}}$ be the irreducible representation with diagram

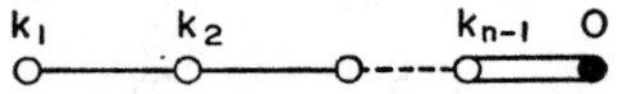

then

$$T_{k_1, \ldots, k_{n-1}} = \overline{\tau_1^{k_1} \otimes \ldots \otimes \tau_{n-1}^{k_{n-1}}}$$

and an arbitrary irreducible representation can be denoted by

$$\overline{T_{k_1, \ldots, k_{n-1}} \otimes \sigma^k}.$$

It is easy to show that

$$\overline{\sigma^2} = \tau_1^{[n]}, \quad \overline{\sigma \quad} = \tau_1^{[n-1]}.$$

11.4. Representations of D_n

Let $n \geqslant 4$. It is known that the roots of D_n are

$$\pm\lambda_i \pm \lambda_k, \qquad i < k,\ i, k = 1, \ldots, n,$$

and

$$\lambda_1 - \lambda_2, \ldots, \lambda_{n-1} - \lambda_n, \lambda_{n-1} + \lambda_n$$

form a fundamental system of roots. The inner product defined on

$$\mathfrak{h}_*^0 = \{a_1\lambda_1 + \ldots + a_n\lambda_n \mid a_i\text{'s are real}\}$$

by

$$(\lambda_1, \lambda_1) = \ldots = (\lambda_n, \lambda_n) = \frac{1}{4n-4},$$

$$(\lambda_p, \lambda_q) = 0, \qquad p \neq q,$$

is positive definite.

If $\omega = a_1\lambda_1 + \ldots + a_n\lambda_n$ is a weight of D_n, then

$$\frac{2(\omega, \pm\lambda_i \pm \lambda_k)}{(\pm\lambda_i \pm \lambda_k, \pm\lambda_i \pm \lambda_k)} = \pm a_i \pm a_k$$

are integers. Therefore, a_i $(1 \leqslant i \leqslant n)$ are all integers or half integers. If ω is the highest weight of some irreducible representation, then

$$a_i - a_{i+1} \geqslant 0 \quad \text{and} \quad a_{n-1} + a_n \geqslant 0,$$

so

$$a_1 \geqslant a_2 \geqslant \ldots \geqslant a_{n-1} \geqslant 0, \quad a_{n-1} \geqslant a_n, \quad a_{n-1} + a_n \geqslant 0.$$

If ω is the highest weight of the fundamental representation $\varrho_{\lambda_i - \lambda_{i+1}}$ $(i = 1, 2, \ldots, n-2)$, then

$$a_i - a_{i+1} = 1,$$
$$a_1 - a_2 = \ldots = a_{i-1} - a_i = a_{i+1} - a_{i+2} = \ldots = a_{n-1} - a_n = a_{n-1} + a_n = 0.$$

Thus the highest weight of $\varrho_{\lambda_i - \lambda_{i+1}}$ $(i = 1, 2, \ldots, n-2)$ is

$$\omega_{\lambda_i - \lambda_{i+1}} = \lambda_1 + \ldots + \lambda_i, \qquad i = 1, 2, \ldots, n-2.$$

If ω is the highest weight of the fundamental representation $\varrho_{\lambda_{n-1} - \lambda_n}$, then

$$a_1 - a_2 = \ldots = a_{n-2} - a_{n-1} = 0, \quad a_{n-1} - a_n = 1, \quad a_{n-1} + a_n = 0.$$

Thus the highest weight of $\varrho_{\lambda_{n-1} - \lambda_n}$ is

$$\omega_{\lambda_{n-1} + \lambda_n} = \tfrac{1}{2}\lambda_1 + \ldots + \tfrac{1}{2}\lambda_{n-1} - \tfrac{1}{2}\lambda_n.$$

If ω is the highest weight of the fundamental representation $\varrho_{\lambda_{n-1} + \lambda_n}$, then

$$a_1 - a_2 = \dots = a_{n-2} - a_{n-1} = a_{n-1} - a_n = 0, \quad a_{n-1} + a_n = 1,$$

so

$$\omega_{\lambda_{n-1}+\lambda_n} = \tfrac{1}{2}\lambda_1 + \tfrac{1}{2}\lambda_2 + \dots + \tfrac{1}{2}\lambda_{n-1} + \tfrac{1}{2}\lambda_n.$$

D_n has three elementary representations $\varrho_{\lambda_1-\lambda_2}$, $\varrho_{\lambda_{n-1}-\lambda_n}$ and $\varrho_{\lambda_{n-1}+\lambda_n}$. For any $X \in D_n$, let

$$X \to X,$$

then we obtain an irreducible representation of D_n with simple weights

$$\pm\lambda_1, \dots, \pm\lambda_n$$

and highest weight λ_1, so this representation is $\varrho_{\lambda_1-\lambda_2}$, denote it by τ_1. Let

$$\tau_k = \overline{\tau_1^{[k]}}, \qquad k = 1, 2, \dots, n-2,$$

then we obtain $n-2$ fundamental representations of D_n. The remaining two elementary representations are denoted by σ_1 and σ_2; these two representations are called the spin representations.

If

$$\tau_{k_1, \dots, k_{n-2}} = \overline{\tau_1^{k_1} \otimes \dots \otimes \tau_{n-2}^{k_{n-2}}}$$

denotes the representation with the diagram

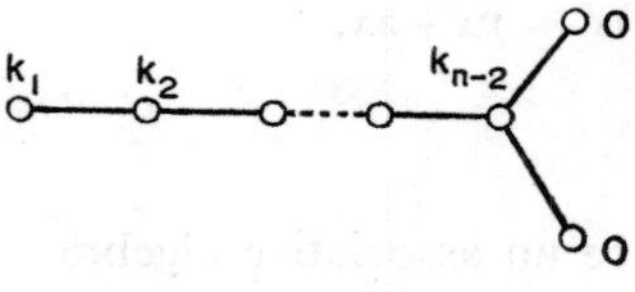

then it can be proved that

$$\overline{\sigma_1 \otimes \sigma_2} \sim \tau_1^{[n-1]},$$

$$\overline{\sigma_1^2 + \sigma_2^2} \sim \tau_1^{[n]},$$

$$\overline{\sigma_1^{[2]}} \sim \overline{\sigma_2^{[2]}} \sim \tau^{[n-2]}.$$

In particular, when $n = 4$, the Dynkin diagram of D_4 is

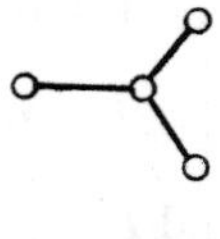

Denote the three elementary representations by σ_1, σ_2 and σ_3, where one of them is $\varrho_{\lambda_1-\lambda_2}$ and the other two are the spin representations; then

$$\sigma_1^{[2]} \sim \sigma_2^{[2]} \sim \sigma_3^{[2]} \sim \varrho_{\lambda_2-\lambda_3}.$$

CHAPTER 12

SPIN REPRESENTATIONS AND THE EXCEPTIONAL LIE ALGEBRAS

12.1. Associative algebras

Let A be a vector space (finite or infinite dimensional) over the complex numbers C. Suppose a binary operation xy ($x, y \in A$) called multiplication is defined on A, i.e. for any $x, y \in A$, there exists a unique $w \in A$ such that $w = xy$. If for any $x, y, z \in A$ and $\lambda \in C$, the conditions:

(1) $x(y+z) = xy+xz, \quad (y+z)x = yx+zx,$
(2) $x(yz) = (xy)z,$
(3) $(\lambda x)y = x(\lambda y) = \lambda(xy)$

are satisfied, then A is said to be an associative algebra over C or simply an *associative algebra.* The dimension of an associative algebra A is defined to be the dimension of A as a vector space.

EXAMPLE 1. The set of all $n\times n$ matrices over C with matrix addition, matrix multiplication and scalar multiplication forms an associative algebra of dimension n^2. This algebra is called the algebra of $n\times n$ matrices. Similarly, the set of all linear transformations on an n-dimensional vector space over C also forms an associative algebra.

EXAMPLE 2. Let V be an r-dimensional vector space over C. Suppose $e_1, e_2, \ldots, e_r$ form a basis of V and T^k is the tensor space of rank k of V, then

$$e_{i_1}\otimes \ldots \otimes e_{i_k}, \qquad 1 \leqslant i_1, \ldots, i_k \leqslant r,$$

form a basis of T^k. In particular, $T^0 = C$ and $T^1 = V$. Consider the direct sum $T = \sum_{k=0}^{\infty} T^k$. Define

$$(e_{i_1}\otimes \ldots \otimes e_{i_k})(e_{j_1}\otimes \ldots \otimes e_{j_l}) = e_{i_1}\otimes \ldots \otimes e_{i_k}\otimes e_{j_1}\otimes \ldots \otimes e_{j_l}$$

for any two elements $e_{i_1}\otimes \ldots \otimes e_{i_k}$ and $e_{j_1}\otimes \ldots \otimes e_{j_l} \in T$. Extend this multiplication by means of (1) and (3) to T, then T becomes an associative algebra; this algebra is called the *tensor algebra* of V. The dimension of T is infinite.

Let A be an associative algebra. An element $1 \in A$ is said to be the identity element if

$$1 \cdot x = x \cdot 1 = x \quad \text{for all } x \in A.$$

A subspace A_1 of A is said to be a subalgebra if it is closed under multiplication. A subspace A_2 of A is said to be an ideal (or two-sided ideal) if for any $x \in A_2$ and $y \in A$, $xy \in A_2$ and $yx \in A_2$. If A_2 is an ideal, then quotient algebra can be defined. A subspace is said to be a left ideal if for any $x \in A_1$ and $y \in A$, $yx \in A_1$. Right ideals can be defined similarly.

Homomorphisms and isomorphisms can also be defined for associative algebras. If a homomorphism maps the identity to the identity, then it is said to be a canonical homomorphism.

Let A be an associative algebra. It is easy to show that the vector space structure of A with commutation defined by $[x, y] = xy - yx$ is a Lie algebra. Denote this Lie algebra by A_L. A mapping from a Lie algebra $\mathfrak{g}$ to A is said to be a representation of $\mathfrak{g}$ if it is a homomorphism from $\mathfrak{g}$ to A_L. Linear representations are special cases of representations.

12.2. Clifford algebra

Let V be an r-dimensional vector space with basis $\{e_1, \ldots, e_r\}$. If T is the tensor algebra tof V, then $T = \sum_{k=0}^{\infty} V^k$ and

$$e_{i_1} \otimes \ldots \otimes e_{i_k}, \quad 1 \leqslant i_1, \ldots, i_k \leqslant r$$

form a basis of T^k. Let

$$T_+ = \sum_{k=0}^{\infty} T^{2k}, \quad T_- = \sum_{k=0}^{\infty} T^{2k+1},$$

then $T = T_+ \dotplus T_-$ and

$$T_+T_+ \subset T_+, \; T_+T_- \subset T_-, \; T_-T_+ \subset T_-, \; T_-T_- \subset T_+.$$

Thus T_+ is a subalgebra of T.

Let I be the ideal of T generated by the elements

$$\left.\begin{aligned} &e_i \otimes e_i - 1, \quad i = 1, \ldots, r \\ &e_i \otimes e_j + e_j \otimes e_i, \quad i \neq j, \; i,j = 1, \ldots, r \end{aligned}\right\} \tag{12.1}$$

and $\mathfrak{C}$ be the quotient algebra T/I, then $\mathfrak{C}$ is said to be the Clifford algebra of V. Since all the generators of I are elements of T_+, we have $I = I_+ \dotplus I_-$ where $I_+ = I \cap T_+$ and $I_- = I \cap T_-$. If $\mathfrak{C}_+ = T_+/I_+$, $\mathfrak{C}_- = T_-/I_-$, then $\mathfrak{C} = \mathfrak{C}_+ \dotplus \mathfrak{C}_-$, and

$$\mathfrak{C}_+\mathfrak{C}_+ \subset \mathfrak{C}_+, \; \mathfrak{C}_+\mathfrak{C}_- \subset \mathfrak{C}_-, \; \mathfrak{C}_-\mathfrak{C}_+ \subset \mathfrak{C}_-, \; \mathfrak{C}_-\mathfrak{C}_- \subset \mathfrak{C}_+.$$

Suppose $x \in \mathfrak{C}$ and $x = y + z$, where $y \in \mathfrak{C}_+$ and $z \in \mathfrak{C}_-$. Define

$$\bar{x} = y - z,$$

then $x \to \bar{x}$ is an automorphism of $\mathfrak{C}$ and $\bar{\bar{x}} = x$. This automorphism is said to be the principal involution.

Let f be the canonical mapping from T onto $\mathfrak{C} = T/I$. For $v \in V$, denote $f(v)$ by $\mathbf{v}$. We have the following:

THEOREM 1. *$\mathfrak{C}$ is a 2^r-dimensional associative algebra and*

$$\mathbf{e}_{i_1}\mathbf{e}_{i_2} \ldots \mathbf{e}_{i_s}, \quad s = 0, 1, \ldots, r, \; 1 \leqslant i_1 < \ldots < i_s \leqslant r,$$

form a basis of $\mathfrak{C}$. Hence $\mathfrak{C}$ is generated by $\mathbf{e}_1, \ldots, \mathbf{e}_r$ and

$$\mathbf{e}_i^2 = 1, \quad i = 1, \ldots, r,$$
$$\mathbf{e}_i\mathbf{e}_j + \mathbf{e}_j\mathbf{e}_i = 0, \quad i \neq j, \quad i, j = 1, 2, \ldots, r.$$

Proof. Since I is generated by $e_i \otimes e_i - 1$ $(i = 1, \ldots, r)$ and $e_i \otimes e_j + e_j \otimes e_i$ $(i \neq j, \; i, j = 1, \ldots, r)$,

$$\mathbf{e}_i^2 = 1, \quad i = 1, \ldots, r,$$

and

$$\mathbf{e}_i\mathbf{e}_j + \mathbf{e}_j\mathbf{e}_i = 0, \quad i \neq j, \quad i, j = 1, \ldots, r.$$

Thus

$$\mathbf{e}_{i_1} \ldots \mathbf{e}_{i_s}, \quad s = 0, 1, \ldots, r, \; 1 \leqslant i_1 < i_2 < \ldots < i_s \leqslant r$$

span $\mathfrak{C}$. To show linear independence of these vectors, we use induction on k to prove the following statement:

(P_k) If $\{i_1, \ldots, i_k\} \subset \{1, \ldots, r\}$ and $i_1 < \ldots < i_k$, then the vectors

$$\mathbf{e}_{j_1} \ldots \mathbf{e}_{j_s}, \quad \text{where} \quad s = 0, \ldots, k, \; \{j_1, \ldots, j_s\} \subseteq \{i_1, \ldots, i_k\}$$
$$\text{and} \quad 1 \leqslant j_1 < \ldots < j_s \leqslant r$$

are linearly independent.

The cases when $k = 0$ and $k = 1$ are obvious.†

Let $k > 1$ and suppose that (P_{k-1}) is true. Let

$$\sum_{\substack{1 \leqslant j_1 < j_2 \ldots \leqslant r \\ \{j_1, \ldots, j_s\} \subset \{i_1, \ldots, i_k\}}} c_{j_1 \ldots j_s} \mathbf{e}_{j_1} \ldots \mathbf{e}_{j_s} = 0 \tag{12.2}$$

be a linear relation. If $\mathbf{e}_{i_k}$ does not appear in (12.2), then by the induction assumption all $c_{j_1, \ldots j_s} = 0$, where $\{j_1, \ldots, j_s\} \subseteq \{i_1, \ldots, i_{k-1}\}$ and $1 \leqslant j_1 < j_2 \ldots < j_s \leqslant r$. Now suppose that $\mathbf{e}_{i_k}$ appears in (12.2). Rewrite (12.2) as

$$x + y\mathbf{e}_{i_k} = 0, \tag{12.3}$$

where x, y are linear combinations of products of elements from $\{\mathbf{e}_{i_1}, \ldots, e_{i_{k-1}}\}$, i.e.

$$x = \sum_{\substack{1 \leqslant j_1 \ldots < j_s \leqslant r \\ \{j_1, \ldots, j_s\} \subset \{i_1, \ldots i_{k-1}\}}} c_{j_1 \ldots j_s} \mathbf{e}_{j_1} \ldots \mathbf{e}_{j_s},$$

$$y = \sum_{\substack{1 \leqslant j_1 < \ldots < j_s \leqslant r \\ \{j_1, \ldots j_{s-1}\} \subset \{i_1, \ldots, i_{k-1}\}}} c_{j_1 \ldots j_{s-1} i_k} \mathbf{e}_{j_1}, \ldots, \mathbf{e}_{j_{s-1}}.$$

† When $k = 0$, we need to show $1 \notin I$; this also follows from Theorem 2.

Multiply (12.3) by $\mathbf{e}_{i_k}$; we get

$$x\mathbf{e}_{i_k}+y = 0. \tag{12.4}$$

Adding (12.3) and (12.4), we get

$$(x+y)(1+\mathbf{e}_{i_k}) = 0. \tag{12.5}$$

Write $x+y$ as

$$x+y = u+v\mathbf{e}_{i_{k-1}},$$

where u, v are linear combinations of products of elements from $\{\mathbf{e}_{i_1}, \ldots, \mathbf{e}_{i_{k-2}}\}$. Thus (12.5) becomes

$$(u+v\,\mathbf{e}_{i_{k-1}})(1+\mathbf{e}_{i_k}) = 0. \tag{12.6}$$

Multiply (12.6) by $\mathbf{e}_{i_{k-1}}$; we get

$$\begin{aligned} 0 &= (u+v\mathbf{e}_{i_{k-1}})(1+\mathbf{e}_{i_k})\,\mathbf{e}_{i_{k-1}} \\ &= (u+v\mathbf{e}_{i_{k-1}})\,\mathbf{e}_{i_{k-1}}(1-\mathbf{e}_{i_k}) = (u\mathbf{e}_{i_{k-1}}+v)(1-\mathbf{e}_{i_k}), \end{aligned}$$

i.e.

$$(u\mathbf{e}_{i_{k-1}}+v)(1-\mathbf{e}_{i_k}) = 0. \tag{12.7}$$

Apply the principal involution of $\mathfrak{C}$ to (12.6), then

$$(\bar{u}+\bar{v}\bar{\mathbf{e}}_{i_{k-1}})(1+\bar{\mathbf{e}}_{i_k}) = 0,$$

i.e.

$$(\bar{u}-\bar{v}\mathbf{e}_{i_{k-1}})(1-\mathbf{e}_{i_k}) = 0.$$

Multiply this equation by $\mathbf{e}_{i_{k-1}}$ and notice that $\mathbf{e}_{i_{k-1}}$ does not appear in $\bar{u}$ or $\bar{v}$; we get

$$0 = \mathbf{e}_{i_{k-1}}(\bar{u}-\bar{v}\mathbf{e}_{i_{k-1}})(1-\mathbf{e}_{i_k}) = (u\mathbf{e}_{i_{k-1}}-v)(1-\mathbf{e}_{i_k}),$$

i.e.

$$(u\mathbf{e}_{i_{k-1}}-v)(1-\mathbf{e}_{i_k}) = 0. \tag{12.8}$$

The sum and difference of (12.7) and (12.8) are

$$2u\mathbf{e}_{i_{k-1}}(1-\mathbf{e}_{i_k}) = 0, \tag{12.9}$$

$$2v(1-\mathbf{e}_{i_k}) = 0. \tag{12.10}$$

Multiply (12.9) by $\mathbf{e}_{i_{k-1}}$; we get

$$2u(1+\mathbf{e}_{i_k}) = 0. \tag{12.11}$$

Now notice that only $k-1$ elements from $\mathbf{e}_{i_1}, \ldots, \mathbf{e}_{i_{k-2}}, \mathbf{e}_{i_k}$ appear in (12.10) and (12.11). By the induction assumption, $u = v = 0$. Hence $x+y = 0$. From the difference of (12.3) and (12.4),

$$(x-y)(1-\mathbf{e}_{i_k}) = 0.$$

Similarly, we can prove that $x-y = 0$. Hence $x = y = 0$. This proves that all coefficients in (12.2) are zero, so (P_k) is also true.

We have proved that

$$\mathbf{e}_{i_1}, \ldots, \mathbf{e}_{i_s}, \quad s = 0, 1, \ldots, r, \qquad 1 \leqslant i_1 < i_2 < \ldots < i_s \leqslant r$$

are linearly independent, so they form a basis of $\mathfrak{C}$.

By Theorem 1, $\mathfrak{C}$ contains a subspace

$$\{c_1\mathbf{e}_1 + \ldots + c_r\mathbf{e}_r \mid c_i\text{'s are complex}\}$$

that is isomorphic to V. In what follows, these two spaces will be identified and $\mathbf{e}_i$ will be written as e_i $(i = 1, \ldots, r)$.

THEOREM 2. *Let $r = 2n$, then $\mathfrak{C}$ is isomorphic to the algebra M_{2^n} of $2^n \times 2^n$ matrices.*

Proof. Let $p_i = e_i$ $(i = 1, \ldots, n)$ and $q_i = e_{n+i}$ $(i = 1, \ldots, n)$, then

$$\left.\begin{array}{l} p_i^2 = q_i^2 = 1, \ i = 1, \ldots, n \\ p_iq_j + q_jp_i = 0, \ i, j = 1, \ldots, n \\ p_ip_j + p_jp_i = q_iq_j + q_jq_i = 0, \qquad i \neq j, \quad i, j = 1, \ldots, n. \end{array}\right\} \tag{12.12}$$

Hence $\mathfrak{C}$ is generated by p_i, q_i $(i = 1, \ldots, n)$ that satisfy (12.12). Let

$$I_2 = \begin{bmatrix} 1 & 0 \\ 0 & 1 \end{bmatrix}, \quad I_2 = \begin{bmatrix} 1 & 0 \\ 0 & -1 \end{bmatrix}, \quad P = \begin{bmatrix} 0 & 1 \\ 1 & 0 \end{bmatrix},$$

$$Q = \begin{bmatrix} 0 & \sqrt{-1} \\ -\sqrt{-1} & 0 \end{bmatrix}$$

and

$$\left.\begin{array}{ll} p_i \to P_i = J_2 \otimes \ldots \otimes J_2 \otimes \underset{(i\text{th position})}{P} \otimes I_2 \otimes \ldots \otimes I_2 & (i = 1, \ldots, n) \\ q_i \to Q_i = J_2 \otimes \ldots \otimes J_2 \otimes \underset{(i\text{th position})}{Q} \otimes I_2 \otimes \ldots \otimes I_2. & \end{array}\right\} \tag{12.13}$$

It is easy to show that P_i, Q_i $(i = 1, \ldots, n)$ also satisfy (12.12). Thus (12.13) defines a homomorphism φ from $\mathfrak{C}$ to M_{2^n}. If φ is onto, then from $\dim \mathfrak{C} = \dim M_{2^n} = 2^{2^n}$ it follows that φ is an isomorphism.

Now

$$u_i = \sqrt{-1}\, p_iq_i \to U_i = \sqrt{-1}\, P_iQ_i = I_2 \otimes \ldots \otimes I_2 \otimes \underset{(i\text{th position})}{J_2} \otimes \ldots \otimes I_2,$$

so

$$U_1 \ldots U_{i-1}P_i = I_2 \otimes \ldots \otimes I_2 \otimes \underset{(i\text{th position})}{P} \otimes I_2 \otimes \ldots \otimes I_2,$$

$$U_1 \ldots U_{i-1}Q_i = I_2 \otimes \ldots \otimes I_2 \otimes Q \otimes \ldots \otimes I_2$$

are in $\varphi(\mathfrak{C})$. We have

$$\tfrac{1}{2}(1+u_i) \to I_2\otimes \dots \otimes I_2\otimes \underset{(i\text{th position})}{\begin{bmatrix}1 & 0\\ 0 & 0\end{bmatrix}}\otimes I_2\otimes \dots \otimes I_2,$$

$$\tfrac{1}{2}(u_1, \dots, u_{i-1})(p_i-\sqrt{-1}\,q_i) \to I_2\otimes \dots \otimes I_2\otimes \begin{bmatrix}0 & 1\\ 0 & 0\end{bmatrix}\otimes I_2 \dots \otimes I_2,$$

$$\tfrac{1}{2}(u_1 \dots u_{i-1})(p_i+\sqrt{-1}\,q_i) \to I_2\otimes \dots \otimes I_2\otimes \begin{bmatrix}0 & 0\\ 1 & 0\end{bmatrix}\otimes I_2 \dots \otimes I_2,$$

$$\tfrac{1}{2}(1-u_i) \to I_2\otimes \dots \otimes I_2\otimes \begin{bmatrix}0 & 0\\ 0 & 1\end{bmatrix}\otimes I_2 \dots \otimes I_2.$$

It follows that $\varphi(\mathfrak{C}) = M_{2^n}$. The proof of Theorem 2 is now completed.

12.3. Spin representations

Let V be an m-dimensional vector space with basis $\{e_1, \dots, e_m\}$, $\mathfrak{C}$ be the Clifford algebra generated by $e_1, \dots, e_m$ and

$$e_i^2 = 1, \qquad i = 1, \dots, m,$$
$$e_ie_j+e_je_i = 0, \qquad i \neq j, \quad i, j = 1, 2, \dots, m.$$

Define commutation on $\mathfrak{C}$ by $[x, y] = xy-yx\ (\forall x, y \in \mathfrak{C})$, then $\mathfrak{C}$ becomes a Lie algebra; denote this Lie algebra by $\mathfrak{C}_L$. Consider all elements in $\mathfrak{C}_L$ of the form

$$\sum_{1\leq i, k\leq m} c_{ik}e_ie_k, \quad c_{ik} \in C.$$

These elements form a subalgebra $\mathfrak{C}_2$ of $\mathfrak{C}_L$.

THEOREM 3. *$\mathfrak{C}_2$ is isomorphic to the Lie algebra $\mathfrak{o}(m, C)$ of all $m\times m$ skew-symmetric matrices and*

$$\sum_{1\leq i<k\leq m} c_{ik}e_ie_k \overset{\tau}{\longleftrightarrow} 2 \sum_{1\leq i<k\leq m} c_{ik}(E_{ik}-E_{ki})$$

is an isomorphism.

Proof. Let

$$\left.\begin{aligned}\tau(e_ie_k) &= 2(E_{ik}-E_{ki}),\\ \tau(e_je_l) &= 2(E_{jl}-E_{lj}).\end{aligned}\right\} \tag{12.14}$$

If no two of i, j, k and l are equal, then

$$[e_ie_k, e_je_l] = e_ie_ke_je_l-e_je_le_ie_k = e_ie_ke_je_l-e_ie_ke_je_l = 0,$$

and

$$[E_{ik}-E_{ki}, E_{jl}-E_{lj}] = 0.$$

If exactly two of i, j, k and l are equal, assume $j = k$. Then

$$[e_ie_k, e_ke_l] = e_ie_ke_ke_l - e_ke_le_ie_k = e_ie_l - e_le_i = 2e_ie_l$$

and

$$[E_{ik}-E_{ki}, E_{kl}-E_{lk}] = E_{il}-E_{li}.$$

If $i = j$, $k = l$, then

$$[e_ie_k, e_ie_k] = 0$$

and

$$[E_{ik}-E_{ki}, E_{ik}-E_{ki}] = 0.$$

This proves that τ is an isomorphism. Thus Theorem 3 is proved.

Now the set $\mathfrak{C}_{12}$ of all elements of the form

$$\sum_1^m a_ie_i + \sum_{1\leq i<k\leq m} c_{ik}e_ie_k, \quad a_i, c_{ik} \in C$$

also form a subalgebra of $\mathfrak{C}_L$. We have the following:

THEOREM 4. *$\mathfrak{C}_{12}$ is isomorphic to $\mathfrak{o}(m+1, C)$ and the mapping*

$$\Sigma a_ie_i + \Sigma c_{ik}e_ie_k \xrightarrow{\tau} \sum_i 2\sqrt{-1}\, a_i(E_{0i}-E_{i0}) + 2\sum_{1\leq i<k\leq m} c_{ik}(E_{ik}-E_{ki})$$

is an isomorphism.

Proof. In this case, we have (12.14) and

$$e_i \xrightarrow{\tau} 2\sqrt{-1}\,(E_{0i}-E_{i0}).$$

If $i \neq k$, then

$$[e_i, e_k] = e_ie_k - e_ke_i = 2e_ie_k,$$
$$[2\sqrt{-1}\,(E_{0i}-E_{i0}), 2\sqrt{-1}\,(E_{0k}-E_{k0})] = -4\{-E_{i0}E_{0k}-(-E_{k0})E_{0i}\}$$
$$= -4(-E_{ik}+E_{ki}) = 4(E_{ik}-E_{ki})$$

and

$$[e_i, e_ie_k] = e_ie_ie_k - e_ie_ke_i = e_k + e_k = 2e_k,$$
$$[2\sqrt{-1}\,(E_{0i}-E_{i0}), 2(E_{ik}-E_{ki})] = 4\sqrt{-1}\,(E_{0k}-E_{k0}).$$

If i, j, k are all different, then

$$[e_i, e_je_k] = e_ie_je_k - e_je_ke_i = 0,$$
$$[2\sqrt{-1}\,(E_{0i}-E_{i0}), 2(E_{jk}-E_{kj})] = 0.$$

This proves Theorem 2.

Now let $m = 2n$ be even. Notice that

$$\begin{bmatrix} \frac{1}{\sqrt{2}} & \frac{1}{\sqrt{2}} \\ \frac{-i}{\sqrt{2}} & \frac{i}{\sqrt{2}} \end{bmatrix} \begin{bmatrix} 0 & 1 \\ 1 & 0 \end{bmatrix} \begin{bmatrix} \frac{1}{\sqrt{2}} & \frac{1}{\sqrt{2}} \\ \frac{-i}{\sqrt{2}} & \frac{i}{\sqrt{2}} \end{bmatrix}' = \begin{bmatrix} 1 & 0 \\ 0 & 1 \end{bmatrix}.$$

If

$$P = \begin{bmatrix} \frac{1}{\sqrt{2}} I_n & \frac{1}{\sqrt{2}} I_n \\ \frac{-i}{\sqrt{2}} I_n & \frac{i}{\sqrt{2}} I_n \end{bmatrix} \quad \text{or} \quad \begin{bmatrix} 1 & & \\ & \frac{1}{\sqrt{2}} I_n & \frac{1}{\sqrt{2}} I_n \\ & \frac{-i}{\sqrt{2}} I_n & \frac{i}{\sqrt{2}} I_n \end{bmatrix},$$

then we have

$$P \begin{bmatrix} 0 & I_n \\ I_n & 0 \end{bmatrix} P' = I_m$$

and

$$P \begin{bmatrix} 1 & & \\ & 0 & I_n \\ & I_n & 0 \end{bmatrix} P' = I_{2n+1}.$$

Hence

$$X \xrightarrow{\psi} P X P^{-1}$$

is an isomorphism from D_n (B_n) onto $\mathfrak{o}(2n, C)$ ($\mathfrak{o}(2n+1, C)$). We consider these two cases separately.

(i) The case of B_n. The element

$$H_{\lambda_1 \ldots \lambda_n} = \begin{bmatrix} 0 & \lambda_1 & & & & & \\ & & \ddots & & & & \\ & & & \lambda_n & & & \\ & & & & -\lambda_1 & & \\ & & & & & \ddots & \\ & & & & & & -\lambda_n \end{bmatrix}$$

is mapped by ψ to

$$\begin{bmatrix} 1 & & \\ & \frac{1}{\sqrt{2}} I & \frac{1}{\sqrt{2}} I \\ & \frac{-i}{\sqrt{2}} I & \frac{i}{\sqrt{2}} I \end{bmatrix} H \begin{bmatrix} 1 & & \\ & \frac{1}{\sqrt{2}} I & \frac{i}{\sqrt{2}} I \\ & \frac{1}{\sqrt{2}} I & \frac{-i}{\sqrt{2}} I \end{bmatrix}$$

$$= \begin{bmatrix} 0 & 0 & \ldots & 0 \\ 0 & & i \begin{bmatrix} \lambda_1 & & \\ & \ddots & \\ & & \lambda_n \end{bmatrix} \\ \vdots & & & \\ 0 & -i \begin{bmatrix} \lambda_1 & & \\ & \ddots & \\ & & \lambda_n \end{bmatrix} & & \end{bmatrix}.$$

Now

$$\begin{bmatrix} 0 & 0 & \ldots & 0 \\ 0 & & i \begin{bmatrix} \lambda_1 & & \\ & \ddots & \\ & & \lambda_n \end{bmatrix} \\ \vdots & & & \\ 0 & -i \begin{pmatrix} \lambda_1 & & \\ & \ddots & \\ & & \lambda_n \end{pmatrix} & & \end{bmatrix} = i \sum_{i=1}^{n} \lambda_i (E_{i,\, n+i} - E_{n+i,\, i}) \xrightarrow{\tau} \tfrac{1}{2} i \sum_{i}^{n} \lambda_i e_i e_{n+i}$$

and

$$\tfrac{1}{2}i\sum_1^n \lambda_i e_i e_{n+i} \xrightarrow{\varphi} \tfrac{1}{2}i\sum_1^n \lambda_i P_i Q_i = \tfrac{1}{2}i\sum_1^n \lambda_i I_2\otimes\ldots\otimes I_2\otimes PQ\otimes\ldots\otimes I_2$$

$$= \sum_1^n I_2\otimes\ldots\otimes I_2\otimes\begin{bmatrix}\frac{1}{2}\lambda_i & \\ & -\frac{1}{2}\lambda_i\end{bmatrix}\otimes I_2\otimes\ldots\otimes I_2.$$

Therefore, $\sigma = \varphi\tau^{-1}\psi$ is an irreducible representation of B_n with simple weights $\frac{1}{2}(\pm\lambda_1\pm\pm\lambda_2\pm\ \ldots\ \pm\lambda_n)$. Hence σ is the spin representation of B_n.

(ii) The case of D_n. The element

$$H_{\lambda_1\ldots\lambda_n} = \begin{bmatrix}\lambda_1 & & & & & \\ & \ddots & & & & \\ & & \lambda_n & & & \\ & & & -\lambda_1 & & \\ & & & & \ddots & \\ & & & & & -\lambda_n\end{bmatrix}$$

is mapped by ψ to

$$\begin{bmatrix} & i\begin{pmatrix}\lambda_1 & & \\ & \ddots & \\ & & \lambda_n\end{pmatrix} \\ -i\begin{pmatrix}\lambda_1 & & \\ & \ddots & \\ & & \lambda_n\end{pmatrix} & \end{bmatrix} = i\sum_{i=1}^n \lambda_i(E_{i,\,n+i}-E_{n+i,\,i}).$$

Now

$$i\sum_{i=1}^n \lambda_i(E_{i,\,n+i}-E_{n+i,\,i}) \xrightarrow{\tau^{-1}} \tfrac{1}{2}i\sum_1^n \lambda_i e_i e_{n+i}$$

and

$$\tfrac{1}{2}i\sum\lambda_i e_i e_{n+i} \xrightarrow{\varphi} \tfrac{1}{2}i\sum_{i=1}^n \lambda_i P_i Q_i = \sum_1^n I_2\otimes\ \ldots\ \otimes I_2\otimes\begin{pmatrix}\frac{1}{2}\lambda_i & \\ & -\frac{1}{2}\lambda_i\end{pmatrix}\otimes\ \ldots\ \otimes I_2.$$

Therefore $\sigma = \varphi\tau^{-1}\psi$ is a representation of D_n with simple weights $\frac{1}{2}(\pm\lambda_1\pm\ \ldots\ \pm\lambda_n)$. σ decomposes into the sum of two irreducible representations with highest weights $\frac{1}{2}(\lambda_1+\ \ldots\ +\lambda_n)$ and $\frac{1}{2}(\lambda_1+\ \ldots\ +\lambda_{n-1}-\lambda_n)$ respectively, so these representations are the spin representation σ_1 and σ_2.

12.4. The exceptional Lie algebras F_4 and E_8

Let $\mathfrak{g}$ be an r-dimensional simple Lie algebra with basis $\{X_1, \ldots, X_r\}$; suppose that

$$[X_i, X_j] = \sum_{k=1}^n c_{ij}^k X_k, \qquad i, j = 1, 2, \ldots, r.$$

Let $\varrho : X \to \varrho(X)$ be an irreducible representation of $\mathfrak{g}$ with representation space V and $\dim V = s$. Suppose that $\{E_1, \ldots, E_s\}$ is a basis of V and

$$\varrho(X_i)E_\beta = \sum_{\alpha=1}^s d_{\alpha\beta}^i E_\alpha, \qquad i = 1, \ldots, r;\ \beta = 1, \ldots, s.$$

Consider the direct sum $\mathfrak{g}_1 = \mathfrak{g} \dotplus V$; the elements $X_1, \ldots, X_r$ and $E_1, \ldots, E_s$ form a basis of $\mathfrak{g}_1$. Define commutation on $\mathfrak{g}_1$ by

$$[X_i, X_j] = \sum_{k=1}^{r} c_{ij}^k X_k, \qquad i, j = 1, 2, \ldots, r$$

$$-[E_\beta, X_i] = [X_i, E_\beta] = \sum_{\alpha=1}^{s} d_{\alpha\beta}^i E_\alpha, \qquad i = 1, 2, \ldots, r;\ \beta = 1, \ldots, s$$

$$[E_\alpha, E_\beta] = \sum_{i=1}^{r} d_{\alpha\beta}^i X_i, \qquad \alpha, \beta = 1, 2, \ldots, s.$$

Suppose that ϱ is not the zero representation and that it is not equivalent to the regular representation of $\mathfrak{g}$. We show that if $\mathfrak{g}_1$ is a Lie algebra, then it is simple.

In fact, $\mathfrak{g}_1$ is the representation space of the representation $f(X)$ of $\mathfrak{g}$ defined by

$$f(X)(Y+E) = \operatorname{ad} XY + \varrho(X)E, \qquad X, Y \in \mathfrak{g}, E \in V.$$

Since ϱ is not equivalent to the regular representation of $\mathfrak{g}$, $\mathfrak{g}_1$ is the direct sum of two representation spaces corresponding to two representations of $\mathfrak{g}$ that are not equivalent. Let $\mathfrak{g}_0$ be an ideal of $\mathfrak{g}_1$, then $\mathfrak{g}_0$ is an invariant subspace of f, so it is the direct sum of irreducible invariant subspaces. There are only three possibilities:

$\mathfrak{g}_0 = \mathfrak{g} \dotplus V$, $\mathfrak{g}_0 = \mathfrak{g}$ or $\mathfrak{g}_0 = V$. Since $\varrho \neq 0$ and ϱ is irreducible, $\mathfrak{g}_0 = \mathfrak{g} \dotplus V$, so $\mathfrak{g}_1$ is simple.

Now we study the conditions under which $\mathfrak{g}_1$ is a Lie algebra.

From the conditions of commutation, we must have

$$d_{\alpha\beta}^i = -d_{\beta\alpha}^i, \qquad i = 1, 2, \ldots, r;\ \alpha, \beta = 1, \ldots, s,$$

i.e. $D_i = \varrho(X_i)$ is skew symmetric ($i = 1, \ldots, r$). In what follows, this condition is assumed.

Now to consider the Jacobi identity, we define

$$\{X, Y, Z\} = [X, [Y, Z]] + [Y, [Z, X]] + [Z, [X, Y]], \quad \text{for } X, Y, Z \in \mathfrak{g}_1.$$

Obviously,

$$\{X_i, X_k, X_l\} = 0.$$

We need to consider the conditions under which the relations

$$\{X_i, X_k, E_\alpha\} = 0,$$
$$\{X_i, E_\alpha, E_\beta\} = 0,$$

and

$$\{E_\alpha, E_\beta, E_\gamma\} = 0$$

are satisfied. Since ϱ is a representation,

$$\begin{aligned} \{X_i, X_k, E_\alpha\} &= [X_i, [X_k, E_\alpha]] + [X_k, [E_\alpha, X_i]] + [E_\alpha, [X_i, X_k]] \\ &= \varrho(X_i)\, \varrho(X_k) E_\alpha - \varrho(X_k)\, \varrho(X_i) E_\alpha - \varrho([X_i, X_k]) E_\alpha \\ &= \{[\varrho(X_i), \varrho(X_k)] - \varrho([X_i, X_k])\} E_\alpha = 0. \end{aligned}$$

Secondly, we have

$$\{X_i, E_\alpha, E_\beta\} = [X_i, [E_\alpha, E_\beta]]+[E_\alpha, [E_\beta, X_i]]+[E_\beta, [X_i, E_\alpha]]$$

$$= \left[X_i, \sum_j d^j_{\alpha\beta}X_j\right]-\left[E_\alpha, \sum_\gamma d^i_{\gamma\beta}E_\gamma\right]+\left[E_\beta, \sum_\gamma d^i_{\gamma\alpha}E_\gamma\right]$$

$$= \sum_{j=1}^{\gamma} d^j_{\alpha\beta}\sum_k c^k_{ij}X_k-\sum_\gamma d^i_{\gamma\beta}\sum_k d^k_{\alpha\gamma}X_k+\sum_\gamma d^i_{\gamma\alpha}\sum_k d^k_{\beta\gamma}X_k$$

$$= \sum_k \left(\sum_{j=1}^{\gamma} d^j_{\alpha\beta}c^k_{ij}-\sum_\gamma d^i_{\gamma\beta}d^k_{\alpha\gamma}+\sum_\gamma d^i_{\gamma\alpha}d^k_{\beta\gamma}\right) X_k. \tag{12.15}$$

On the other hand,

$$[\varrho(X_i), \varrho(X_j)]E_\alpha = \sum_k c^k_{ij}\varrho(X_k)E_\alpha$$

and

$$[\varrho(X_i), \varrho(X_j)]E_\alpha = \varrho(X_i)\,\varrho(X_j)E_\alpha-\varrho(X_j)\,\varrho(X_i)E_\alpha$$

$$= \varrho(X_i)\sum_\gamma d^j_{\gamma\alpha}E_\gamma-\varrho(X_j)\sum_\gamma d^i_{\gamma\alpha}E_\gamma = \sum_\beta\sum_\gamma d^j_{\gamma\alpha}d^i_{\beta\gamma}E_\beta-\sum_\beta\sum_\gamma d^i_{\gamma\alpha}d^j_{\beta\gamma}E_\beta$$

$$= \sum_\beta\sum_\gamma (d^j_{\gamma\alpha}d^i_{\beta\gamma}-d^i_{\gamma\alpha}d^j_{\beta\gamma})E_\beta,$$

$$\sum_k c^k_{ij}\varrho(X_k)E_\alpha = \sum_k c^k_{ij}\sum_\beta d^k_{\beta\alpha}E_\beta.$$

Thus

$$\sum_\gamma (d^j_{\gamma\alpha}d^i_{\beta\gamma}-d^i_{\gamma\alpha}d^j_{\beta\gamma}) = \sum_k c^k_{ij}d^k_{\beta\alpha},$$

i.e.

$$\sum_\gamma (d^i_{\gamma\beta}d^k_{\alpha\gamma}-d^i_{\gamma\alpha}d^k_{\beta\gamma}) = \sum_j c^j_{ik}d^j_{\beta\alpha}.$$

Substituting this equation in (12.15), we get

$$\{X_i, E_\alpha, E_\beta\} = \sum_k \left(\sum_j d^i_{\alpha\beta}c^k_{ij}-\sum_j d^j_{\beta\alpha}c^j_{ik}\right) X_k$$

$$= \sum_k \left\{\sum_j d^j_{\alpha\beta}(c^k_{ij}+c^j_{ik})\right\} X_k.$$

If we assume that

$$\operatorname{Tr} D_iD_k = -s\delta_{ik}, \tag{12.16}$$

then

$$0 = \operatorname{Tr}[D_i, D_jD_k] = \operatorname{Tr}[D_i, D_j]D_k+\operatorname{Tr} D_j[D_i, D_k]$$

$$= \operatorname{Tr}\sum_l c^l_{ij}D_lD_k+\operatorname{Tr} D_j\sum_l c^l_{ik}D_l = -(c^k_{ij}+c^j_{ik})s.$$

Therefore

$$\{X_i, E_\alpha, E_\beta\} = 0.$$

Finally,

$$\begin{aligned}\{E_\alpha, E_\beta, E_\gamma\} &= [E_\alpha, [E_\beta, E_\gamma]]+[E_\beta, [E_\gamma, E_\alpha]]+[E_\gamma, [E_\alpha, E_\beta]] \\ &= \Big[E_\alpha, \sum_i d^i_{\beta\gamma}X_i\Big]+\Big[E_\beta, \sum_i d^i_{\gamma\alpha}X_i\Big]+\Big[E_\gamma, \sum_i d^i_{\alpha\beta}X_i\Big] \\ &= -\sum_i d^i_{\beta\gamma}\sum_\delta d^i_{\delta\alpha}E_\delta-\sum_i d^i_{\gamma\alpha}\sum_\delta \delta^i_{\delta\beta}E_\delta-\sum_i d^i_{\alpha\beta}\sum_\delta d^i_{\delta\gamma}E_\delta \\ &= \sum_{i,\delta}(d^i_{\beta\gamma}d^i_{\alpha\delta}+d^i_{\gamma\alpha}d^i_{\beta\delta}+d^i_{\alpha\beta}d^i_{\gamma\delta})E_\delta.\end{aligned}$$

Set

$$p_{\alpha\beta\gamma\delta} = \sum_i (d^i_{\beta\gamma}d^i_{\alpha\delta}+d^i_{\gamma\alpha}d^i_{\beta\delta}+d^i_{\alpha\beta}d^i_{\gamma\delta}),$$

then for $\{E_\alpha, E_\beta, E_\gamma\}$ to be zero, it suffices to have $p_{\alpha\beta\gamma\delta} = 0$. Notice that if π is a permutation of $\alpha, \beta, \gamma, \delta$, then

$$p_{\alpha\beta\gamma\delta} = \begin{cases} p_{\pi(\alpha)\pi(\beta)\pi(\gamma)\pi(\delta)}, & \text{if } \pi \text{ is even} \\ -p_{\pi(\alpha)\pi(\beta)\pi(\gamma)\pi(\delta)}, & \text{if } \pi \text{ is odd.}\end{cases}$$

We have

$$\operatorname{Tr}(D_iD_k)^2 = \Big(\sum_\beta d^i_{\alpha\beta}d^k_{\beta\alpha}\Big)^2 = \sum_{\alpha,\beta,\gamma,\delta} d^i_{\alpha\beta}d^k_{\beta\alpha}d^i_{\gamma\alpha}d^k_{\delta\gamma},$$

$$\operatorname{Tr}(D_iD_k)^2 = \sum_{\alpha,\beta,\gamma,\delta} d^i_{\alpha\beta}d^k_{\beta\gamma}d^i_{\gamma\delta}d^k_{\delta\alpha}.$$

Hence

$$\begin{aligned}\sum_{i,k}(\operatorname{Tr} D_iD_k)^2-2\sum_{i,k}\operatorname{Tr}(D_iD_k)^2 &= \sum_{i,k}\sum_{\alpha,\beta,\gamma,\delta} d_{\alpha\beta}d^i_{\gamma\delta}(d^k_{\beta\alpha}d^k_{\delta\gamma}-2d^k_{\beta\gamma}d^k_{\delta\alpha}) \\ &= \sum_{\alpha,\beta,\gamma,\delta} p_{\alpha\beta\gamma\delta}\sum_i d^i_{\alpha\beta}d^i_{\gamma\delta} = \tfrac{1}{3}\sum_{\alpha,\beta,\gamma,\delta} p^2_{\alpha\beta\gamma\delta}.\end{aligned}$$

On the other hand, by (12.16)

$$\sum_{i,k}(\operatorname{Tr} D_iD_k)^2 = rs^2.$$

Therefore, if

$$\sum_{i,k}\operatorname{Tr}(D_iD_k)^2 = \tfrac{1}{2}rs^2$$

and $D_i = \varrho(X_i)$ $(1 \leq i \leq r)$ are real matrices, then $p_{\alpha\beta\gamma\delta} = 0$ and

$$\{E_\alpha, E_\beta, E_\gamma\} = 0.$$

We have proved the following:

THEOREM 5 *(E. Witt).*† *Let $\mathfrak{g}$ be an r-dimensional simple Lie algebra with basis $\{X_1, \ldots, X_r\}$. Suppose ϱ is an irreducible representation of $\mathfrak{g}$ with representation space V and dim $V = s$. Let*

$$D_i = (d^i_{\alpha\beta})_{1\leq\alpha,\beta\leq s}, \quad i = 1, \ldots, r,$$

† E. Witt, Spiegelungsgruppen und Aufzählung halbeinfacher Liescher Ringe, *Abh. Math. Sem. Univ. Hamburg* **14** (1941), 289–337.

be the matrix of $\varrho(X_i)$ *relative to a basis* $\{E_1, \ldots, E_s\}$ *of* V. *Assume that* ϱ *is not the zero representation,* ϱ *is not equivalent to the regular representation of* $\mathfrak{g}$ *and*

(i) *Every* $D_i = \varrho(X_i)$ *is real and skew symmetric,*
(ii) $Tr\, D_i D_k = -s\delta_{ik}$,
(iii) $Tr \sum_{i,k} (D_i D_k)^2 = \frac{1}{2} rs^2$.

Then $\mathfrak{g}_1 = \mathfrak{g} \dot{+} V$ *is a simple Lie algebra with commutation defined by*

$$[X_i, X_j] = \sum_{k=1}^{r} c_{ij}^k X_k, \quad i, j = 1, \ldots, r$$

$$-[E_\beta, X_i] = [X_i, E_\beta] = \sum_{\alpha=1}^{s} d_{\alpha\beta}^i E_\alpha, \quad i = 1, \ldots, r; \beta = 1, \ldots, s$$

$$[E_\alpha, E_\beta] = \sum_{i=1}^{r} d_{\alpha\beta}^i X_i, \quad \alpha, \beta = 1, \ldots, s.$$

With proved the existence of F_4 and E_8 by Theorem 5.

We consider first the case of F_4. Let V_8 be an eight-dimensional vector space with basis $\{e_1, \ldots, e_8\}$. Consider the Clifford algebra $\mathfrak{C}$ generated by $e_1, \ldots, e_8$ satisfying

$$e_i^2 = 1, \quad i = 1, 2, \ldots, 8$$

and

$$e_i e_j + e_j e_i = 0, \quad i \neq j, \quad i, j = 1, \ldots, 8.$$

Let

$$e_0 = e_1, \ldots, e_8.$$

It can be proved that

$$e_0^2 = 1,$$

$$e_0 e_i + e_i e_0 = 0, \quad i = 1, 2, \ldots, 8.$$

Let $\mathfrak{C}_2'$ be the set of elements of the form

$$\sum_{0 \leq i < k \leq 8} c_{ik} e_i e_k, \quad c_{ik} \in C.$$

in $\mathfrak{C}$; as in the proof of Theorem 3, it can be shown that

$$\sum_{0 \leq i < k \leq 8} c_{ik} e_i e_k \xrightarrow{\tau} 2 \sum_{0 \leq i < k \leq 8} c_{ik} (E_{ik} - E_{ki})$$

is an isomorphism between $\mathfrak{o}(9, C)$ and $\mathfrak{C}_2'$. By Theorem 2, $\mathfrak{C}$ is isomorphic to the ring of $2^4 \times 2^4$ matrices; under the isomorphism φ defined in Theorem 2, $e_1, \ldots, e_8$ are mapped to $P_1\ P_2, P_3, P_4, Q_1, Q_2, Q_3$ and Q_4, respectively. Now there exists T such that

$$TP_1T^{-1} = E_1 = \begin{bmatrix} I_8 & \\ & -I_8 \end{bmatrix},$$

$$TP_2T^{-1} = E_2 = \begin{bmatrix} & I_8 \\ I_8 & \end{bmatrix},$$

$$TP_3T^{-1} = E_3 = \left[\begin{array}{cc|cc} & & 0 & I_4 \\ & & -I_4 & 0 \\ \hline 0 & -I_4 & & \\ I_4 & 0 & & \end{array}\right],$$

$$TP_4T^{-1} = E_4 = \left[\begin{array}{c|c} 0 & \begin{array}{c|c} \begin{matrix} 0 & I_2 \\ -I_2 & 0 \end{matrix} & 0 \\ \hline 0 & \begin{matrix} 0 & -I_2 \\ I_2 & 0 \end{matrix} \end{array} \\ \hline \begin{array}{c|c} \begin{matrix} 0 & -I_2 \\ I_2 & 0 \end{matrix} & 0 \\ \hline 0 & \begin{matrix} 0 & I_2 \\ -I_2 & 0 \end{matrix} \end{array} & 0 \end{array}\right]$$

$$TQ_1T^{-1} = E_5 = \left[\begin{array}{c|c} 0 & \begin{array}{c|c} \begin{array}{c|c} \begin{matrix} 0 & 1 \\ -1 & 0 \end{matrix} & 0 \\ \hline 0 & \begin{matrix} 0 & -1 \\ 1 & 0 \end{matrix} \end{array} & 0 \\ \hline 0 & \begin{array}{c|c} \begin{matrix} 0 & -1 \\ 1 & 0 \end{matrix} & 0 \\ \hline 0 & \begin{matrix} 0 & 1 \\ -1 & 0 \end{matrix} \end{array} \end{array} \\ \hline \begin{array}{c|c} \begin{array}{c|c} \begin{matrix} 0 & -1 \\ 1 & 0 \end{matrix} & 0 \\ \hline 0 & \begin{matrix} 0 & 1 \\ -1 & 0 \end{matrix} \end{array} & 0 \\ \hline 0 & \begin{array}{c|c} \begin{matrix} 0 & 1 \\ -1 & 0 \end{matrix} & 0 \\ \hline 0 & \begin{matrix} 0 & -1 \\ 1 & 0 \end{matrix} \end{array} \end{array} & 0 \end{array}\right],$$

$$TQ_2T^{-1} = E_6 = \left[\begin{array}{c|c} 0 & \begin{array}{c|c} 0 & \begin{array}{c|c} \begin{matrix} 0 & 1 \\ -1 & 0 \end{matrix} & 0 \\ \hline 0 & \begin{matrix} 0 & 1 \\ -1 & 0 \end{matrix} \end{array} \\ \hline \begin{array}{c|c} \begin{matrix} 0 & 1 \\ -1 & 0 \end{matrix} & 0 \\ \hline 0 & \begin{matrix} 0 & 1 \\ -1 & 0 \end{matrix} \end{array} & 0 \end{array} \\ \hline \begin{array}{c|c} 0 & \begin{array}{c|c} \begin{matrix} 0 & -1 \\ 1 & 0 \end{matrix} & 0 \\ \hline 0 & \begin{matrix} 0 & -1 \\ 1 & 0 \end{matrix} \end{array} \\ \hline \begin{array}{c|c} \begin{matrix} 0 & -1 \\ 1 & 0 \end{matrix} & 0 \\ \hline 0 & \begin{matrix} 0 & -1 \\ 1 & 0 \end{matrix} \end{array} & 0 \end{array} & 0 \end{array}\right],$$

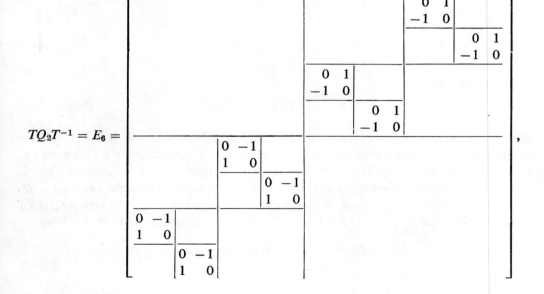

$$TQ_3T^{-1} = E_7 = \left[\begin{array}{cccccccccccccccc}
&&&&&&&&&&&1 \\
&&&&&&&&&&-1 \\
&&&&&&&&&1 \\
&&&&&&&&-1 \\
&&&&&&&&&&&&&&&-1 \\
&&&&&&&&&&&&&&1 \\
&&&&&&&&&&&&&-1 \\
&&&&&&&&&&&&1 \\
&&&-1 \\
&&1 \\
&-1 \\
1 \\
&&&&&&&1 \\
&&&&&&-1 \\
&&&&&1 \\
&&&&-1
\end{array}\right],$$

$$TQ_4T^{-1} = E_8 = \left[\begin{array}{cccccccccccccccc}
&&&&&&&&&&&&&&&1 \\
&&&&&&&&&&&&&&1 \\
&&&&&&&&&&&&&-1 \\
&&&&&&&&&&&&-1 \\
&&&&&&&&&&&1 \\
&&&&&&&&&&1 \\
&&&&&&&&&-1 \\
&&&&&&&&-1 \\
&&&&&&&-1 \\
&&&&&&-1 \\
&&&&&1 \\
&&&&1 \\
&&&-1 \\
&&-1 \\
&1 \\
1
\end{array}\right].$$

If ϱ_0 is defined by

$$X \xrightarrow{\varrho_0} TXT^{-1}, X \in M_{16}(C)$$

then $\varrho = \varrho_0\varphi$ is also an isomorphism from $\mathfrak{C}$ to $M_{16}(C)$. Now $\varrho(e_i) = E_i$ $(i = 1, \ldots, 8)$ are real orthogonal matrices, thus $\varrho(e_0) = E_0$ is also a real orthogonal matrix. Order the e_ie_k $(j = 1, \ldots, 36)$ as X_j $(j = 1, \ldots, 36)$ and denote the representation induced by ϱ on $\mathfrak{C}_2'$ also by ϱ. Since $\varrho(e_i)$ can be written as products of the $\varrho(X_j)$, ϱ is irreducible. Let $\varrho(X_j) = D_j$, then D_j is real orthogonal. Since $X_j = e_\lambda e_\mu$, $X_j^2 = -1$. Hence $\varrho(X_j)^2 = -I$. From $\varrho(X_j)\,\varrho(X_j)' = I$, it follows that $D_j = \varrho(X_j)$ $(j = 1, \ldots, 36)$ are skew symmetric matrices. We have

$$\begin{aligned}
\operatorname{Tr} D_j^2 &= \operatorname{Tr} \varrho(X_j)\,\varrho(X_j) = \operatorname{Tr} \varrho(e_\lambda e_\mu e_\lambda e_\mu) \\
&= \operatorname{Tr} \varrho(-1) = -16.
\end{aligned}$$

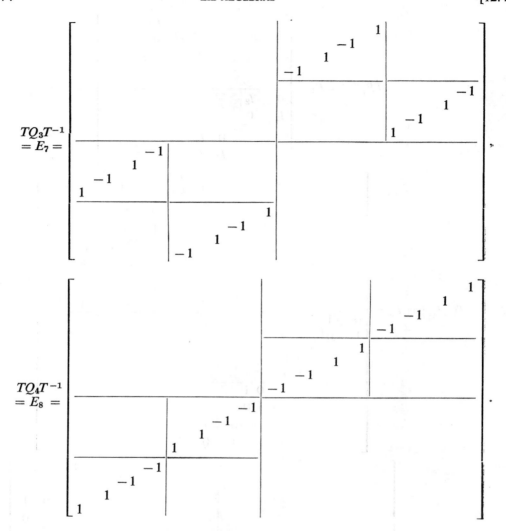

Now let $X_k = e_\nu e_k$. If $\mu = \nu$, then

$$\operatorname{Tr} D_j D_k = \operatorname{Tr} \varrho(e_\lambda e_\mu e_\mu e_k) = \operatorname{Tr} \varrho(e_\lambda e_k) = 0.$$

If no two of λ, μ, ν, k are equal, then $(e_\lambda e_\mu e_\nu e_k)^2 = 1$ and

$$\operatorname{Tr} D_j D_k = 0.$$

We have proved that

$$\operatorname{Tr} D_j D_k = -\delta_{jk}(16).$$

Finally, we compute $\operatorname{Tr} \sum_{j,k} (D_j D_k)^2$ for a fixed j. When $k = j$, $\operatorname{Tr} (D_j D_j)^2 = 16$. Let $D_j = E_\lambda E_\mu$, then there are 14 $D_k = E_\nu E_k$ such that one of ν, k is equal to one of λ, μ and $\operatorname{Tr} (D_j D_k)^2 = -16$; there are 21 $D_k = E_\nu E_k$ such that none of ν and k is equal to λ or μ and

$$\operatorname{Tr} (D_j D_k)^2 = \operatorname{Tr} I = 16.$$

Therefore,

$$\operatorname{Tr} \sum_{j,k} (D_j D_k)^2 = 36(16-16\cdot 14+16\cdot 21)$$
$$= 36\cdot 16\cdot 8 = \tfrac{1}{2}\cdot 36\cdot 16^2.$$

Therefore, the representation ϱ satisfies the conditions of Theorem 5 and a simple Lie algebra $\mathfrak{g}_1$ is obtained. Clearly the Cartan subalgebra of $\mathfrak{C}_2$ is the Cartan subalgebra of $\mathfrak{g}_1$, so the rank of $\mathfrak{g}_1$ is 4. Now $\dim \mathfrak{g}_1 = 36+16 = 52$, $\dim A_4 = 5^2-1 = 24$, $\dim B_4 = \dim C_4 = 36$ and $\dim D_4 = 28$, so $\mathfrak{g}_1$ is not one of the classical Lie algebras of rank 4. Hence $\mathfrak{g}_1$ is a Lie algebra with fundamental system of roots $\Pi(F_4)$. This proves the existence of F_4.

Clearly, the root system of F_4 is the union of the root system of $\mathfrak{o}(9, C)$ and the set of all weights of the representation ϱ. Now ϱ is the spin representation of $\mathfrak{o}(9, C)$, so the roots of F_4 are

$$\pm\lambda_i \quad (i = 1, 2, 3, 4), \qquad \pm\lambda_i\pm\lambda_k, \quad (i \neq k, \quad i, k = 1, 2, 3, 4),$$
$$\tfrac{1}{2}(\pm\lambda_1\pm\lambda_2\pm\lambda_3\pm\lambda_4).$$

We now consider E_8. Let V_{14} be a 14-dimensional vector space with basis $\{e_1, \ldots, e_{14}\}$ and T_{14} be the tensor algebra of V_{14}. Consider the ideal I of T_{14} generated by the elements

$$\left.\begin{array}{ll} e_i^2-1, & i = 1, 2, \ldots, 14 \\ e_i e_j + e_j e_i, & i, j = 1, \ldots, 14, i \neq j. \end{array}\right\} \qquad (12.17)$$

As in the proof of Theorem 1, it can be shown that T/I is a 2^{14}-dimensional association algebra $\mathfrak{C}'$ generated by $e_1, \ldots, e_{14}$ satisfying the relations (12.17). As in the proof of Theorem 2, it can be shown that $\mathfrak{C}'$ is isomorphic to the algebra of $2^7\times 2^7$ matrices; in this case, we let

$$P = \begin{bmatrix} 0 & 1 \\ -1 & 0 \end{bmatrix}, \quad Q = \begin{bmatrix} 0 & \sqrt{-1} \\ \sqrt{-1} & 0 \end{bmatrix}.$$

Let ϱ be this isomorphism. Rewrite $e_1, \ldots, e_7$ as $e_2, \ldots, e_8$ and $e_8, e_9, \ldots, e_{14}$ as $e_{10}, e_{11}, \ldots, e_{16}$. Let

$$e_9 = e_2 e_3 \ldots . e_8 e_{10} e_{11} \ldots . e_{16},$$

then

$$e_9^2 = -1,$$

$$e_9 e_i + e_i e_9 = 0, \qquad i = 2, 3, \ldots, 8, 10, 11, \ldots, 16.$$

Let $\mathfrak{C}'_{12}$ be the set of elements of the form

$$\sum_{i=2}^{16} a_i e_i + \sum_{2 \leq i < k \leq 16} c_{ik} e_i e_k, \qquad a_i, c_{ik} \text{ are complex};$$

then $\mathfrak{C}'_{12}$ is a subalgebra of $\mathfrak{C}'_2$. As in Theorem 4, it can be shown that

$$\sum_{i=2}^{16} a_i e_i + \sum_{2 \leq i < k \leq 16} c_{ik} e_i e_k \xrightarrow{\tau} 2 \sum_{i=2}^{16} a_i (E_{1i} - E_{i1}) - 2 \sum_{2 \leq i < k \leq 16} c_{ik} (E_{ik} - E_{ki})$$

is an isomorphism between $\mathfrak{C}'_{12}$ and $\mathfrak{o}(16, C)$. Let

$$P = \begin{bmatrix} \frac{1}{\sqrt{2}} I_8 & \frac{1}{\sqrt{2}} I_8 \\ \frac{-i}{\sqrt{2}} I_8 & \frac{i}{\sqrt{2}} I_8 \end{bmatrix}$$

and ψ be the isomorphism from D_8 to $\mathfrak{o}(16, C)$ defined by

$$X \xrightarrow{\psi} PXP^{-1},$$

then $\varphi \tau^{-1} \psi$ is a representation of D_8. We now consider the weights of this representation. The element

$$H_{\lambda_1 \ldots \lambda_8} = \begin{bmatrix} \lambda_1 & & & & & \\ & \ddots & & & & \\ & & \lambda_8 & & & \\ & & & -\lambda_1 & & \\ & & & & \ddots & \\ & & & & & -\lambda_8 \end{bmatrix}$$

is mapped by ψ to

$$\begin{bmatrix} & \sqrt{-1} \begin{pmatrix} \lambda_1 & & \\ & \ddots & \\ & & \lambda_8 \end{pmatrix} \\ -\sqrt{-} \begin{pmatrix} \lambda_1 & & \\ & \ddots & \\ & & \lambda_8 \end{pmatrix} & \end{bmatrix} = \sqrt{-1} \sum_{i=1}^{8} \lambda_i (E_{i, 8+i} - E_{8+i, i}).$$

Moreover,

$$\sqrt{-1} \sum_{i=1}^{8} \lambda_i (E_{i, 8+i} - E_{8+i, i}) \xrightarrow{\tau^{-1}} \frac{\sqrt{-1}}{2} \lambda_1 e_9 - \frac{\sqrt{-1}}{2} \sum_{i=2}^{8} \lambda_i e_i e_{8+i}$$

$$\xrightarrow{\varphi} \frac{\sqrt{-1}}{2} \lambda_1 \varphi(e_9) - \frac{\sqrt{-1}}{2} \sum_{i=2}^{8} \lambda_i P_{i-1} Q_{i-1},$$

where $P_{i-1} = \varphi(e_i)$ and $Q_{i-1} = \varphi(e_{8+i})$ $(i = 2, \ldots, 8)$. We have

$$\frac{\sqrt{-1}}{2}\lambda_1\varphi(e_9) - \frac{\sqrt{-1}}{2}\sum_{i=2}^{8}\lambda_i P_i Q_i = \frac{\sqrt{-1}}{2}\lambda_1 P_1 \ldots P_7 Q_1 \ldots Q_7 - \frac{\sqrt{-1}}{2}$$

$$\times \sum_{i=2}^{8}\lambda_i I_2 \otimes \ldots \otimes I_2 \otimes PQ \otimes \ldots \otimes I_2 \quad \big((i-1)\text{th position}\big)$$

$$= \frac{\sqrt{-1}}{2}\lambda_1 \begin{bmatrix} \sqrt{-1} & 0 \\ 0 & -\sqrt{-1} \end{bmatrix} \otimes \begin{bmatrix} \sqrt{-1} & 0 \\ 0 & -\sqrt{-1} \end{bmatrix} \otimes \ldots \otimes \begin{bmatrix} \sqrt{-1} & 0 \\ 0 & -\sqrt{-1} \end{bmatrix}$$

$$- \frac{\sqrt{-1}}{2}\sum_{i=2}^{8}\lambda_i I_2 \otimes \ldots \otimes I_2 \otimes \begin{bmatrix} \sqrt{-1} & 0 \\ 0 & -\sqrt{-1} \end{bmatrix} \otimes I_2 \ldots \otimes I_2$$

$$= -\frac{1}{2}\lambda_1 \begin{bmatrix} 1 & 0 \\ 0 & -1 \end{bmatrix} \otimes \begin{bmatrix} 1 & 0 \\ 0 & -1 \end{bmatrix} \otimes \ldots \otimes \begin{bmatrix} 1 & 0 \\ 0 & -1 \end{bmatrix}$$

$$+ \frac{1}{2}\sum_{i=2}^{8}\lambda_i I_2 \otimes \ldots \otimes I_2 \otimes \begin{bmatrix} 1 & 0 \\ 0 & -1 \end{bmatrix} \otimes I_2 \otimes \ldots \otimes I_2.$$

Therefore, $\varphi\tau^{-1}\psi$ is an irreducible representation of D_8 with highest weight $\frac{1}{2}(\lambda_1 + \ldots + \lambda_7 - \lambda_8)$.

Now from

$$P_i^2 = Q_i^2 = -I \quad (i = 1, \ldots, 7)$$

and the definitions of $P_1, \ldots, P_7, Q_1, \ldots, Q_6$ and Q_7, there exists a $2^7 \times 2^7$ non-singular T such that†

$$TP_1T^{-1} = \begin{bmatrix} & I_{64} \\ -I_{64} & \end{bmatrix},$$

$$TP_2T^{-1} = \left[\begin{array}{c|c} \begin{matrix} & I_{32} \\ -I_{32} & \end{matrix} & \\ \hline & \begin{matrix} & -I_{32} \\ I_{32} & \end{matrix} \end{array}\right],$$

$$TP_3T^{-1} = \left[\begin{array}{c|c} \begin{array}{c|c} \begin{matrix} & I_{16} \\ -I_{16} & \end{matrix} & \\ \hline & \begin{matrix} & -I_{16} \\ I_{16} & \end{matrix} \end{array} & \\ \hline & \begin{array}{c|c} \begin{matrix} & -I_{16} \\ I_{16} & \end{matrix} & \\ \hline & \begin{matrix} & I_{16} \\ -I_{16} & \end{matrix} \end{array} \end{array}\right],$$

† The explicit forms of the fourteen matrices listed here were obtained by Gen-Dao Li.

$$TP_4T^{-1} = \left[\begin{array}{cccc|cccc} & & & I_{16} & & & & \\ & & -I_{16} & & & & & \\ & I_{16} & & & & & & \\ -I_{16} & & & & & & & \\ \hline & & & & & & & -I_{16} \\ & & & & & & I_{16} & \\ & & & & & -I_{16} & & \\ & & & & I_{16} & & & \end{array}\right],$$

$$TP_5T^{-1} = \left[\begin{array}{cccc|cccc} & & & & & & I_{16} & \\ & & & & & & & -I_{16} \\ & & & & -I_{16} & & & \\ & & & & & I_{16} & & \\ \hline & & I_{16} & & & & & \\ & & & -I_{16} & & & & \\ -I_{16} & & & & & & & \\ & I_{16} & & & & & & \end{array}\right],$$

$$TP_6T^{-1} = \left[\begin{array}{cccc|cccc} & & & & & I_{16} & & \\ & & & & -I_{16} & & & \\ & & & & & & & I_{16} \\ & & & & & & -I_{16} & \\ \hline & I_{16} & & & & & & \\ -I_{16} & & & & & & & \\ & & & I_{16} & & & & \\ & & -I_{16} & & & & & \end{array}\right],$$

$$TP_7T^{-1} = \left[\begin{array}{cccc|cccc} & & & & & & E_8 & \\ & & & & & & & -E_8 \\ & & & & -E_8 & & & \\ & & & & & E_8 & & \\ \hline & & E_8 & & & & & \\ & & & -E_8 & & & & \\ -E_8 & & & & & & & \\ & E_8 & & & & & & \end{array}\right],$$

$$TQT_i^{-1} = \left[\begin{array}{cccc|cccc} & & & & & & E_i & \\ & & & & & & & -E_i \\ & & & & -E_i & & & \\ & & & & & E_i & & \\ \hline & & E_i & & & & & \\ & & & -E_i & & & & \\ -E_i & & & & & & & \\ & E_i & & & & & & \end{array}\right],$$

$(i = 1, \ldots, 7)$,

where $E_i^2 = I$ and $E_iE_j = -E_jE_i$. Therefore, $E_1, \ldots, E_8$ can be identified with the matrices defined for the case of F_4. Hence TP_iT^{-1} and TQ_iT^{-1} $(i = 1, 2, \ldots, 7)$ are real orthogonal matrices. If ϱ_0 is defined by

$$X \xrightarrow{\varrho_0} TXT^{-1}, \quad X \in M_{2^7}(C)$$

then $\varrho = \varrho_0\varphi$ is also an isomorphism from $\mathfrak{C}'$ to $M_{128}(C)$ and $\varrho(e_i)$ $(2 \leqslant i \leqslant 16)$ are real orthogonal matrices. $\varrho(e_ie_k)$ $(2 \leqslant i < k \leqslant 16)$ can also be shown to be real orthogonal matrices. Order the e_i, e_ie_k as X_j $(j = 1, \ldots, 120)$ and let $\varrho(X_j) = D_j$, then D_j $(j = 1, \ldots, 120)$ are all real orthogonal and skew symmetric matrices. As in the case of F_4, we have

$$\operatorname{Tr} D_j^2 = -2^7,$$
$$\operatorname{Tr} D_jD_k = 0,$$
$$\operatorname{Tr} \sum_{j,k} (D_jD_k)^2 = \tfrac{1}{2}\cdot 120\cdot(2^7)^2.$$

Thus ϱ satisfies the conditions of Theorem 5 and a simple Lie algebra $\mathfrak{g}_1$ is obtained. Clearly, the Cartan subalgebra of $\mathfrak{o}(16, C)$ is a Cartan subalgebra of $\mathfrak{g}_1$, so the rank of $\mathfrak{g}_1$ is 8. Now $\dim \mathfrak{g}_1 = 120+2^7 = 248$, $\dim A_8 = 9^2-1 = 80$, $\dim B_8 = \dim C_8 = 136$ and $\dim D_8 = 120$, so $\mathfrak{g}_1$ is not isomorphic to any of the classical Lie algebras of rank 8. Hence $\mathfrak{g}_1$ is a simple Lie algebra with fundamental system of roots $\Pi(E_8)$; this proves the existence of E_8. Clearly, the root system of E_8 is the union of the root system of D_8 and the set of weights of ϱ, so the roots of E_8 are

$$\pm\lambda_i\pm\lambda_k \qquad (i \neq k, \ , i, k = 1, 2, \ldots, 8)$$
$$\tfrac{1}{2}(\pm\lambda_1\pm \ \ldots \ \pm\lambda_8) \text{ (with an odd number of negative signs).}$$

CHAPTER 13

POINCARÉ–BIRKHOFF–WITT THEOREM AND ITS APPLICATIONS TO REPRESENTATION THEORY OF SEMISIMPLE LIE ALGEBRAS

13.1. Enveloping algebras of Lie algebras

Let $\mathfrak{g}$ be a Lie algebra over C and A be an associative algebra with identity 1 over C. If f is a representation from $\mathfrak{g}$ to A and $\{f(\mathfrak{g}), 1\}$ is a set of generators of A, then A is said to be an enveloping algebra (relative to f) of $\mathfrak{g}$. If $\{X_1, \ldots, X_r\}$ is a basis of $\mathfrak{g}$, then A is an enveloping algebra of $\mathfrak{g}$ iff $\{f(X_1), \ldots, f(X_r)\}$ generate A.

An enveloping algebra U (relative to some representation ϱ) of $\mathfrak{g}$ is said to be a universal enveloping algebra (relative to ϱ) of $\mathfrak{g}$ if for any enveloping algebra A (relative to some f), there exists a homomorphism $\tilde{f}$ from U onto A such that

$$\tilde{f}\circ\varrho = f, \qquad \tilde{f}(1) = 1. \qquad \begin{array}{ccc} \mathfrak{g} & \xrightarrow{\varrho} & U \\ & {\scriptstyle f}\searrow \quad \swarrow{\scriptstyle \tilde{f}} & \\ & A & \end{array}$$

Since $\tilde{f}(1) = 1$, $\tilde{f}$ is a canonical homomorphism.

THEOREM 1. *Let $\mathfrak{g}$ be a Lie algebra, then there exists a universal enveloping algebra of $\mathfrak{g}$. Moreover, any two universal enveloping algebras U and U' (relative to ϱ and ϱ') are isomorphic with respect to an isomorphism satisfying*

$$\varrho(X) \to \varrho'(X), \quad \textit{for all} \quad X \in \mathfrak{g}.$$

Proof. We show first the uniqueness of the universal enveloping algebra (up to isomorphisms). By definition, there exists a homomorphism $\tilde{\varrho}'$ from U onto U' such that

$$\tilde{\varrho}'\circ\varrho = \varrho', \qquad \tilde{\varrho}'(1) = 1. \qquad \begin{array}{ccc} \mathfrak{g} & \xrightarrow{\varrho} & U \\ & {\scriptstyle \varrho'}\searrow \quad \swarrow{\scriptstyle \tilde{\varrho}'} & \\ & U' & \end{array}$$

Similarly, there exists $\tilde{\varrho}$ from U' onto U such that

$$\tilde{\varrho}\circ\varrho' = \varrho, \qquad \mathfrak{g}\xrightarrow{\varrho'} U'$$
$$\tilde{\varrho}(1) = 1. \qquad \varrho\searrow \ \swarrow\tilde{\varrho}$$
$$U$$

If $\{X_1, \ldots, X_r\}$ is a basis of $\mathfrak{g}$, then

$$\tilde{\varrho}\circ\tilde{\varrho}'\big(\varrho(X_i)\big) = \tilde{\varrho}\circ\tilde{\varrho}'\circ\varrho(X_i) = \tilde{\varrho}\circ\varrho'(X_i) = \varrho(X_i)$$

and $\tilde{\varrho}\circ\tilde{\varrho}'(1) = 1$.

Now U is generated by 1 and $\varrho(X_1), \ldots, \varrho(X_r)$, so $\tilde{\varrho}\circ\tilde{\varrho}' = I$ on U. Similarly, $\tilde{\varrho}'\circ\tilde{\varrho} = I$ on U'. Thus $\tilde{\varrho}'$ is an isomorphism between U and U' and

$$\tilde{\varrho}'\big(\varrho(X)\big) = \varrho'(X), \quad X\in\mathfrak{g}.$$

We now show the existence of the universal enveloping algebra. Choose any basis $\{X_1, \ldots, X_r\}$ of $\mathfrak{g}$ and consider the tensor algebra T of $\mathfrak{g}$. The elements in T of the form

$$X_i\otimes X_j - X_j\otimes X_i - [X_i, X_j], \quad 1\leqslant i, j\leqslant r$$

generate an ideal J of T. The mapping

$$\sum a_iX_i \to \sum a_iX_i$$

from $\mathfrak{g}$ to T induces a mapping ϱ from $\mathfrak{g}$ to the quotient algebra T/J:

$$\sum a_iX_i \xrightarrow{\ \varrho\ } \sum a_iX_i + J.$$

We now show that T/J (relative to ϱ) is a universal enveloping algebra. Clearly, ϱ is a representation of $\mathfrak{g}$ (to T/J). Since $\{1, X_1, \ldots, X_r\}$ is a set of generators of T, $\{1, \varrho(X_1), \ldots, \varrho(X_r)\}$ is a set of generators of T/J.) This proves that T/J (relative to ϱ) is an enveloping algebra of $\mathfrak{g}$.

Now let A be any enveloping algebra (relative to some f). Extend f to a canonical homomorphism f^0 from T to A:

$$f^0(X_{i_1}\otimes\ldots\otimes X_{i_k}) = f(X_{i_1})\ldots f(X_{i_k}), \quad f^0(1) = 1.$$

Since f is a representation,

$$f^0(X_i\otimes X_j - X_j\otimes X_i - [X_i, X_j]) = f(X_i)f(X_j) - f(X_j)f(X_i) - f([X_i, X_j]) = 0.$$

Thus $f^0(J) = \{0\}$ and f^0 induces a homomorphism $\tilde{f}$ from T/J to A:

$$\tilde{f}(X_{i_1}\otimes\ldots\otimes X_{i_k} + J) = f(X_{i_1})\ldots f(X_{i_k}).$$

Naturally,

$$\tilde{f}\circ\varrho = f, \quad \tilde{f}(1) = 1.$$

Since $\{f(X_1), \ldots, f(X_r), 1\}$ is a set of generators of A, $\tilde{f}$ is onto. This proves that T/J is a universal enveloping algebra of $\mathfrak{g}$. The proof of Theorem 1 is now completed.

Based on Theorem 1, we always consider T/J as the unique universal enveloping algebra of $\mathfrak{g}$.

13.2. Poincaré–Birkhoff–Witt theorem

Let $\mathfrak{g}$ be a Lie algebra and $U = T/J$ be the universal enveloping algebra of $\mathfrak{g}$. For any $X \in \mathfrak{g}$, let $\varrho(X) = \mathbf{X}$. Suppose that $\{X_1, \ldots, X_r\}$ is a basis of $\mathfrak{g}$, then we have the following:

THEOREM 2 *(Poincaré–Birkhoff–Witt*†*). The elements*

$$\mathbf{X}_{i_1}, \ldots, \mathbf{X}_{i_p}, \quad 1 \leqslant i_1 \leqslant \ldots \leqslant i_p \leqslant r, \quad p = 0, 1, 2, \ldots, \tag{13.1}$$

form a basis of U.

Proof (Iwasawa). (1) We show first that the elements (13.1) span U. We call scalar multiples of an element of the form (13.1) a standard monomial of order p; linear combinations of standard monomials are called standard polynomials. The order of a standard polynomial is defined to be the highest order of the monomial summands. Let F_p be the image of the subspace $\sum_{i \leqslant p} T^i$ under the canonical mapping $T \to U = T/J$. To show that the elements (13.1) generate U, it suffices to show that $F_p (p = 0, 1, 2, \ldots)$ is equal to the set of all standard polynomials. This can be proved by induction. Obviously, F_p is spanned by the elements

$$\mathbf{X}_{i_1}, \ldots, \mathbf{X}_{_k}, \quad 1 \leqslant i_1, i_2, \ldots, i_k \leqslant r, \quad k = 0, 1, 2, \ldots, p.$$

Hence it suffices to prove the following:

LEMMA 1. *If π is a permutation of 1, 2, ..., p, then*

$$\mathbf{X}_{i_1}, \ldots, \mathbf{X}_{i_p} - \mathbf{X}_{i_{\pi(1)}}, \ldots, \mathbf{X}_{i_{\pi(p)}} \in F_{p-1}.$$

Proof. It suffices to consider the case when $\pi = (j, j+1)$. We have

$$X_{i_j} \otimes X_{i_{j+1}} - X_{i_{j+1}} \otimes X_{i_j} \equiv [X_{i_j}, X_{i_{j+1}}] = \sum c^k_{i_j, i_{j+1}} X_k \pmod{J},$$

so

$$\mathbf{X}_{i_j}\mathbf{X}_{i_{j+1}} - \mathbf{X}_{i_{j+1}}\mathbf{X}_{i_j} = \sum_k c^k_{i_j, i_{j+1}} \mathbf{X}_k,$$

and

$$\mathbf{X}_{i_1} \ldots, \mathbf{X}_{i_{j-1}}\mathbf{X}_{i_j}\mathbf{X}_{i_{j+1}}\mathbf{X}_{i_{j+2}} \ldots, \mathbf{X}_{i_p} - \mathbf{X}_{i_1} \ldots, \mathbf{X}_{i_{j-1}}\mathbf{X}_{i_{j+1}}\mathbf{X}_{i_j}\mathbf{X}_{i_{j+2}} \ldots \mathbf{X}_{i_p}$$
$$= \sum c^k_{i_j i_{j+1}} \mathbf{X}_{i_1} \ldots \mathbf{X}_{i_{j+1}} \mathbf{X}_k \mathbf{X}_{i_{j+2}} \ldots \mathbf{X}_{i_p} \in F_{p-1}.$$

(2) We now show that the elements (13.1) are linearly independent. Let $P = C[x_1, \ldots, x_r]$ be the ring of polynomials of r independent variables, then P is infinite dimensional over C. Let I be the sequence $i_1, \ldots, i_p$ and $x_1 = x_{i_1} \ldots x_{i_p}$, $\mathbf{X}_I = \mathbf{X}_{i_1}, \ldots, \mathbf{X}_{i_p}$. We have the following:

LEMMA 2. *There exists a representation f of $\mathfrak{g}$ with representation space P and*

$$f(X_i)\, x_I = x_i x_I, \quad \textit{for } i \leqslant I, \quad \textit{i.e.} \quad i \leqslant k \textit{ for all } k \in I.$$

† H. Poincaré, Quelques remarques sur les groupes finis et continus, *Œuvres complètes*, Tome III, Paris, 1954. G. Birkhoff, Representations of Lie algebras and Lie groups by matrices, *Ann. Math.* **38** (1937), 526–32. E. Witt, Treue Darstellung Liescher Ringe, *J. reine angew. Math.* **177** (1937), 152–60.

We show first that Theorem 2 follows from Lemma 2.

By definition, there exists a representation $\tilde{f}$ of U such that

$$\tilde{f} \circ \varrho(X) = f(X) \quad \text{for all} \quad X \in \mathfrak{g},$$
$$\tilde{f}(1) = 1.$$

If the elements in $I = \{i_1, \ldots, i_p\}$ satisfy $i_1 \leqslant i_2 \leqslant \ldots \leqslant i_p$, then by Lemma 2 and induction on p, we have

$$\tilde{f}(\mathbf{X}) \cdot 1 = x_I,$$

where 1 is the identity of P. Now let I denote any finite non-decreasing sequence consisting of elements from $\{1, 2, \ldots, r\}$. Since the elements x_I (for all possible I) are independent in P, the elements X_r are also independent.

(3) We now prove Lemma 2. Let P_p be the subspace spanned by all homogeneous polynomials of degree p and $Q_p = \sum_{i \leqslant p} P_i$. Then Lemma 2 is a consequence of the following statement:

(Γ_p) For any non-negative integer p, there exists a homomorphism (as vector spaces) ψ from $\mathfrak{g} \otimes Q_p$ to P such that

$$\psi(X_i \otimes x_I) = x_i x_I, \quad i \leqslant I, \quad x_I \in Q_p, \tag{13.2}$$
$$\psi(X_i \otimes x_I) \in Q_{q+1}, \quad x_I \in Q_q, \quad q < p, \tag{13.3}$$
$$\psi(X_i \otimes \psi(X_j) \otimes x_J)) = \psi(X_j \otimes \psi(X_i \otimes x_J)) + \psi([X_i, X_j] \otimes x_J), \quad x_J \in Q_{p-1}, \tag{13.4}$$
$$\psi(X_i \otimes x_I) - x_i x_I \in Q_q, \quad x_I \in Q_q, \quad q \leqslant p. \tag{13.5}$$

If (Γ_p) is true for $p = 0, 1, 2, \ldots$, then by Lemma 3, the mapping f defined by

$$f(X) \Sigma a_I x_I = \Sigma c_i a_I \psi(X_i \otimes x_I)$$

for any $X = \Sigma_i c_i X_i \in \mathfrak{g}$ and any $\Sigma a_I x_I$ $(a_I \in C)$ in P is a representation of $\mathfrak{g}$ with representation space P. By (13.2), f satisfies the condition of Lemma 2.

(4) We now prove (Γ_p). Notice that (13.3) is a consequence of (13.5).

Let $p = 0$. From (13.2), $\psi(X_i \otimes 1) = x_i$, so (13.3), (13.4) and (13.5) are satisfied.

Suppose (Γ_{p-1}) is true for some $p > 0$. We show that the mapping ψ satisfying (Γ_{p-1}) has a unique extension that satisfies (Γ_p). We must define $\psi(X_i \otimes x_I)$ for a sequence I of length p. Let I denote any increasing sequence of length p. Since P is abelian, the elements Z_I (for all possible I) span P_p. Thus it suffices to consider increasing sequences of length p. Suppose $i \leqslant I$. From (13.2), $\psi(X_i \otimes x_I)$ must be defined as $x_i x_I$. If $i \leqslant I$ is not true, then let j (j can appear in I several times) be the first element of I. Delete j from I and call the new series J, then $i > j \leqslant J$. Hence $x_I = x_j x_J = \psi(X_j \otimes x_J)$ and the left-hand side of (13.4) is $\psi(X_i \otimes x_I)$. To define $\psi(X_i \otimes x_I)$ by (13.4), it is necessary to define the two terms on the right-hand side of (13.4). By (13.5), write

$$\psi(X_i \otimes x_J) = x_i x_J + w, \quad w \in Q_{p-1},$$

then the right-hand side of (13.4) becomes

$$x_j x_i x_J + \psi(X_j \otimes w) + \psi([X_\alpha, X_\beta] \otimes x_J),$$

where each term is defined. Now ψ is defined for all cases such that (13.2), (13.3) and (13.5) are satisfied. Moreover, (13.4) is satisfied for $i > j \leqslant J$. From the anti-symmetry of $[X_i, X_j]$, it follows that (13.4) is also satisfied for $j > i \leqslant J$; (13.4) is obviously satisfied for $i = j$. We have seen that (13.4) is true for $i \leqslant J$ or $j \leqslant J$; we now show that (13.4) is also true for other cases.

Suppose $i \leqslant J$ and $j \leqslant J$ are not satisfied, then the length of J is not zero. Let k be the first element of J and L be the sequence obtained by deleting k from J, then $k \leqslant L$, $k < i$ and $k < j$. By the induction assumption,

$$\begin{aligned}\psi(X_j \otimes X_J) &= \psi(X_j \otimes \psi(X_k \otimes x_L)) \\ &= \psi(X_k \otimes \psi(X_j \otimes x_L)) + \psi([X_j, X_k] \otimes x_L) \\ &= \psi(X_k \otimes x_j x_L) + \psi(X_k \otimes w) + \psi([X_j, X_k] \otimes x_L),\end{aligned}$$

where $w = \psi(X_j \otimes x_L) - x_j x_L \in Q_{p-2}$. Therefore,

$$\begin{aligned}\psi(X_i \otimes \psi(X_j \otimes x_J)) = {}& \psi(X_i \otimes \psi(X_k \otimes x_j x_L)) + \psi(X_i \otimes \psi(X_k \otimes w)) \\ &+ \psi(X_i \otimes \psi([X_j, X_k] \otimes x_L)).\end{aligned}$$

Since $k \leqslant j$ and $k \leqslant L$, (13.4) can be applied to the first term on the right-hand side of this equation; by the induction assumption, (13.4) can be applied to the other two terms of this equation. Therefore

$$\begin{aligned}\psi(X_i \otimes \psi(X_j \otimes x_J)) = {}& \psi\big(X_k \otimes \psi(X_i \otimes \psi(X_j \otimes x_L))\big) \\ &+ \psi([X_i, X_k] \otimes (X_j \otimes x_L)) + \psi([X_j, X_k] \otimes \psi(X_i \otimes x_L)) + \psi([X_i, [X_j, X_k]] \otimes x_L).\end{aligned} \tag{13.6}$$

Exchanging i and j, we also have

$$\begin{aligned}\psi(X_j \otimes \psi(X_i \otimes x_J)) = {}& \psi\big(X_k \otimes \psi(X_j \otimes \psi(X_i \otimes x_L))\big) \\ &+ \psi([X_j, X_k] \otimes \psi(X_i \otimes x_L)) + \psi([X_i, X_k] \otimes \psi(X_j \otimes x_L)) + \psi([X_j, [X_i, X_k]] \otimes x_L).\end{aligned} \tag{13.7}$$

From the difference of (13.6) and (13.7), we have

$$\begin{aligned}&\psi(X_i \otimes \psi(X_j \otimes x_J)) - \psi(X_j \otimes \psi(X_i \otimes x_J)) \\ &= \psi\Big(X_k \otimes \Big\{ \psi(X_i \otimes \psi(X_j \otimes x_L)) + \psi(X_j \otimes \psi(X_i \otimes x_L)) \Big\}\Big) \\ &\quad + \psi([X_i, [X_j, X_k]] \otimes x_L) - \psi([X_j, [X_i, X_k]] \otimes x_L).\end{aligned} \tag{13.8}$$

By (13.4),

$$\begin{aligned}&\psi\Big(X_k \otimes \Big\{\psi(X_i \otimes \psi(X_j \otimes x_L)) - \psi(X_j \otimes \psi(X_i \otimes x_L))\Big\}\Big) \\ &= \psi(X_k \otimes \psi([X_i, X_j] \otimes x_L)) \\ &= \psi([X_i, X_j] \otimes \psi(X_k \otimes x_L)) + \psi([X_k, [X_i, X_j]] \otimes x_L).\end{aligned}$$

Substituting this into (13.7) and using the Jacobi identity,

$$\psi(X_i \otimes \psi(X_j \otimes x_J)) - \psi(X_j \otimes \psi(X_i \otimes x_J)) = \psi([X_i, X_j] \otimes x_J).$$

Thus (13.4) is true for other cases. The proof of Theorem 2 is now completed.

COROLLARY. *If* $\mathfrak{g}$ *is a Lie algebra and* U *is the universal enveloping algebra of* $\mathfrak{g}$ *(relative to* ϱ*), then* ϱ *is one–one.*

In what follows, $\varrho(X)$ $(\forall X \in \mathfrak{g})$ will be denoted by X and ϱ by i. i is said to be the embedding of $\mathfrak{g}$ into U.

13.3. Applications to representations of semisimple Lie algebras

Let $\mathfrak{g}$ be a semisimple Lie algebra over C, $\mathfrak{h}$ be a Cartan subalgebra of $\mathfrak{g}$, Σ be the system of roots of $\mathfrak{h}$ and $\Pi = \{\alpha_1, \ldots, \alpha_n\}$ be a fundamental system of roots. Define a partial ordering on $\mathfrak{h}$ by means of Π and let Σ_+ be the set of positive roots. Let

$$\mathfrak{g} = \mathfrak{h} \dotplus \sum_{\alpha \in \Sigma} \mathfrak{g}^{\alpha}$$

be the Cartan decomposition of $\mathfrak{g}$ and

$$\mathfrak{g}_+ = \sum_{\alpha>0} \mathfrak{g}^{\alpha}, \quad \mathfrak{g}_- = \sum_{\alpha<0} \mathfrak{g}^{\alpha}, \quad \mathfrak{g}_0 = \mathfrak{h} + \mathfrak{g}_+.$$

Then $\mathfrak{g}_+$ and $\mathfrak{g}_-$ are nilpotent, $\mathfrak{g}_0$ is solvable and

$$[\mathfrak{g}_0, \mathfrak{g}_0] = [\mathfrak{h}, \mathfrak{g}_0] = \mathfrak{g}_+.$$

Also,

$$\mathfrak{g} = \mathfrak{h} \dotplus \mathfrak{g}_+ \dotplus \mathfrak{g}_- = \mathfrak{g}_- + \mathfrak{g}_0.$$

Obviously, $\mathfrak{g}_+$ is generated by $\mathfrak{g}^{\alpha_i}$ $(i = 1, \ldots, n)$ and $\mathfrak{g}_-$ is generated by $\mathfrak{g}^{-\alpha_i}$ $(i = 1, \ldots, n)$.

For any $\alpha \in \Sigma$, let $H_\alpha \in \mathfrak{h}$ be such that

$$(H, H_\alpha) = \alpha(H), \quad \text{for all } H \in \mathfrak{h}.$$

If

$$H_i = \frac{2H_{\alpha_i}}{(\alpha_i, \alpha_i)}, \quad i = 1, \ldots, n,$$

then

$$\alpha_i(H_i) = 2.$$

Let

$$\alpha_i(H_j) = \frac{2(H_{\alpha_i}, H_{\alpha_j})}{(\alpha_j, \alpha_j)} = -a_{ji}, \quad i, j = 1, \ldots, n,$$

then all a_{ji} are integers (in fact, $a_{ji} = q-p$, where q, p are largest non-negative integers such that $\alpha_i - p\alpha_j, \ldots, \alpha_i - \alpha_j, \alpha_i, \alpha_i + \alpha_j, \ldots, \alpha_i + q\alpha_j$ are roots and $\alpha_i - (p+1)\alpha_j$, $\alpha_i - (q+1)\alpha_j$ are not roots). Naturally,

$$a_{ii} = -2.$$

$\{a_{ij}\}$ $(i, j = 1, \ldots, n)$ is said to be a set of Cartan integers. Obviously, two fundamental systems of roots are equivalent iff they have the same set of Cartan integers. Therefore, we have the following:

THEOREM 3. *Two complex semisimple Lie algebras are isomorphic iff they have the same set of Cartan integers.*

Choose $X_i \in \mathfrak{g}^{\alpha_i}$, $Y_i \in \mathfrak{g}^{-\alpha_i}$ such that

$$[X_i, Y_i] = H_i,$$

then $\{H_1, \ldots, H_n, X_1, \ldots, X_n, Y_1, \ldots, Y_n\}$ is a set of generators of $\mathfrak{g}$; these generators are said to be *the canonical generators*. The canonical generators satisfy the following relations:

$$\begin{aligned} [H_i, H_j] &= 0, \\ [H_i, X_j] &= -a_{ij}X_j, \\ [H_i, Y_j] &= a_{ij}Y_j, \\ [X_i, Y_j] &= \delta_{ij}H_i. \end{aligned}$$

If ω is a weight of some representation of $\mathfrak{g}$, then

$$\omega(H_i) = \frac{2\omega(H_{\alpha_i})}{(\alpha_i, \alpha_i)}, \qquad (i = 1, 2, \ldots, n)$$

are integers. If ω_0 is the highest weight of some representation, then $\omega_0(H_i)\,(i = 1, \ldots, n)$ are non-negative integers. If $\omega_0 \in \mathfrak{h}_0^*$ and $\omega_0(H_i)\,(i = 1, \ldots, n)$ are non-negative integers, then ω_0 is said to be a dominant integral linear function. We now prove the following:

THEOREM 4. *If ω_0 is a dominant integral linear function of a complex semisimple Lie algebra $\mathfrak{g}$, then ω_0 is the highest weight of an irreducible representation of $\mathfrak{g}$.*

Proof (Harish-Chandra†). The proof consists of the following steps.

(1) We have

$$\mathfrak{g} = \mathfrak{g}_- \dotplus \mathfrak{g}_0.$$

Let U, U_- and U_0 be the universal enveloping algebras of $\mathfrak{g}$, $\mathfrak{g}_-$ and $\mathfrak{g}_0$, respectively. For every $\alpha \in \Sigma_+$, choose $X_\alpha \in \mathfrak{g}^\alpha$, $X_{-\alpha} \in \mathfrak{g}^{-\alpha}$ (when $\alpha = \alpha_i$, choose $X_\alpha = X_i$ and $X_{-\alpha} = Y_i$). Then $X_\alpha, X_{-\alpha}\,(\alpha \in \Sigma)$ and $H_1, \ldots, H_n$ form a basis of $\mathfrak{g}$. Order this basis as $X_{-\alpha_{i_1}}, \ldots, X_{-\alpha_{i_r}}$ $(\alpha_{i_1}, \ldots, \alpha_{i_r} \in \Sigma_+)$, $H_1, \ldots, H_n$, $X_{\alpha_{j_1}}, \ldots, X_{\alpha_{j_r}} (\alpha_{j_1}, \ldots, \alpha_{j_r} \in \Sigma_+)$ where r is the number of positive roots and the orderings on the negative and positive roots are arbitrary. Then U is the set of all standard polynomials of these elements, U_- is the set of standard polynomials of $X_{-\alpha_{i_1}}, \ldots, X_{-\alpha_{i_r}}$ and U_0 is the set of standard polynomials of $H_1, \ldots, H_n, X_{\alpha_{j_1}}, \ldots, X_{\alpha_{j_r}}$. By the Poincaré–Birkhoff–Witt theorem, the vector space U is isomorphic to the Kronecker product of the vector spaces U_- and U_0. Therefore, U can be identified with $U_- \otimes U_0$, i.e. we may assume $U = U_- \times U_0$.

(2) Let I_{ω_0} be the left ideal generated by $\mathfrak{g}_+$ and $H - \omega_0(H)$ $(H \in \mathfrak{h})$ in U. We show that $I_{\omega_0} \neq U$, i.e. $U/I_\omega \neq \{0\}$.

Let I'_{ω_0} be the left ideal generated by $\mathfrak{g}_+$ and $H - \omega_0(H)$ $(H \in \mathfrak{h})$ in U_0. Define a one-dimensional representation θ by

$$\begin{aligned} \theta(H) &= \omega_0(H), \qquad && H \in \mathfrak{h}, \\ \theta(X) &= 0, && X \in \mathfrak{g}_+. \end{aligned}$$

† Harish-Chandra, Some applications of the universal enveloping algebra of a semisimple Lie algebra, *Trans. Amer. Math. Soc.* **70** (1951), 28–99.

θ can be considered as a representation of $\mathfrak{g}_0$ with representation space C. Since U_0 is the universal enveloping algebra of $\mathfrak{g}_0$ (relative to i), there exists a homomorphism $\tilde{\theta}$ from U_0 to C such that

$$\tilde{\theta} \circ i = \theta,$$
$$\tilde{\theta}(1) = 1.$$

By identifying $i(X)$and X, $\mathfrak{g}_0$ can be considered as a subspace of U_0. Hence

$$\tilde{\theta}(X) = \theta(X), \quad \text{for all } X \in \mathfrak{g}$$

and θ is extended to a representation $\tilde{\theta}$ of U_0. For simplicity, $\tilde{\theta}$ will also be denoted by θ. The kernel of θ contains I'_{ω_0}. Since θ does not map every element to zero, $I'_{\omega_0} \neq U_0$. By Theorem 2, the vector space $U_- \otimes U_0$, so $I_{\omega_0} = U_- \otimes I'_{\omega_0}$. Since $I'_{\omega_0} \neq U_0$, $I_{\omega_0} \neq U$.

(3) We now show that there exists a unique maximal left ideal of U containing I_{ω_0}.

Consider U as a vector space with subspace I_{ω_0}. For any $\xi \in U$ and $\eta + I\omega_0 \in V = U/I_{\omega_0}$, define

$$\varrho(\xi)\,(\eta + I_{\omega_0}) = \xi\eta + I_{\omega_0}.$$

Since I_{ω_0} is a left ideal of U, it can be shown that

$$\xi \to \varrho(\xi)$$

is a representation (possibly infinite-dimensional) of U and the restriction of ϱ to $\mathfrak{g}$ is a representation of $\mathfrak{g}$; denote $\varrho \mid \mathfrak{g}$ also by ϱ.

Let $v = 1 + I_{\omega_0} \in V$, then

$$\varrho(H)\,v = H + I_{\omega_0} = \omega_0(H) + I_{\omega_0} = \omega_0(H)\,v$$

and v is of weight ω_0. The vector

$$\varrho(Y_{i_1}) \ldots \varrho(Y_{i_m})\,v$$

is of weight

$$\omega_0 - \alpha_{i_1} - \ldots - \alpha_{i_m}.$$

Clearly, all such vectors span V and $V = \dot{\Sigma}_\omega V_\omega$; V^0_ω is one-dimensional.

Let W be an invariant subspace of V; it will be shown that $W = \dot{\sum}_\omega (W \cap V_\omega)$. Obviously, $\dot{\sum}_\omega (W \cap V_\omega) \subset W$. Let $u \in W$ and $u = \dot{\sum}_\omega u_\omega$ be the decomposition of u with respect to $V = \dot{\sum} V_\omega$. Now the quotient space V/W is also a representation space. Since V is generated by weight vectors, so is V/W, i.e. $V/W = \dot{\sum}_\omega (V/W)_\omega$, where $(V/W)_\omega$ is the weight space of ω. From $\dot{\sum}_\omega u_\omega = 0 \pmod{W}$, it follows that $u_\omega = 0 \pmod{W}$, i.e. $u_\omega \in W$. This proves the assertion.

Since V_{ω_0} is one-dimensional, $W \cap V_{\omega_0} = 0$ or $W \cap V_{\omega_0} = V_{\omega_0}$. If $W \cap V_{\omega_0} = V_{\omega_0}$, then $v \in W$. Hence from the invariance of W, we have $W = V$. If $W \cap V_{\omega_0} = \{0\}$, then from $W = \dot{\sum}_\omega (W \cap V_\omega)$ it follows that $W \subset V^+ = \dot{\sum}_{\omega < \omega_0} V_\omega$. Thus the union of all the invariant subspaces contained in V^+ is the unique maximal invariant subspace of V. The elements that are mapped to this maximal invariant subspace by the homomorphism

$$U \to U/I_{\omega_0} = V$$

is the unique maximal left ideal in U containing I_{ω_0}.

(4) Let I be the unique maximal ideal containing I_{ω_0}, then the quotient space U/I is the representation space of the representation ϱ of $\mathfrak{g}$ (and U) defined by

$$\varrho(\xi)(\eta+I) = \xi\eta+I, \quad \text{for all} \quad \xi \in U, \eta+I \in U/I.$$

Since I is maximal, U/I is irreducible. As in (3), it can be shown that the weights of ϱ are of the form

$$\omega_0-\alpha_{i_1}-\ldots-\alpha_{i_m},$$

U/I is generated by weight vectors, $v = 1+I$ is of weight ω_0 and ω_0 is simple. Therefore, to prove Theorem 4, it suffices to show that U/I is finite-dimensional.

Now let $V = U/I = \sum^{\cdot} V_\omega$, where V_ω is the weight space of ω and $\dim V_{\omega_0} = 1$. Clearly, every V_ω is finite-dimensional. It will be shown that ϱ has only a finite number of weights.

(5) Let $\mathfrak{g}_i$ be the three-dimensional simple algebra generated by X_i, Y_i, and H_i. Let T_i be the subspace of V generated by

$$v_k = \varrho(Y_i)^k v, \quad k = 0, 1, 2, \ldots$$

then T_i is invariant under $\varrho(\mathfrak{g}_i)$. It will be shown that T_i is a finite-dimensional irreducible invariant subspace under $\varrho(\mathfrak{g}_i)$.

Let $v_k \in V$ be of weight $\omega_0-k\alpha_i$ $(k=0, 1, 2, \ldots)$ and $\alpha_i(H_i)=2$, then every v_k is a weight vector in T_i of weight $\omega_0(H_i)-2k$. Let $\omega_k = \omega_0(H_i)-2k$, then $T_i = \sum_k^{\cdot} V_{\omega_k}$, where V_{ω_k} is the weight space of ω_k and $\dim V_{\omega_k} = 1$.

Let U_i be an invariant subspace under $\varrho(\mathfrak{g}_i)$ contained in T_i. By repeating the argument in (3), it can be shown that $U_i = \sum_k^{\cdot} U_i \cap V_{\omega_k}$. Suppose $U_i \neq T_i$, then $U_i \subset T_i^+ = \sum_{k\geq 1} V_{\omega_k}$. We want to show that $U_i = 0$.

If $j \neq i$, then $[X_j, Y_i] = 0$. Hence

$$\varrho(X_j)\, v_k = \varrho(X_j)\, \varrho(Y_i)^k\, v = \varrho(Y_i)^k\, \varrho(X_j)\, v = 0$$

and any $\varrho(X_j) (j \neq i)$ annihilates T_i. In particular, any $\varrho(X_j)$ $(j \neq i)$ annihilates U_i. Since U_i is invariant under $\varrho(\mathfrak{g}_i)$, it is also invariant under $\varrho(\mathfrak{g}_+)$. On the other hand, from $U_i = \sum_k^{\cdot} U_i \cap V_{\omega_k}$, it follows that U_i is also invariant under $\varrho(\mathfrak{h})$. Therefore, U_i is invariant under $\varrho(\mathfrak{g}_0)$ and $\varrho(U_0)$, so

$$\varrho(U)\, U_i = \varrho(U_-)\, \varrho(U_0)\, U_i \subset \varrho(U_-)\, U_i \subset \varrho(U_-)\, T_i^+ \subset \varrho(U_-) V^+ \subset V^+.$$

Since ϱ is an irreducible representation of $\mathfrak{g}$ (and U) and $\varrho(U)U_i$ is an invariant subspace, so $\varrho(U)U_i = 0$. Hence $U_i = 0$. This proves the irreducibility of T_i.

By Theorem 3 of Chapter 9, $\mathfrak{g}_i$ has a finite-dimensional irreducible representation with highest weight $\omega_0(H_i)$. By Theorem 2 of Chapter 10, T_i is finite-dimensional.

(6) Let $\mathcal{F}_i$ be the collection of all invariant subspaces of $\varrho(\mathfrak{g}_i)$ and W_i be the union of all the subspaces in $\mathcal{F}_i$. We claim that $W_i = V$.

It suffices to show that W_i is invariant under $\varrho(\mathfrak{g})$. Notice that for any $M, N \in \mathcal{F}_i$, we have $M+N \in \mathcal{F}_i$, and that $\varrho(\mathfrak{g})M$ is finite-dimensional. Moreover,

$$\varrho(\mathfrak{g}_i)\, \varrho(\mathfrak{g})M \subset \varrho([\mathfrak{g}, \mathfrak{g}_i])\, M+\varrho(\mathfrak{g})\, \varrho(\mathfrak{g}_i)\, M \subset \varrho(\mathfrak{g})\, M,$$

so $\varrho(\mathfrak{g})M \subset \mathcal{F}_i$. It follows that W_i is invariant under $\varrho(\mathfrak{g})$ and $W_i = V$.

(7) We now prove that if ω is a weight of ϱ, then so is $S(\omega)$ ($\forall S \in W$).

Let x ($\neq 0$) of weight ω. By (6), x is contained in a finite-dimensional subspace U of V which is invariant under $\varrho(\mathfrak{g}_i)$. Since V is spanned by weight vectors, V is invariant under $\varrho(\mathfrak{h})$. By Lemma 2 of Chapter 10,

$$\omega-\omega(H_i)\alpha_i = \omega-\frac{2(\omega, \alpha_i)}{(\alpha_i, \alpha_i)}\alpha_i$$

is also a weight of ϱ. Let

$$S_i : \omega \to \omega-\frac{2(\omega, \alpha_i)}{(\alpha_i, \alpha_i)}\alpha_i.$$

We have proved that if ω is a weight, then $S_i(\omega)$ is also a weight. Since W is generated by S_i ($i = 1, \ldots, n$), so any $S(\omega)$ ($S \in W$) is also a weight of ϱ.

(8) Recall that a weight ω of ϱ is said to be dominant if $\omega(H_i)$ ($i = 1, \ldots, n$) are non-negative integers. Two weights ω and ω' of ϱ are said to be equivalent if $\omega = S(\omega')$ for some $S \in W$; clearly, this is an equivalent relation. Any equivalence class of weights contains a highest weight ω. Since $\omega \geqslant S_i(\omega) = \omega-\omega(H_i)\alpha_i$, we have $\omega(H_i) \geqslant 0$ ($i = 1, \ldots, n$). Hence ω is a dominant weight. Since every equivalence class only contains a finite number of weights, to show that ϱ has finitely many weights, it suffices to show that it has finitely many dominant weights.

If ω is a dominant weight, then it is of the form

$$\omega = \omega_0-\beta = \omega_0-\sum m_i\alpha_i, \quad m_i \geqslant 0.$$

Thus

$$(\omega_0, \omega_0) = (\omega+\beta, \omega+\beta) = (\omega, \omega)+(\beta, \beta)+2(\omega, \beta).$$

Since ω is a dominant weight, $\omega(H_i) \geqslant 0$ ($i = 1, \ldots, n$), so

$$(\omega, \beta) = \sum m_i(\omega, \alpha_i) = \tfrac{1}{2}\sum m_i\omega(H_i)(\alpha_i, \alpha_i) \geqslant 0$$

and

$$(\omega_0, \omega_0) \geqslant (\omega, \omega).$$

This proves that ω is contained in the sphere with radius $\sqrt{\{(\omega_0, \omega_0)\}}$ centered at the origin. Since ω is of the form $\omega_0-\sum m_i\alpha_i$ and there are only a finite number of such points in the sphere (contained in $\mathfrak{h}_0^*$) mentioned above, ϱ can only have a finite number of weights. The proof of Theorem 4 is now completed.

CHAPTER 14

CHARACTERS OF IRREDUCIBLE REPRESENTATIONS OF SEMISIMPLE LIE ALGEBRAS

14.1. A recursion formula for the multiplicity of a weight of an irreducible representation

Let $\mathfrak{g}$ be a semisimple Lie algebra, $\mathfrak{h}$ be a Cartan subalgebra, Σ be the system of roots of $\mathfrak{g}$ with respect to $\mathfrak{h}$ and $\Pi = \{\alpha_1, \ldots, \alpha_n\}$ be a fundamental system of roots. For every $\alpha \in \Sigma$, choose $E_{\pm\alpha} \in \mathfrak{g}^{\pm\alpha}$ such that $[E_\alpha, E_{-\alpha}] = H_\alpha$.

Let ϱ be an irreducible representation of $\mathfrak{g}$ with highest weight ω_0 and representation space V. V is decomposed as

$$V = \sum_{\omega}{}^{\cdot}\, V_\omega,$$

where V_ω is the weight space of ω. Let $\dim V_\omega = m_\omega$, then m_ω is said to be the multiplicity of ω. When a linear function ω on $\mathfrak{h}$ is not a weight of ϱ, let $m_\omega = 0$.

For a fixed root α, let

$$N(\alpha) = \{v \in V | \varrho(E_\alpha)v = 0\},$$

then $N(\alpha)$ is a subspace of V. Let $v \in N(\alpha)$ and $H \in \mathfrak{h}$, then

$$\begin{aligned}\varrho(E_\alpha)\big(\varrho(H)\, v\big) &= \varrho(H)\, \varrho(E_\alpha)\, v - [\varrho(H), \varrho(E_\alpha)]\, v \\ &= -\alpha(H)\, \varrho(E_\alpha)\, v = 0.\end{aligned}$$

Thus $N(\alpha)$ is invariant under $\varrho(\mathfrak{h})$ and

$$N(\alpha) = \sum_{\omega}{}^{\cdot}\, N(\alpha)_\omega,$$

where $N(\alpha)_\omega = N(\alpha) \cap V_\omega$.

LEMMA 1. *Let* $N(\alpha)_\omega \neq 0$, *then*

1° $2(\omega, \alpha)/(\alpha, \alpha)$ *is a non-negative integer.*

2° $\dim \varrho(E_{-\alpha})^k N(\alpha)\omega =$

$$= \begin{cases} \dim N(\alpha)_\omega, & 0 \leqslant k \leqslant \dfrac{2(\omega, \alpha)}{(\alpha, \alpha)}, \\ 0, & k > \dfrac{2(\omega, \alpha)}{(\alpha, \alpha)}. \end{cases}$$

3° If $0 \leqslant k \leqslant 2(\omega, \alpha)/(\alpha, \alpha)-1$, *then* $\varrho(E_\alpha)\ \varrho(E_{-\alpha})$ *is a one–one linear transformation from* $\varrho(E_{-\alpha})_k N(\alpha)_\omega$ *onto itself.*

Proof. Let $u_0 \in N(\alpha)_\omega$ and $u_0 \neq 0$. Set

$$u_{-k} = \varrho(E_{-\alpha})^k u_0, \qquad k = 0, 1, 2, \ldots.$$

If p is the smallest non-negative integer such that

$$e_p \neq 0 \quad \text{and} \quad e_{-p-1} = \varrho(E_{-\alpha})e_{-p} = 0,$$

then by Lemma 2 of Chapter 10,

$$p = \frac{2(\omega, \alpha)}{(\alpha, \alpha)}$$

and

$$\varrho(E_\alpha)u_{-k} = \varrho(E_\alpha)\ \varrho(E_{-\alpha})u_{-k+1} = k\left\{(\omega, \alpha) - \frac{k-1}{2}(\alpha, \alpha)\right\} u_{-k+1}, \tag{14.1}$$

$$\varrho(H)u_{-k} = (\omega - k\alpha, \alpha)u_{-k}, \qquad k = 0, 1, \ldots, p, \tag{14.2}$$

where $u_1 = 0$. Therefore, p is independent of the choice of u_0 in $N(\alpha)_\omega$ and $2(\omega, \alpha)/(\alpha, \alpha)$ is a non-negative integer. This proves 1°.

When $0 \leqslant k \leqslant p$, since $u_{-k} = \varrho(E_{-\alpha})^k u_0 \neq 0$ for any non-zero $u_0 \in N(\alpha)$, the linear mapping

$$u_0 \to u_{-k} = \varrho(E_{-\alpha})^k u_0, \qquad 0 \leqslant k \leqslant p$$

from $N(\alpha)_\omega$ to $\varrho(E_{-\alpha})^k N(\alpha)_\omega$ is one–one and onto. Hence

$$\dim \varrho(E_{-\alpha})^k N(\alpha)_\omega = \dim N(\alpha)_\omega, \qquad 0 \leqslant k \leqslant p.$$

Obviously,

$$\dim \varrho(E_{-\alpha})^k N(\alpha)_\omega \neq 0, \qquad k > p.$$

Thus 2° is also true.

Finally, from (14.1), it follows that $\varrho(E_\alpha)\ \varrho(E_{-\alpha})$ is one–one linear mapping from $\varrho(E_{-\alpha})^k N(\alpha)_\omega$ onto itself $(0 \leqslant k \leqslant p-1)$. This proves 3°.

LEMMA 2. *Suppose* ω *is a weight of* ϱ *and* $\omega+\alpha$ *is not a weight of* ϱ. *If* k *is an integer satisfying* $0 \leqslant k \leqslant (\omega, \alpha)/(\alpha, \alpha)$, *then*

$$V_{\omega-k\alpha} = \sum_{j=0}^{k}{}^{\cdot}\, \varrho(E_{-\alpha})^{k-j} N(\alpha)_{\omega-j\alpha} \tag{14.3}$$

Proof. We use induction on k. When $k = 0$, since $\omega+\alpha$ is not a weight of ϱ, $V_\omega \subset N(\alpha)$. Hence $V_\omega = N(\alpha)_\omega$ and (14.3) is true.

Suppose (14.3) holds for $k-1$. We show first that the right-hand side of (14.3) is a direct sum. If

$$y_j \in \varrho(E_{-\alpha})^{k-j} N(\alpha)_{\omega-j\alpha} \quad \text{and} \quad \sum_{j=0}^{k} y_j = 0, \quad \text{then} \quad \sum_{j=0}^{k} \varrho(E_\alpha) y_j = 0.$$

When $j < k$, we have $j \leqslant k-1 < p/2$, so

$$0 \leqslant k-1-j \leqslant 2(p/2-j)+1 = \frac{2(\omega-j\alpha, \alpha)}{(\alpha, \alpha)} - 1.$$

By 3° of Lemma 1,

$$\varrho(E_\alpha) y_j \in \varrho(E_\alpha)\, \varrho(E_{-\alpha})^{k-j} N(\alpha)_{\omega-j\alpha} = \varrho(E_{-\alpha})^{k-1-j} N(\alpha)_{\omega-j\alpha}.$$

When $j = k$, since $y_k = N(\alpha)_{\omega-k\alpha} \subset N(\alpha)$, we have $\varrho(E_\alpha) y_k = 0$. Hence by the induction assumption, $\varrho(E_\alpha) y_j = 0\,(0 \leqslant j \leqslant k-1)$. On the other hand, $y_i \in \varrho(E_{-\alpha})^{k-j} N(\alpha)_{\omega-j\alpha} = \varrho(E_{-\alpha}) \times \varrho(E_\alpha)^{k-1-j} N(\alpha)_{\omega-j\alpha}\,(1 \leqslant j \leqslant k-1)$, so when $1 \leqslant j \leqslant k-1$, there exists $x_j \in \varrho(E_{-\alpha})^{k-1-} \times N(\alpha)_{\omega-j\alpha}$ such that $y_j = \varrho(E_{-\alpha}) x_j$. Thus $\varrho(E_\alpha)\, \varrho(E_{-\alpha}) x_j = \varrho(E_\alpha) y_j = 0$. By 3° of Lemma 1, $x_j = 0$. Therefore, for j satisfying $0 \leqslant j \leqslant k-1$, $y_j = 0$, so $y_k = 0$. Hence the right-hand side of (14.3) is a direct sum.

We now show that the right-hand side of (14.3) is $V_{\omega-k\alpha}$. Let $x \in V_{\omega-k\alpha}$. Since $\varrho(E_\alpha) x \in V_{\omega-(k-1)\alpha}$, by the induction assumption,

$$\varrho(E_\alpha) x \in \sum_{j=0}^{k-1} \varrho(E_{-\alpha})^{k-1-j} N(\alpha)_{\omega-k\alpha}.$$

Now $j \leqslant k-1 < p/2$, so by 3° of Lemma 1, $\varrho(E_\alpha)\varrho(E_{-\alpha})$ is a one-one linear transformation from

$$\sum_{j=0}^{k-1} \varrho(E_{-\alpha})^{k-1-j} N(\alpha)_{\omega-k\alpha}$$

onto itself. Thus there exists

$$y \in \sum_{j=0}^{k-1} \varrho(E_{-\alpha})^{k-1-j} N(\alpha)_{\omega-k\alpha}$$

such that $\varrho(E_\alpha) x = \varrho(E_\alpha)\varrho(E_{-\alpha}) y$, i.e. $\varrho(E_\alpha)\,(x - \varrho(E_{-\alpha}) y) = 0$. Hence $x - \varrho(E_{-\alpha}) y \in N(\alpha)$. Let $z = x - \varrho(E_\alpha) y$. Since $x - \varrho(E_{-\alpha}) y \in V_{\omega-k\alpha}$, $z \in N(\alpha)_{\omega-k\alpha}$. On the other hand,

$$\varrho(E_{-\alpha}) y \in \sum_{j=0}^{k-1} \varrho(E_{-\alpha})^{k-j} N(\alpha)_{\omega-j\alpha},$$

so

$$x = z + \varrho(E_{-\alpha}) y \in \sum_{j=0}^{k} \varrho(E_{-\alpha})^{k-j} N(\alpha)_{\omega-j\alpha}.$$

The proof of Lemma 2 is now completed.

COROLLARY. *Under the assumptions of Lemma* 2, *if* $n_\omega = \dim N(\alpha)_\omega$ *then*

$$m_{\omega-k\alpha} = \sum_{j=0}^{k} n_{\omega-j\alpha}, \tag{14.4}$$

so

$$m_{\omega-k\alpha}-m_{\omega-(k-1)\alpha} = n_{\omega-j\alpha}. \tag{14.5}$$

Proof. From Lemma 2,

$$m_{\omega-k\alpha} = \sum_{j=0}^{k} \dim \varrho(E_{-\alpha})^{k-j} N(\alpha)_{\omega-j\alpha}.$$

When $0 \leqslant j \leqslant k \leqslant (\omega, \alpha)/(\alpha, \alpha)$, $0 \leqslant k-j \leqslant 2(\omega, \alpha)/(\alpha, \alpha)$. By 2° of Lemma 1,

$$\dim \varrho(E_{-\alpha})^{k-j} N(\alpha)_{\omega-j\alpha} = \dim N(\alpha)_{\omega-j\alpha}.$$

Hence (14.4) is true. (14.5) is clearly a consequence of (14.4).

LEMMA 3. *If ω is a weight of ϱ and α is a root of $\mathfrak{g}$, then*

$$\sum_{j=-\infty}^{\infty} m_{\omega+j\alpha}(\omega+j\alpha, \alpha) = 0. \tag{14.6}$$

Proof. Recall that if $\omega+j\alpha$ is not a weight of ϱ, then $m_{\omega+j\alpha} = 0$. Hence the sum in (14.6) is a finite sum. Let p, q be non-negative integers such that $\omega+j\alpha$ $(-p \leqslant j \leqslant q)$ are weights of ϱ and all other $\omega+j\alpha$ are not weights of ϱ. If

$$W = \sum_{j=-p}^{q} V_{\omega+j\alpha},$$

then W is invariant under $\{\varrho(H_\alpha), \varrho(E_\alpha), \varrho(E_{-\alpha})\}$. Thus

$$\mathrm{Tr}_w\, \varrho(H_\alpha) = \mathrm{Tr}_w\, (\varrho(E_\alpha)\, \varrho(E_{-\alpha}) - \varrho(E_{-\alpha})\, \varrho(E_\alpha)) = 0$$

and

$$\mathrm{Tr}_w\, \varrho(H_\alpha) = \sum_{j=-p}^{q} m_{\omega+j\alpha}(\omega+j\alpha, \alpha).$$

If $-p \not\leqslant j \not\leqslant q$, then $m_{\omega+j\alpha} = 0$. Hence Lemma 3 is true.

LEMMA 4. *Suppose ω is a weight of ϱ, α is a root of $\mathfrak{g}$ and $\omega+\alpha$ is not a weight of ϱ. If $0 \leqslant k \leqslant 2(\omega, \alpha)/(\alpha, \alpha)$, then*

$$\mathrm{Tr}_{V_{\omega-k\alpha}}\, \varrho(E_\alpha)\, \varrho(E_{-\alpha}) = \sum_{j=0}^{k} (\omega-j\alpha, \alpha) m_{\omega-j\alpha}.$$

Proof. We assume first that $0 \leqslant k \leqslant (\omega, \alpha)/(\alpha, \alpha)$ and compute

$$\mathrm{Tr}_{V_{\omega-k\alpha}}\, \varrho(E_\alpha)\, \varrho(E_{-\alpha}).$$

By Lemma 2,

$$V_{\omega-k\alpha} = \sum_{j=0}^{k} \varrho(E_{-\alpha})^{k-j} N(\alpha)_{\omega-j\alpha}.$$

By (14.1), $\varrho(E_\alpha)^{k-j} N(\alpha)_{\omega-j\alpha}$ is invariant under $\varrho(E_\alpha)\, \varrho(E_{-\alpha})$, so

$$\mathrm{Tr}_{V_{\omega-k\alpha}} \varrho(E_\alpha)\, \varrho(E_{-\alpha}) = \sum_{j=0}^{k} \mathrm{Tr}_{\varrho(E_{-\alpha})^{k-j} N(\alpha)_{\omega-j\alpha}} \varrho(E_\alpha)\, \varrho(E_{-\alpha}).$$

By (14.5) and (14.1),

$$\mathrm{Tr}_{V_{\omega-k\alpha}}\varrho(E_\alpha)\,\varrho(E_{-\alpha}) = \sum_{j=0}^{k}(m_{\omega-j\alpha}-m_{\omega-(j-1)\alpha})\,(k-j+1)\left(\omega-j\alpha-\frac{k-j}{2}\alpha,\alpha\right)$$

$$= \sum_{j=0}^{k}(\omega-j\alpha,\alpha)\,m_{\omega-j\alpha}.$$

Now let $(\omega, \alpha)/(\alpha, \alpha) < k \leqslant 2(\omega, \alpha)/(\alpha, \alpha)$. Let $\omega' = S_\alpha(\omega)$ and $\alpha' = S_\alpha(\alpha)$, then

$$\alpha' = S_\alpha(\alpha) = -\alpha,$$

$$\omega' = S_\alpha(\omega) = \omega-\frac{2(\omega,\alpha)}{(\alpha,\alpha)}\alpha = \omega-p\alpha,$$

and

$$\frac{2(\omega',\alpha')}{(\alpha',\alpha')} = \frac{2(\omega,\alpha)}{(\alpha,\alpha)} = p.$$

If $0 \leqslant k' < (\omega', \alpha')/(\alpha', \alpha')$, then by the previous argument,

$$\mathrm{Tr}_{V_{\omega'-k\alpha'}}\varrho(E_{\alpha'})\,\varrho(E_{-\alpha'}) = \sum_{j=0}^{k'}(\omega'-j\alpha',\alpha')\,m_{\omega'-j\alpha'}.$$

Notice that

$$\omega'-k'\alpha' = \omega-p\alpha+k'\alpha = \omega-(p-k')\alpha,$$

$$\omega'-j\alpha' = \omega-p\alpha+j\alpha = \omega-(p-j)\alpha,$$

thus

$$\mathrm{Tr}_{V_{\omega-(p-k')\alpha}}\varrho(E_{-\alpha})\,\varrho(E_\alpha) = -\sum_{j=0}^{k'}(\omega-(p-j)\alpha,\alpha)\,m_{\omega-(p-j)\alpha}$$

$$= -\sum_{j=p-k'}^{p}(\omega-j\alpha,\alpha)\,m_{\omega-j\alpha}.$$

Since

$$0 \leqslant k' \leqslant \frac{(\omega',\alpha')}{(\alpha',\alpha')} = \frac{(\omega,\alpha)}{(\alpha,\alpha)}, \quad \frac{(\omega,\alpha)}{(\alpha,\alpha)} \leqslant p-k' \leqslant \frac{2(\omega,\alpha)}{(\alpha,\alpha)}.$$

Let $p-k' = k$, then

$$\mathrm{Tr}_{V_{\omega-k\alpha}}\varrho(E_{-\alpha})\,\varrho(E_\alpha) = -\sum_{j=k}^{p}(\omega-j\alpha,\alpha)\,m_{\omega-j\alpha}$$

and $(\omega, \alpha)/(\alpha, \alpha) \leqslant k \leqslant 2(\omega, \alpha)/(\alpha, \alpha)$. Now

$$\mathrm{Tr}_{V_{\omega-k\alpha}}\varrho(E_\alpha)\,\varrho(E_{-\alpha}) = \mathrm{Tr}_{V_{\omega-k\alpha}}\,\varrho(E_{-\alpha})\varrho(E_\alpha)+\mathrm{Tr}_{V_{\omega-k\alpha}}\varrho(H_\alpha)$$

and

$$\mathrm{Tr}_{V_{\omega-k\alpha}}\varrho(H_\alpha) = (\omega-k\alpha,\alpha)\,m_{\omega-k\alpha}.$$

By Lemma 3 and noticing that $q = 0$ because $\omega+\alpha$ is not a weight of ϱ, we have

$$\mathrm{Tr}_{V_{\omega-k\alpha}}\varrho(E_\alpha)\,\varrho(E_{-\alpha}) = -\sum_{j=k}^{p}(\omega-j\alpha,\alpha)\,m_{\omega-j\alpha}+(\omega-k\alpha,\alpha)\,m_{\omega-k\alpha}+\sum_{j=-p}^{0}(\omega+j\alpha,\alpha)\,m_{\omega+j\alpha}$$
$$= \sum_{j=0}^{k}(\omega-j\alpha,\alpha)\,m_{\omega-j\alpha}.$$

The proof of Lemma 4 is now completed.

LEMMA 5. *If ω is any weight of ϱ, then*

$$\mathrm{Tr}_{V_\omega}\varrho(E_\alpha)\,\varrho(E_{-\alpha}) = \sum_{j=0}^{\infty}(\omega+j\alpha,\alpha)\,m_{\omega+j\alpha}.$$

Proof. Notice first that the right-hand side is a finite sum. Let p, q be non-negative integers such that $\omega+j\alpha$ $(-p \leqslant j \leqslant q)$ are weights of ϱ and all other $\omega+j\alpha$ are not weights of ϱ. If $\omega' = \omega+q\alpha$, then $\omega'+\alpha$ is not a weight of ϱ. Now $\omega = \omega'-q\alpha$ and $0 \leqslant q \leqslant 2(\omega',\alpha)/(\alpha,\alpha) = p+q$, so by Lemma 4,

$$\mathrm{Tr}_{V_\omega}\varrho(E_\alpha)\,\varrho(E_{-\alpha}) = \sum_{j=0}^{q}(\omega'-j\alpha,\alpha)\,m_{\omega'-j\alpha} = \sum_{j=0}^{q}(\omega+j\alpha,\alpha)\,m_{\omega+j\alpha}.$$

If $j > q$, then $\omega+j\alpha$ is not a weight of σ, so $m_{\omega+j\alpha} = 0$. Hence this equation can be written as

$$\mathrm{Tr}_{V_\omega}\varrho(E_\alpha)\,\varrho(E_{-\alpha}) = \sum_{j=0}^{\infty}(\omega+j\alpha,\alpha)\,m_{\omega+j\alpha}.$$

THEOREM 1 *(H. Freudenthal†). Let ϱ be an irreducible representation of a semisimple Lie algebra $\mathfrak{g}$ with highest weight ω_0, $\varrho(G)$ be the Casimir operator of ϱ and*

$$\varrho(G) = rI.$$

Then for any $\omega \in \mathfrak{h}^*$,

$$rm_\omega = m_\omega(\omega,\omega)+\sum_{\alpha\in\Sigma}\sum_{j=1}^{\infty}m_{\omega+j\alpha}(\omega+j\alpha,\alpha). \tag{14.7}$$

If $\delta = \frac{1}{2}\sum_{\alpha>0}\alpha$, then

$$rm_\omega = m_\omega(\omega+2\delta,\omega)+2\sum_{\alpha>0}\sum_{j=1}^{\infty}m_{\omega+j\alpha}(\omega+j\alpha,\alpha). \tag{14.8}$$

In particular, for the highest weight ω_0,

$$r = (\omega_0+2\delta,\omega_0) = (\omega_0+\delta,\omega_0+\delta)-(\delta,\delta). \tag{14.9}$$

Therefore,

$$m_\omega\{(\omega_0+\delta,\omega_0+\delta)-(\omega+\delta,\omega+\delta)\} = 2\sum_{\alpha>0}\sum_{j=1}^{\infty}m_{\omega+j\alpha}(\omega+j\alpha,\alpha). \tag{14.10}$$

† H. Freudenthal, Zur Berechnung des Charakters der halbeinfachen Lieschen Gruppen, 1, *Indag. Math.* **16** (1954), 369–76.

Proof. Let $\{H_1, \ldots, H_n\}$ be a basis of $\mathfrak{h}$ and

$$((H_i, H_j))^{-1}_{1\le i,j\le n} = (q^{ij})_{1\le i,j\le n},$$

then

$$\varrho(G) = \sum_{i,j=1}^{n} q^{ij}\varrho(H_i)\,\varrho(H_j) + \sum_{\alpha} \varrho(E_\alpha)\,\varrho(E_{-\alpha}).$$

If ω is a weight of ϱ, then

$$\mathrm{Tr}_{V\omega} \sum_{i,j=1}^{n} q^{ij}\varrho(H_i)\,\varrho(H_j) = \sum_{i,j=1}^{n} m_\omega q^{ij}\omega(H_i)\,\omega(H_j) = m_\omega(\omega, \omega).$$

By Lemma 5,

$$\mathrm{Tr}_{V_\omega}\, \varrho(E_\alpha)\,\varrho(E_{-\alpha}) = \sum_{j=0}^{\infty} (\omega+j\alpha, \alpha)m_{\omega+j\alpha},$$

thus

$$\mathrm{Tr}_{V_\omega}\, \varrho(G) = m_\omega(\omega, \omega) + \sum_{\alpha\in\Sigma}\sum_{j=0}^{\infty} (\omega+j\alpha, \alpha)m_{\omega+j\alpha}.$$

Since α and $-\alpha$ are roots of $\mathfrak{g}$, $\sum_\alpha m_\omega(\omega, \alpha) = 0$. Hence

$$rm_\omega = m_\omega(\omega, \omega) + \sum_{\alpha\in\Sigma}\sum_{j=1}^{\infty} m_{\omega+j\alpha}(\omega+j\alpha, \alpha).$$

This proves that (14.7) is true if ω is a weight of ϱ.

Suppose ω is not a weight of ϱ, then $m_\omega = 0$. Hence to show (14.7), it suffices to show that

$$\sum_{\alpha\in\Sigma}\sum_{j=1}^{\infty} m_{\omega+j\alpha}(\omega+j\alpha, \alpha) = 0;$$

if the $\omega+j\alpha$ $(j = 1, 2, \ldots)$ are not weights of ϱ, this is obvious. Suppose $\omega' = \omega+k\alpha$ is a weight of ϱ, $\omega'+j\alpha(-p \le j \le q)$ are weights of ϱ and any other $\omega+j\alpha$ is not a weight. Since ω is not a weight of ϱ, $k > p$. By Lemma 3,

$$\begin{aligned} 0 &= \sum_{j=-p}^{q} m_{\omega'+j\alpha}(\omega'+j\alpha, \alpha) \\ &= \sum_{j=-1}^{q} m_{\omega+(j+k)\alpha}\big(\omega+(j+k)\alpha, \alpha\big) = \sum_{j=k-p}^{k+q} m_{\omega+j\alpha}(\omega+j\alpha, \alpha) \\ &= \sum_{j=p}^{\infty} m_{\omega+j\alpha}(\omega+j\alpha, \alpha). \end{aligned}$$

This proves that (14.7) is true for any $\omega \in \mathfrak{h}^*$.

By Lemma 3,

$$\sum_{j=1}^{\infty} m_{\omega+j\alpha}(\omega+j\alpha, \alpha) = -\sum_{j=0}^{\infty} m_{\omega-j\alpha}(\omega-j\alpha, \alpha).$$

Hence

$$\sum_{\alpha}\sum_{j=1}^{\infty} m_{\omega+j\alpha}(\omega+j\alpha, \alpha) = \sum_{\alpha>0}\sum_{j=1}^{\infty} m_{\omega+j\alpha}(\omega+j\alpha,\alpha)+\sum_{\alpha<0}\sum_{j=0}^{\infty} m_{\omega+j\alpha}(\omega+j\alpha, \alpha)$$

$$= 2\sum_{\alpha>0}\sum_{j=1}^{\infty} m_{\omega+j\alpha}(\omega+j\alpha, \alpha)+\sum_{\alpha>1} m_{\omega}(\omega, \alpha).$$

Since $\delta = \frac{1}{2}\sum_{\alpha>0}\alpha$, $\sum_{\alpha>0}(\omega, \alpha) = (\omega, 2\delta)$. Thus

$$rm_{\omega} = m_{\omega}(\omega+2\delta, \omega)+2\sum_{\alpha>0}\sum_{j=1}^{\infty} m_{\omega+j\alpha}(\omega+j\alpha, \alpha).$$

This is (14.8). In particular, when $\omega = \omega_0$, $m_{\omega_0} = 1$ and the $\omega_0+j\alpha$ ($j = 1, 2, \ldots$) are not weights of ϱ if $\alpha > 0$, thus (14.8) becomes

$$r = (\omega_0+2\delta, \omega_0) = (\omega_0+\delta, \omega_0+\delta)-(\delta, \delta).$$

Finally, from (14.8) and (14.9), we have

$$m_{\omega}\{(\omega_0+\delta, \omega_0+\delta)-(\omega+\delta, \omega+\delta)\} = 2\sum_{\alpha>0}\sum_{j=1}^{\infty} m_{\omega+j\alpha}(\omega+j\alpha, \alpha).$$

The proof of Theorem 1 is now completed.

14.2. Half of the sum of all the positive roots

In this section, some properties of $\delta = \frac{1}{2}\sum_{\alpha>0}\alpha$ are obtained; these properties will be used to derive formulas for the characters of irreducible representations of $\mathfrak{g}$.

Let $\Pi = \{\alpha_1, \ldots, \alpha_n\}$ be a fundamental system of roots of $\mathfrak{g}$.

LEMMA 6. *Let* $\delta = \frac{1}{2}\sum_{\alpha>0}\alpha$, *then*

$$\frac{2(\delta, \alpha_i)}{(\alpha_i, \alpha_i)} = 1, \qquad i = 1, 2, \ldots, n,$$

thus δ *is a dominant linear function. Moreover, for any* $S \in W$, $\delta - S(\delta)$ *is a sum of some positive roots.*

Proof. Let $S_i = S_{\alpha_i}$ and $\alpha \in \Sigma_+$. If $\alpha = \alpha_i$, then $S_i(\alpha_i) = -\alpha_i$. If $\alpha \neq \alpha_i$, then by Lemma 6 of Chapter 6,

$$S_i(\alpha) = \alpha - \frac{2(\alpha, \alpha_i)}{(\alpha_i, \alpha_i)}\alpha > 0.$$

Therefore,

$$S_i(\delta) = \frac{1}{2}\sum_{\substack{\alpha>0\\ \alpha\neq\alpha_i}}\alpha - \frac{1}{2}\alpha_i$$

and $\delta - S_i(\delta) = \alpha_i$. Now

$$S_i(\delta) = \delta - \frac{2(\delta, \alpha_i)}{(\alpha_i, \alpha_i)}\alpha_i,$$

thus

$$\frac{2(\delta, \alpha_i)}{(\alpha_i, \alpha_i)} = 1.$$

By Lemma 3 of Chapter 10, δ is a dominant integral linear function on $\mathfrak{h}$.

Now let $S \in W$ and $S \neq 1$, then S cannot transform all the positive roots to all the positive roots. Suppose $\alpha \in \Sigma_+$ and $S(\alpha) < 0$, then $-S(\alpha) = \beta \in \Sigma_+$ and $\delta - S(\delta) = \Sigma\beta$, where every β is in Σ_+.

LEMMA 7. *Let* $\mu = \frac{1}{2}\sum_{\alpha>0} \varepsilon_\alpha \alpha$ *and* $\varepsilon_\alpha = \pm 1$. *If* $\mu > S(\mu)$ *for all* $S\ (\neq 1)$ *in* W, *then* $\mu = \delta$.

Proof. Suppose $\mu \neq \delta$. Let ν be the sum of the roots α with $\varepsilon_\alpha = -1$, then $\mu = \delta - \nu$. Since $\mu \neq \delta$, $\nu \neq 0$. Let $\nu = \sum_i m_i \alpha_i$, where all m_i are non-negative integers. Since $\nu \neq 0$, we have

$$(\nu, \nu) = \sum m_i(\nu, \alpha_i) > 0,$$

so for at least one i_0, $m_{i_0} > 0$ and $(\nu, \alpha_{i_0}) > 0$. Thus $2(\nu, \alpha_{i_0})/(\alpha_{i_0}, \alpha_{i_0})$ is a positive integer. By assumption,

$$\mu > S_{i_0}(\mu) = \mu - \frac{2(\mu, \alpha_{i_0})}{(\alpha_{i_0}, \alpha_{i_0})}\alpha_{i_0},$$

so $2(\mu, \alpha_{i_0})/(\alpha_{i_0}, \alpha_{i_0}) > 0$. Now we also have

$$\frac{2(\mu, \alpha_{i_0})}{(\alpha_{i_0}, \alpha_{i_0})} = \frac{2(\delta, \alpha_{i_0})}{(\alpha_{i_0}, \alpha_{i_0})} - \frac{2(\nu, \alpha_{i_0})}{(\alpha_{i_0}, \alpha_{i_0})} = 1 - \frac{2(\nu, \alpha_{i_0})}{(\alpha_{i_0}, \alpha_{i_0})} \leqslant 0,$$

this is a contradiction. Therefore, $\mu = \delta$.

LEMMA 8. *Let* ω_0 *be the highest weight of an irreducible representation* ϱ *of a semisimple Lie algebra* $\mathfrak{g}$, ω *be a weight of* ϱ *and* $\omega \neq \omega_0$, *then*

$$(\omega + \delta, \omega + \delta) < (\omega_0 + \delta, \omega_0 + \delta).$$

Proof. Since $\omega_0 \neq \omega$, for some positive root α, $\omega + \alpha$ is also a weight of ϱ.

Suppose ω is a dominant weight, i.e. $\omega \geqslant S(\omega)\ (\forall S \in W)$. Let α be the highest root such that $\omega' = \omega + \alpha$ is also a weight of ϱ. We show first that $(\omega' + \delta, \omega' + \delta) > (\omega + \delta, \omega + \delta)$. We have

$$(\omega' + \delta, \omega' + \delta) - (\omega + \delta, \omega + \delta) = 2(\omega + \delta, \alpha) + (\alpha, \alpha). \tag{14.11}$$

Since ω is dominant,

$$\omega \geqslant S_\alpha(\omega) = \omega - \frac{2(\omega, \alpha)}{(\alpha, \alpha)}\alpha, \quad \text{so} \quad \frac{2(\omega, \alpha)}{(\alpha, \alpha)}$$

is a non-negative integer and $(\omega, \alpha) \geqslant 0$. Similarly, since δ is a dominant linear function, $(\delta, \alpha) \geqslant 0$. From (14.11), we get

$$(\omega + \delta, \omega + \delta) < (\omega' + \delta, \omega' + \delta).$$

If $\omega' = \omega_0$, then the assertion is true. Suppose $\omega' \neq \omega_0$; we claim that ω' is also a dominant weight. Otherwise, $2(\omega', \alpha_j)/(\alpha_j, \alpha_j) < 0$ for some α_j, i.e.

$$\frac{2(\omega', \alpha_j)}{(\alpha_j, \alpha_j)} = \frac{2(\omega, \alpha_j)}{(\alpha_j, \alpha_j)} + \frac{2(\alpha, \alpha_j)}{(\alpha_j, \alpha_j)} < 0.$$

Since ω is a dominant weight, $2(\omega, \alpha_j)/(\alpha_j, \alpha_j) \geqslant 0$, so

$$\frac{2(\alpha, \alpha_j)}{(\alpha_j, \alpha_j)} < \frac{2(\omega', \alpha_j)}{(\alpha_j, \alpha_j)} < 0.$$

Therefore,

$$\beta = \alpha - \frac{2(\omega', \alpha_j)}{(\alpha_j, \alpha_j)}\alpha_j$$

is also a root and $\beta > \alpha$, but

$$S_j(\omega') = \omega' - \frac{2(\omega', \alpha_j)}{(\alpha_j, \alpha_j)}\alpha_j = \omega + \beta$$

is also a weight and this contradicts the choice of α, so ω' is necessarily a dominant weight. As before, let α' be the highest root such that $\omega'' = \omega' + \alpha'$ is a weight of ϱ, then

$$(\omega' + \delta, \omega' + \delta) < (\omega'' + \delta, \omega'' + \delta)$$

and ω'' is a dominant weight. If $\omega'' = \omega_0$, then the assertion is true. Otherwise, by repeating the same argument, we obtain a sequence of dominant weights satisfying $\omega < \omega' < \omega'' < \ldots$; ω_0 must finally be reached. Moreover,

$$(\omega + \delta, \omega + \delta) < (\omega' + \delta, \omega' + \delta) < (\omega'' + \delta, \omega'' + \delta) < \ldots < (\omega_0 + \delta, \omega_0 + \delta).$$

Hence Lemma 8 is true if ω is a dominant weight.

Suppose ω is not a dominant weight, then for some S_i, $\omega' = S_i(\omega) > \omega$. Let $g = -2(\omega, \alpha_i)/(\alpha_i, \alpha_i)$, then $g > 0$ and $\omega' = \omega + g\alpha_i$. We have

$$(\omega' + \delta, \omega' + \delta) = (\omega + \delta, \omega + \delta) + 2g(\omega + \delta, \alpha_i) + g^2(\alpha_i, \alpha_i).$$

Since $2(\omega, \alpha_i) = -g(\alpha_i, \alpha_i)$, $2g(\omega, \alpha_i) = -g^2(\alpha_i, \alpha_i)$. From $2(\delta, \alpha_i)/(\alpha_i, \alpha_i) = 1$, it follows that $2g(\delta, \alpha_i) = 2g(\alpha_i, \alpha_i)$, so $2g(\delta, \alpha_i) > 0$. Therefore,

$$(\omega + \delta, \omega + \delta) < (\omega' + \delta, \omega' + \delta).$$

If $\omega' = \omega_0$, then Lemma 8 is true. Suppose $\omega' \neq \omega_0$. If ω' is a dominant weight, then $(\omega' + \delta, \omega' + \delta) < (\omega_0 + \delta, \omega_0 + \delta)$ and Lemma 8 is also true. If ω' is not a dominant weight, then by repeating the previous argument, we obtain a sequence of weights satisfying $\omega < \omega' < \omega'' < \ldots$; a dominant weight ω_1 must finally be reached. Moreover,

$$(\omega + \delta, \omega + \delta) < (\omega' + \delta, \omega' + \delta) < \ldots < (\omega_1 + \delta, \omega_1 + \delta).$$

If $\omega_1 = \omega_0$, then Lemma 8 is true. Otherwise, $(\omega_1 + \delta, \omega_1 + \delta) < (\omega_0 + \delta, \omega_0 + \delta)$ and Lemma 8 is also true.

14.3. Alternating functions

Let $\mathfrak{h}^*$ be the dual space of the Cartan subalgebra $\mathfrak{h}$ of a semisimple Lie algebra $\mathfrak{g}$ and F be a complex-valued function defined on $\mathfrak{h}^*$. For any $S \in W$, define SF by

$$SF(\lambda) = F(S^{-1}(\lambda)), \qquad \lambda \in \mathfrak{h}^*.$$

If $SF = F\,(\forall S \in W)$, then F is said to be a symmetric function. If $SF = \det S \cdot F\,(\forall S \in W)$, then F is said to be an alternating function. Since W is generated by $S_1, \ldots, S_n$ F is symmetric iff $S_iF = F\,(i = 1, \ldots, n)$ and F is alternating iff $S_iF = -F\,(i = 1, \ldots, n)$.

In what follows, the exponential function e^t will be denoted by exp t.

LEMMA 9. *Let $\mu \in \mathfrak{h}_0^*$. For any $\lambda \in \mathfrak{h}^*$, let*

$$A_\mu(\lambda) = \sum_{S \in W} \det S \exp\,(S(\mu), \lambda), \tag{14.12}$$

then $A_\mu(\lambda)$ is alternating and $A_{S(\mu)}(\lambda) = \det S A_\mu(\lambda)$. Moreover, if $S(\mu) = \mu$ for some $S\,(\neq 1)$ in W, then $A_\mu = 0$.

Proof. Clearly,

$$\begin{aligned} SA_\mu(\lambda) = A_\mu(S^{-1}(\lambda)) &= \sum_{S_1 \in W} \det S_1 \exp\,(S_1(\mu), S^{-1}(\lambda)) \\ &= \sum_{S_1 \in W} \det S_1 \exp\,(SS_1(\mu), \lambda) = \det S A_\mu(\lambda), \end{aligned}$$

.e. $A_\mu(\lambda)$ is alternating. We also have

$$A_{S(\mu)}(\lambda) = \sum_{S_1 \in W} \det S_1 \exp\,(S_1S(\mu), \lambda) = \det SA_\mu(\lambda).$$

Now suppose $S \neq 1$ and $S(\mu) = \mu$. We show first that μ is not contained in any Weyl chamber. Suppose μ is in some Weyl chamber C, then from $S(\mu) = \mu$, it follows that $S(C) \cap C \neq \phi$. Now $S(C)$ is also a Weyl chamber, so $S(C) = C$ and $S = 1$ —— contradiction. Thus for some root α, $(\alpha, \mu) = 0$ and $S_\alpha(\mu) = \mu$. Now $\det S_\alpha = -1$, so

$$A_{S_\alpha(\mu)}(\lambda) = \det S_\alpha A_\mu(\lambda) = -A_\mu(\lambda)$$

and $A_\mu(\lambda) = 0$.

LEMMA 10. *If*

$$\mu_i(t_1, \ldots, t_n) = \sum_{j=1}^{n} a_{ij}t_j, \qquad i = 1, \ldots, m,$$

are m distinct linear functions of the n complex variables $t_1, \ldots, t_n$, then the functions exp $\mu_1(t_1, \ldots, t_n), \ldots,$ exp $\mu_m(t_1, \ldots, t_n)$ *are linearly independent over C.*

Proof. Since the μ_i $(i = 1, \ldots, m)$ are distinct, there exist complex numbers $a_1, \ldots, a_n$ such that $\mu_1(a_1, \ldots, a_n), \ldots, \mu_m(a_1, \ldots, a_n)$ are distinct. Let $b_i = \mu_i\,(a_1, \ldots, a_n)$, then $b_1, \ldots, b_m$ are distinct.

Suppose

$$c_1 \exp \mu_1(t_1, \ldots, t_n) + \ldots + c_m \exp \mu_m(t_1, \ldots, t_n) = 0, \tag{14.13}$$

is a linear relation among the $\exp \mu_i$ $(i = 1, \ldots, m)$. Let t be a new variable and substitute $t_1 = a_1 t, \ldots, t_n = a_n t$ into (14.13), then

$$c_1 \exp b_1 t + \ldots + c_m \exp b_m t = 0.$$

Differentiating this equation k times with respect to t, we get

$$c_1 b_1^k \exp b_1 t + \ldots + c_m b_m^k \exp b_m t = 0,$$
$$(k = 0, 1, 2, \ldots, m-1).$$

Since $b_1, \ldots, b_m$ are distinct, the Vandermonde determinant

$$\begin{vmatrix} 1 & 1 & \ldots & 1 \\ b_1 & b_2 & \ldots & b_m \\ \cdots & \cdots & \cdots & \cdots \\ \cdots & \cdots & \cdots & \cdots \\ b_1^{m-1} & b_2^{m-1} & \ldots & b_m^{m-1} \end{vmatrix} \neq 0.$$

Hence $c_1 = c_2 = \ldots = c_m = 0$. This proves Lemma 10.

Lemma 11. *Suppose*

$$F(\lambda) = \sum_\mu c_\mu \exp(\mu, \lambda) \ (c_\mu \in C \text{ and the summation is over finitely many } \mu \in \mathfrak{h}_0^*)$$

is an alternating function. Then

$$F(\lambda) = \sum_{\mu'} c_{\mu'} A_{\mu'}(\lambda), \tag{14.14}$$

where each $A_{\mu'}(\lambda)$ *has the form* (14.12), $\mu' > S(\mu')(\forall S \in W)$ *and* $c_{\mu'} \in C$.

Proof. For any $S \in W$, we have

$$(SF)(\lambda) = \sum_\mu c_\mu \exp\big(\mu, S^{-1}(\lambda)\big) = \sum_\mu c_\mu \exp\big(S(\mu), \lambda\big).$$

Since F is alternating, i.e. $SF(\lambda) = \det SF(\lambda)$, so

$$F(\lambda) = \sum_\mu c_\mu \det S \exp\big(S(\mu), \lambda\big).$$

By Lemma 10, the functions $\exp(\mu, \lambda)$ are linearly independent. Therefore, if $\exp(\mu, \lambda)$ is a summand of F with coefficient c_μ and $\nu = S(\mu)$, then $\exp(\nu, \lambda)$ is also a summand of F with coefficient $\det S c_\mu$. If in the right-hand side of (14.4) we first sum over all weights that are equivalent to a dominant weight and then sum over all dominant weights, then we get

$$F(\lambda) = \sum_{\mu'} c_{\mu'} A_{\mu'}(\lambda) \quad (\mu' \text{ dominant}).$$

By Lemma 9, it can be assumed that $\mu < S(\mu)$ for all $S\ (\neq 1)$ in W.

LEMMA 12. *Let $\mu_1, \ldots, \mu_m$ be m distinct dominant integral linear functions on $\mathfrak{h}$ satisfying $\mu_i > S(\mu_i)$ $(i = 1, \ldots, m)$, where $S \neq 1$ and $S \in W$. Then $A_{\mu_1}, \ldots, A_{\mu_m}$ are linearly independent over C.*

Proof. Notice first that if $i \neq j$, then there exist no $S, T \in W$ such that $S(\mu_i) = T(\mu_j)$. For otherwise, $\mu_i > T^{-1}S(\mu_i) = \mu_j$ and $\mu_j > S^{-1}T(\mu_j) = \mu_i$——contradiction.

Let $\beta_i = 2\alpha_i/(\alpha_i, \alpha_i)$ $(i = 1, 2, \ldots, n)$, then $\beta_1, \ldots, \beta_n$ form a basis of $\mathfrak{h}^*$ and $\mu \in \mathfrak{h}^*$ is an integral linear function iff every (μ, β_i) $(i = 1, \ldots, n)$ is an integer. Let

$$\xi = 2\pi\sqrt{-1}\sum_{i=1}^{n} x_i\beta_i \in \mathfrak{h}^*,$$

where $x_1, \ldots, x_n$ are real variables. Let

$$A\mu_j(\xi) = P_j(x_1, \ldots, x_n),$$

then it can be shown that

$$\int_0^1 \cdots \int_0^1 P_j(x_1, \ldots, x_n)\overline{P_k(x_1, \ldots, x_n)}\,dx_1 \ldots dx_n = \begin{cases} 0, & j \neq k \\ W:1, & j = k \end{cases} \tag{14.15}$$

where $W:1$ is the order of W. In fact,

$$P_j(x_1, \ldots x_n)\overline{P_k(x_1, \ldots, x_n)} = \sum_{S, T \in W} \det ST \exp(S(\mu_j), \xi) \exp(T(\mu_k), \xi),$$

$$\exp(S(\mu_j), \xi) = \exp\left\{2\pi\sqrt{-1}\sum_{i=1}^{n}(S(\mu_j), \beta_i)\,x_i\right\},$$

$$\exp(T(\mu_k), \xi) = \exp\left\{2\pi\sqrt{-1}\sum_{i=1}^{n}(T(\mu_k), \beta_i)\,x_i\right\}.$$

Since $S(\mu_j)$ and $T(\mu_k)$ are equivalent to μ_j and μ_k respectively, $(S(\mu_j), \beta_i)$ and $(T(\mu_k), \beta_i)$ $(i = 1, \ldots, n)$ are integers. Now for $j \neq k$, there exist no $S, T \in W$ such that $S(\mu_j) = T(\mu_k)$, so $(S(\mu_j), \beta_i) \neq (T(\mu_k), \beta_i)$ for at least one i. Hence

$$\int_0^1 \cdots \int_0^1 P_j\bar{P}_k\,dx_1 \ldots dx_n = 0.$$

When $j = k$, since for $S \neq T$, we have $S(\mu_j) \neq T(\mu_j)$. Thus

$$\int_0^1 \cdots \int_0^1 P_j\bar{P}_j\,dx_1 \ldots dx_n$$

$$= \sum_{S \in W}\int_0^1 \cdots \int_0^1 |\exp 2\pi\sqrt{-1}\sum_1^n (S(\mu_j), \beta_i)\,x_i|^2\,dx_1 \ldots dx_n = W:1.$$

Hence (14.15) holds.

Now suppose $\sum_{j=1}^m c_jA_{\mu_j}(\lambda) = 0$, where $c_j \in C$, then $\sum_{j=1}^m c_jP_j = 0$. Multiplying this equation by $\overline{P_k(x_1, \ldots, x_n)}$ and integrating both sides, we get $c_k(W:1) = 0$, i.e. $c_j = 0$ $(j = 1, \ldots, m)$. This proves that $A_{\mu_1}, \ldots, A_{\mu_m}$ are linearly independent.

14.4. Formula of the character of an irreducible representation

Let $\mathfrak{g}$ be a semisimple Lie algebra, $\mathfrak{h}$ be a Cartan subalgebra, Σ be the system of roots and $\Pi = \{\alpha_1, \ldots, \alpha_n\}$ be a fundamental system of roots. Suppose ϱ is an irreducible representation of $\mathfrak{g}$. For $H \in \mathfrak{h}$, define

$$\chi(H) = \operatorname{Tr} \exp \varrho(H),$$

where

$$\exp \varrho(H) = I+\varrho(H)+\frac{\varrho(H)^2}{2!}+\frac{\varrho(H)^3}{3!}+\ldots,$$

then χ is a function defined on $\mathfrak{h}$; χ is said to be the character of ϱ. Let $m_\omega = \dim V_\omega$, then

$$\chi(H) = \sum_{\omega} m_\omega \exp_\omega \omega(H),$$

where summation is over all weights of ϱ.

THEOREM 2 *(H. Weyl*†*). Let ϱ be an irreducible representation of a semisimple Lie algebra $\mathfrak{g}$ with highest weight ω_0. If $\delta = \frac{1}{2}\sum_{\alpha>0} \alpha$, then for any $H \in \mathfrak{h}$,*

$$\chi(H) = \frac{\sum_{S\in W} \det S \exp [(S(\omega_0+\delta))(H)]}{\sum_{S\in W} \det S \exp [(S(\delta))(H)]} \tag{14.16}$$

and

$$\sum_{S\in W} \det S \exp [(S(\delta))(H)] = \prod_{\alpha>0} [\exp (\tfrac{1}{2}\alpha(H))-\exp (-\tfrac{1}{2}\alpha(H))]. \tag{14.17}$$

Proof. For any $H \in \mathfrak{h}$, there exists $\lambda \in \mathfrak{h}^*$ such that $\mu(H) = (\mu, \lambda)$ for any $\mu \in \mathfrak{h}^*$. If $F(H)$ is a function defined on $\mathfrak{h}$, let $F(\lambda) = F(H)$. Therefore, we have

$$\chi(\lambda) = \sum_{\omega} m_\omega \exp (\omega, \lambda)$$

and (14.16), (14.17) can be rewritten as

$$\chi(\lambda) = \frac{\sum_{S\in W} \det S \exp (S(\omega_0+\delta), \lambda)}{\sum_{S\in W} \det S \exp (S(\delta), \lambda)} \tag{14.16$'$}$$

and

$$\sum_{S\in W} \det S \exp (S(\delta), \lambda) = \prod_{\alpha>0} [\exp (\tfrac{1}{2}(\alpha, \lambda))-\exp (-\tfrac{1}{2}(\alpha, \lambda))]. \tag{14.17$'$}$$

We show first that (14.17′) holds. Let

$$Q(\lambda) = \prod_{\alpha>0} [\exp (\tfrac{1}{2}(\alpha, \lambda))-\exp (-\tfrac{1}{2}(\alpha, \lambda))]. \tag{14.18}$$

† See footnote for Theorem 6 in Section 5.3.

Now if $S_i = S_{\alpha_i}, \alpha > 0$ and $\alpha = \alpha_i$ then $S_i(\alpha_i) = -\alpha_i$; if $\alpha > 0$ and $\alpha \neq \alpha_i$, then by Lemma 6 of Chapter 6,

$$S_i(\alpha) = \alpha - \frac{2(\alpha, \alpha_i)}{(\alpha_i, \alpha_i)}\alpha_i > 0.$$

Therefore,

$$\begin{aligned}(S_iQ)(\lambda) &= Q(S_i^{-1}(\lambda)) = Q(S_i(\lambda))\\ &= \prod_{\alpha>0}\left[\exp\left(\tfrac{1}{2}(\alpha, S_i(\lambda))\right) - \exp\left(-\tfrac{1}{2}(\alpha, S_i(\lambda))\right)\right]\\ &= \prod_{\alpha>0}\left[\exp\left(\tfrac{1}{2}(S_i(\alpha), \lambda)\right) - \exp\left(-\tfrac{1}{2}(S_i(\alpha), \lambda)\right)\right]\\ &= \prod_{\substack{\alpha>0\\ \alpha\neq\alpha_i}}\left[\exp\left(\tfrac{1}{2}(\alpha, \lambda)\right) - \exp\left(-\tfrac{1}{2}(\alpha, \lambda)\right)\right]\cdot\left[\exp\left(-\tfrac{1}{2}(\alpha_i, \lambda)\right) - \exp\left(\tfrac{1}{2}(\alpha_i, \lambda)\right)\right]\\ &= -Q(\lambda).\end{aligned}$$

This proves that $Q(\lambda)$ is alternating.

Expanding the right-hand side of (14.18), we get

$$Q(\lambda) = \sum_{\mu} \pm\exp(\mu, \lambda),$$

where $\mu = \frac{1}{2}\sum_{\alpha>0}\varepsilon_\alpha\alpha$ and $\varepsilon_\alpha = \pm 1$. By Lemma 11,

$$Q(\lambda) = \sum_{\mu} \pm A_\mu(\lambda),$$

where $\mu > S(\mu)$ for all $S\ (\neq 1) \in W$. By Lemma 7,

$$Q(\lambda) = \pm A_\delta(\lambda).$$

Comparing the coefficient of exp (δ, λ) on both sides of this equation, we get

$$Q(\lambda) = A_\delta(\lambda) = \sum_{S\in W} \det S \exp(S(\delta), \lambda).$$

This proves (14.17′).

We now show that (14.16′) holds. The proof here is due to H. Freudenthal† and consists of the following steps:

1. Multiplying both sides of (14.8) by $\exp(\omega, \lambda)$ and summing over all integral linear functions ω (when ω is not a weight of ϱ, $m_\omega = 0$), we get

$$\begin{aligned} r\chi(\lambda) &= \sum_{\omega} m_\omega(\omega+2\delta, \omega)\exp(\omega, \lambda) + 2\sum_{\omega}\sum_{\alpha>0}\sum_{j=1}^{\infty} m_{\omega+j\alpha}(\omega+j\alpha, \alpha)\exp(\omega, \lambda)\\ &= \sum_{\omega} m_\omega(\omega+2\delta, \omega)\exp(\omega, \lambda) + 2\sum_{\omega}\sum_{\alpha>0}\sum_{j=1}^{\infty} m_\omega(\omega, \alpha)\exp(\omega - j\alpha, \lambda).\end{aligned}$$

† See footnote for Theorem 1 in Section 14.1.

Suppose $\tau \in \mathfrak{h}^*$ and $|\exp(\alpha, \tau)| > 1$ for all $\alpha > 0$, then

$$\sum_{\omega}\sum_{\alpha>0}\sum_{j=1}^{\infty} m_\omega(\omega, \alpha)\exp(\omega - j\alpha, \tau) = \sum_{\omega}\sum_{\alpha>0} m_\omega(\omega, \alpha)\exp(\omega, \tau)\sum_{j=1}^{\infty}[\exp(-\alpha, \tau)]^j$$

$$= \sum_{\omega}\sum_{\alpha>0} m_\omega(\omega, \alpha)\exp(\omega, \tau)\frac{\exp(-\alpha, \tau)}{1-\exp(-\alpha, \tau)}.$$

Therefore,

$$r\chi(\tau) = \sum_{\omega} m_\omega(\omega, \omega)\exp(\omega, \lambda) + \sum_{\omega} m_\omega(2\delta, \omega)\exp(\omega, \lambda)$$

$$+2\sum_{\omega}\sum_{\alpha>0} m_\omega(\omega, \alpha)\exp(\omega, \tau)\frac{\exp(-\alpha, \tau)}{1-\exp(-\alpha, \tau)}. \tag{14.19}$$

2. Choose a basis $\{\varepsilon_1, \ldots, \varepsilon_n\}$ of $\mathfrak{h}^*$ such that $(\varepsilon_i, \varepsilon_j) = \delta_{ij}$. We can assume that $\varepsilon_1, \ldots, \varepsilon_n \in \mathfrak{h}_0^*$. If $\lambda \in \mathfrak{h}^*$, write $\lambda = t_1\varepsilon_1 + \ldots + t_n\varepsilon_n$, then any complex-valued function $F(\lambda)$ can be considered as a function of $t_1, \ldots, t_n$. If F is differentiable, define a mapping grad F by

$$(\text{grad } F)(\lambda) = \sum_1^n \left(\frac{\partial F}{\partial t_i}\right)(\lambda)\varepsilon_i;$$

define the Laplace operator Δ on F by

$$(\Delta F)(\lambda) = \sum_1^n \frac{\partial^2 F(\lambda)}{\partial t_i^2}.$$

By simple computations, it can be shown that

$$\left.\begin{aligned} \text{grad} \exp(\omega, \lambda) &= \exp(\omega, \lambda)\cdot\omega, \\ \text{grad} \log(1-\exp(-\alpha, \tau)) &= \frac{\exp(-\alpha, \tau)}{1-\exp(-\alpha, \tau)}\alpha, \\ \Delta \exp(\omega, \lambda) &= (\omega, \omega)\exp(\omega, \lambda). \end{aligned}\right\} \tag{14.20}$$

3. Using (14.20) and $\chi(\lambda) = \sum_\omega m_\omega \exp(\omega, \lambda)$, we get

$$\text{grad } \chi(\lambda) = \sum_{\omega} m_\omega \exp(\omega, \lambda)\cdot\omega,$$

$$\Delta\chi(\lambda) = \sum_{\omega} m_\omega(\omega, \omega)\exp(\omega, \lambda).$$

Hence (14.18) can be written as

$$r\chi(\tau) = \Delta\chi(\tau) + (2\delta, \text{grad } \chi(\tau)) + 2\left(\text{grad} \log \prod_{\alpha>0}((1-\exp(-\alpha, \tau)), \text{grad } \chi(\tau)\right)$$

$$= \Delta\chi(\tau) + 2\left(\delta + \text{grad} \log \prod_{\alpha>0}(1-\exp(-\alpha, \tau)), \ \text{grad } \ \chi(\tau)\right). \tag{14.21}$$

On the other hand, from (14.18) it follows that

$$Q(\tau) = \prod_{\alpha>0} [\exp(\tfrac{1}{2}(\alpha, \tau)) - \exp(-\tfrac{1}{2}(\alpha, \tau))]$$
$$= \exp(\delta, \tau) \prod_{\alpha>0} (1-\exp(-\alpha, \tau)),$$

so

$$\text{grad } Q(\tau) = \text{grad} \exp(\delta, \tau) \cdot \prod_{\alpha>0} (1-\exp(-\alpha, \tau)) + \exp(\delta, \tau) \text{ grad} \prod_{\alpha>0} (1-\exp(-\alpha, \tau))$$
$$= \delta \exp(\delta, \tau) \prod_{\alpha>0} (1-\exp(-\alpha, \tau)) + \exp(\delta, \tau) \prod_{\alpha>0} (1-\exp(-\alpha, \tau))$$
$$\times \text{ grad log} \prod_{\alpha>0} (1-\exp(-\alpha, \tau))$$
$$= \delta Q(\tau) + Q(\tau) \text{ grad log} \prod_{\alpha>0} (1-\exp(-\alpha, \tau)). \tag{14.22}$$

Multiplying (14.21) by $Q(\tau)$ and using (14.22), we get

$$r\chi(\tau)\, Q(\tau) = Q(\tau)\, \Delta\chi(\tau) + 2\,(\text{grad } Q(\tau), \text{grad } \chi(\tau)). \tag{14.23}$$

Let

$$\varphi(\lambda) = Q(\lambda)\, \chi(\lambda);$$

then from simple computation,

$$\Delta\varphi(\lambda) = \Delta Q(\lambda) \cdot \chi(\lambda) + Q(\lambda) \cdot \Delta\chi(\lambda) + 2(\text{grad } Q(\lambda), \text{grad } \chi(\lambda)),$$

so (14.23) becomes

$$r\varphi(\tau) = \Delta\varphi(\tau) - \Delta Q(\tau) \cdot \chi(\tau). \tag{14.24}$$

4. We now show that $\chi(\lambda)$ is symmetric. By definition,

$$\chi(\lambda) = \sum_{\omega} m_{\omega} \exp(\omega, \lambda).$$

Therefore, from $m_{S(\omega)} = m_{\omega}$, we have

$$S\chi(\lambda) = \sum_{\omega} m_{\omega} \exp(\omega, S^{-1}(\lambda)) = \sum_{\omega} m_{\omega} \exp(S(\omega), \lambda)$$
$$= \sum_{\omega} m_{S(\omega)} \exp(S(\omega), \lambda) = \chi(\lambda).$$

Hence $\chi(\lambda)$ is symmetric. Therefore, $\varphi(\lambda) = Q(\lambda)\, \chi(\lambda)$ is alternating.

By (14.17′) and the definition of $\chi(\lambda)$,

$$\varphi(\lambda) = Q(\lambda)\, \chi(\lambda) = \sum_{S \in W} \det S \exp(S(\delta), \lambda) \sum_{\omega} m_{\omega} \exp(\omega, \lambda)$$
$$= \sum_{\omega, S} \det S m_{\omega} \exp(\omega + S(\delta), \lambda).$$

By Lemma 11,

$$\varphi(\lambda) = \sum_{\mu} c_{\mu} A_{\mu}(\lambda),$$

where $\mu = \omega + S(\delta)$ and $\mu > T(\mu)$ for any $T\ (\neq 1) \in W$. Now

$$\begin{aligned}\Delta A_\mu(\lambda) &= \sum_{S \in W} \det S\, \Delta \exp(S(\mu), \lambda)\\ &= \sum_{S \in W} \det S (S(\mu), S(\mu)) \exp(S(\mu), \lambda)\\ &= (\mu, \mu) A_\mu(\lambda),\end{aligned}$$

so

$$\Delta\varphi(\lambda) = \Sigma c_\mu (\mu, \mu) A_\mu(\lambda). \tag{14.25}$$

From (14.24),

$$\Delta\varphi(\tau) = r\varphi(\tau) + \Delta Q(\tau)\cdot\chi(\tau), \tag{14.26}$$

where

$$r = (\omega_0 + \delta, \omega_0 + \delta) - (\delta, \delta).$$

Since $Q(\lambda) = A_\delta(\lambda)$,

$$\Delta Q(\lambda) = (\delta, \delta) Q(\lambda).$$

Hence

$$\Delta Q(\tau)\cdot\chi(\tau) = (\delta, \delta) Q(\tau)\chi(\tau) = (\delta, \delta)\varphi(\tau) \tag{14.27}$$

Substituting (14.27) into (14.26) and then comparing (14.25) and (14.26), we get

$$\sum_\mu c_\mu (\mu, \mu) A_\mu(\tau) = (\omega_0 + \delta, \omega_0 + \delta) \sum_\mu c_\mu A_\mu(\tau).$$

Hence

$$\sum_\mu c_\mu ((\mu, \mu) - (\omega_0 + \delta, \omega_0 + \delta)) A_\mu(\tau) = 0.$$

Since $\{\tau \in \mathfrak{h}^* \mid |\exp(\alpha, \tau)| > 1 \text{ for all } \alpha > 0\}$ is an open set of $\mathfrak{h}^*$ and all $A_\mu(\tau)$ are analytic functions on $\mathfrak{h}^*$,

$$\sum_\mu c_\mu ((\mu, \mu) - (\omega_0 + \delta, \omega_0 + \delta)) A_\mu(\lambda) = 0, \quad \text{for all} \quad \lambda \in \mathfrak{h}^*.$$

Notice that μ is of the form $\omega + S(\delta)$ and $\mu > T(\mu)$ for all $T (\neq 1) \in W$. Clearly, all $\mu = \omega + S(\delta)$ are dominant integral linear functions; by Lemma 12, all $A_\mu(\lambda)$ are linearly independent over C, so

$$(\mu, \mu) = (\omega_0 + \delta, \omega_0 + \delta).$$

Now

$$(\mu, \mu) = \big(S^{-1}(\mu), S^{-1}(\mu)\big) = \big(S^{-1}(\omega) + \delta, S^{-1}(\omega) + \delta\big),$$

so

$$(\omega_0 + \delta, \omega_0 + \delta) = \big(S^{-1}(\omega) + \delta, S^{-1}(\omega) + \delta\big).$$

By Lemma 8, $\omega_0 = S^{-1}(\omega)$, i.e. $\omega = S(\omega_0)$, thus $\mu = S(\omega_0 + \delta)$. If $S \neq 1$, then $S^{-1}(\mu) = \omega_0 + \delta < \mu = \omega + S(\delta)$, but $\delta > S(\delta)$, so $\omega_0 < \omega$ —— contradiction. Hence $S = 1$ and $\mu = \omega_0 + \delta$. Therefore,

$$\varphi(\lambda) = c A_{\omega_0 + \delta}(\lambda).$$

Comparing the coefficients of $\exp(\omega_0 + \delta, \lambda)$ on both sides of this equation, we get $c = 1$. Hence

$$\varphi(\lambda) = A_{\omega_0 + \delta}(\lambda) = \sum_{S \in W} \det S \exp(S(\omega_0 + \delta), \lambda).$$

Consequently,

$$\chi(\lambda) = \frac{\varphi(\lambda)}{Q(\lambda)} = \frac{\sum_{S\in W} \det S \exp\,(S(\omega_0+\delta), \lambda)}{\sum_{S\in W} \det S \exp\,(S(\delta), \lambda)}.$$

This proves that (14.17′) holds. The proof of Theorem 2 is now completed.

By means of the character formula, H. Weyl also obtained the formula for the dimension of the corresponding representation space.

THEOREM 3. *If ϱ is an irreducible representation of a semisimple Lie algebra $\mathfrak{g}$ with highest weight ω_0, then*

$$\dim \varrho = \frac{\prod_{\alpha>0} (\alpha, \omega_0+\delta)}{\prod_{\alpha>0} (\alpha, \delta)}. \tag{14.28}$$

Proof. We have

$$\dim \varrho = \chi(0).$$

Let t be a real variable, then $\chi(\sqrt{-1}t\delta)$ is a continuous function of t, so

$$\chi(0) = \lim_{t\to 0} \chi(\sqrt{-1}t\delta).$$

We show first that if ε is sufficiently small and $0 < |t| < \varepsilon$, then $Q(\sqrt{-1}t\delta) \neq 0$. We have

$$Q(\sqrt{-1}t\delta) = \prod_{\alpha>0} [\exp\,(\tfrac{1}{2}\sqrt{-1}t(\alpha, \delta)) - \exp\,(-\tfrac{1}{2}\sqrt{-1}t(\alpha, \delta))]$$

$$= \prod_{\alpha>0} 2\sqrt{-1} \sin \frac{t}{2} (\alpha, \delta) = (2\sqrt{-1})^m \prod_{\alpha>0} \sin \frac{t}{2} (\alpha, \delta),$$

where m is the number of positive roots. Now for any $S\ (\neq 1) \in W$, $\delta > S(\delta)$, in particular,

$$\delta > S_\alpha(\delta) = \delta - \frac{2(\delta, \alpha)}{(\alpha, \alpha)}\alpha.$$

Hence for any positive root α, $(\delta, \alpha) > 0$. Consequently, there exists $\varepsilon > 0$ such that when $0 < |t| < \varepsilon$, we have $\prod_{\alpha>0} \sin t/2\,(\alpha, \delta) \neq 0$. Hence $Q(\sqrt{-1}t\delta) \neq 0$.

Now suppose an $\varepsilon > 0$ is chosen so that for t satisfying $0 < |t| < \varepsilon$ we have $Q(\sqrt{-1}t\delta) \neq 0$. From $\chi(\sqrt{-1}t\delta)Q(\sqrt{-1}t\delta) = \varphi(\sqrt{-1}t\delta)$, it follows that

$$\dim \varrho = \chi(0) = \lim_{t\to 0} \frac{\varphi(\sqrt{-1}t\delta)}{Q(\sqrt{-1}t\delta)},$$

We also have

$$\varphi(\sqrt{-1}t\delta) = \sum_{S\in W} \det S \exp\,(S(\omega_0+\delta), \sqrt{-1}t\delta)$$

$$= \sum_{S\in W} \det S \exp\,(S^{-1}(\delta), \sqrt{-1}t\,(\omega_0+\delta))$$

$$= \sum_{S\in W} \det S \exp\,(S^{-1}(\delta), \sqrt{-1}t(\omega_0+\delta))$$

$$= A_\delta(\sqrt{-1}t\,(\omega_0+\delta)) = Q(\sqrt{-1}t\,(\omega_0+\delta))$$

$$= (2\sqrt{-1})^m \prod_{\alpha>0} \sin t/2\,(\alpha, \omega_0+\delta).$$

Thus

$$\dim \varrho = \chi(0) = \lim_{t\to 0} \frac{\varphi(\sqrt{-1}t\delta)}{Q(\sqrt{-1}t\delta)}$$

$$= \lim_{t\to 0} \frac{(2\sqrt{-1})^m \prod_{\alpha>0} (\sin t/2)(\alpha, \omega_0+\delta)}{(2\sqrt{-1})^m \prod_{\alpha>0} (\sin t/2)(\alpha, \delta)}$$

$$= \lim_{t\to 0} \frac{\prod_{\alpha>0} (t/2)(\alpha, \omega_0+\delta)}{\prod_{\alpha>0} (t/2)(\alpha, \delta)} = \frac{\prod_{\alpha>0} (\alpha, \omega_0+\delta)}{\prod_{\alpha>0} (\alpha, \delta)}.$$

The proof of Theorem 3 is now completed.

By means of the character formula, B. Kostant† derived a formula for the multiplicity of a weight of ϱ.

THEOREM 4. *Let ϱ be an irreducible representation of a semisimple Lie algebra $\mathfrak{g}$ with highest weight ω_0. For any $\mu \in \mathfrak{h}^*$, let $P(\mu)$ be the number of ways of expressing μ as a sum of positive roots, then*

$$m_\omega = \sum_{S\in W} \det S P\big(S(\omega_0+\delta)-(\omega+\delta)\big) \qquad (14.29)$$

Proof (P. Cartier‡). We have

$$Q(\lambda) = \exp(\delta, \lambda) \prod_{\alpha>0} (1-\exp(-\alpha, \lambda)),$$

so

$$Q(\tau)^{-1} = \exp(-\delta, \tau) \sum_{\mu} P(\mu) \exp(-\mu, \tau),$$

for any $\tau \in \mathfrak{h}^*$ satisfying $|\exp(\alpha, \tau)| > 1$ $(\forall \alpha \in \Sigma_+)$. From Weyl's formula, we have

$$\chi(\tau) = \sum_{\omega} m_\omega \exp(\omega, \tau) = \varphi(\tau) Q(\tau)^{-1} = \sum_{\mu} \sum_{S\in W} \det S \exp\big(S(\omega_0+\delta)-(\mu+\delta), \tau\big) P(\mu)$$

$$= \sum_{\omega} \sum_{S\in W} \det S P\big(S(\omega_0+\delta)-(\omega_0+\delta)\big) \exp(\omega, \tau).$$

Therefore, for any $\lambda \in \mathfrak{h}^*$,

$$\sum_{\omega} m_\omega \exp(\omega, \lambda) = \sum_{\omega} \sum_{S\in W} \det S P\big(S(\omega_0+\delta)-(\omega+\delta)\big) \exp(\omega, \lambda).$$

By Lemma 10 and comparing the coefficients of $\exp(\omega, \lambda)$ on both sides of this equation, we get (14.29).

† B. Kostant, A formula for the multiplicity of a weight, *Trans. Amer. Math. Soc.* **93** (1959), 53–73.
‡ P. Cartier, On H. Weyl's character formula, *Bull. Amer. Math. Soc.* **67** (1961), 228–30.

CHAPTER 15

REAL FORMS OF COMPLEX SEMISIMPLE LIE ALGEBRAS

15.1. Complex extension of real Lie algebras and real forms of complex Lie algebras

Let V be an n-dimensional real vector space with basis $\{e_1, \ldots, e_n\}$. The set of linear combinations of $e_1, \ldots, e_n$ with complex coefficients is an n-dimensional complex vector space; denote this space by $[V]$. $[V]$ is said to be the complex extension of V. Clearly, V is a subset of $[V]$ and $\dim V = \dim [V]$.

Let $\mathfrak{g}$ be an r-dimensional real Lie algebra with basis $\{e_1, \ldots, e_r\}$ and

$$[e_i, e_j] = c_{ij}^k e_k, \tag{15.1}$$

then the structure of $\mathfrak{g}$ is determined by the r^3 real numbers c_{ij}^k. Now $\{e_1, \ldots, e_r\}$ is also a basis of the vector space $[\mathfrak{g}]$. If commutation is defined on $[\mathfrak{g}]$ by means of (15.1), then $[\mathfrak{g}]$ becomes a complex Lie algebra; this Lie algebra is said to be the *complex extension* of $\mathfrak{g}$. Clearly, $\dim [\mathfrak{g}] = \dim \mathfrak{g}$.

Suppose $\mathfrak{g}_1$ is a complex Lie algebra. If $\mathfrak{g}$ is a subset of $\mathfrak{g}_1$, $\mathfrak{g}$ is a real Lie algebra and $[\mathfrak{g}] = \mathfrak{g}_1$, then $\mathfrak{g}$ is said to be a real form of $\mathfrak{g}_1$. A complex Lie algebra does not necessarily have a real form. *Every real Lie algebra is a real form of its complex extension.*

Let $\mathfrak{g}_0$ be a real Lie algebra and $\mathfrak{g}_1 = [\mathfrak{g}_0]$ be the complex extension of $\mathfrak{g}_0$, then every element in $\mathfrak{g}_1$ can be expressed uniquely as

$$X+iY, \quad X, Y \in \mathfrak{g}_0.$$

Define a mapping σ on $\mathfrak{g}_1$ by

$$X+iY \xrightarrow{\sigma} X-iY,$$

then σ is one-one from $\mathfrak{g}_1$ onto itself and

$$\left.\begin{aligned} \sigma^2 &= 1, \\ \sigma(Z_1+Z_2) &= \sigma(Z_1)+\sigma(Z_2), \\ \sigma(\lambda Z) &= \bar{\lambda}\sigma(Z), \\ \sigma([Z_1, Z_2]) &= [\sigma(Z_1), \ \sigma(Z_2)], \quad (Z_1, Z_2, Z \in \mathfrak{g}_1, \lambda \in C). \end{aligned}\right\} \tag{15.2}$$

A one-one mapping from $\mathfrak{g}_1$ onto itself that satisfies (15.2) is said to be a *semi-involution*. Clearly, $\mathfrak{g}_0$ is the set of fixed points of σ, i.e.

$$\mathfrak{g}_0 = \{Z \in \mathfrak{g}_1 | \sigma(Z) = Z\}.$$

Conversely, suppose $\mathfrak{g}_1$ is a complex Lie algebra with a semi-involution σ. If $\mathfrak{g}_0 = \{Z \in \mathfrak{g}_1 | \sigma(Z) = Z\}$, then it can be shown that $\mathfrak{g}_0$ is a real Lie algebra. In fact, if $X, Y \in \mathfrak{g}_0$ and λ is real, then $\sigma(X+Y) = \sigma(X)+\sigma(Y) = X+Y$, $\sigma(\lambda Y) = \lambda\sigma(X) = \lambda X$ and $\sigma([X, Y]) = [\sigma(X), \sigma(Y)] = [X, Y]$, thus $X+Y$, λX and $[X, Y]$ are in $\mathfrak{g}_0$. Moreover, if $Z \in \mathfrak{g}_1$, then Z can be written as

$$Z = \frac{1}{2}(Z+\sigma(Z))+i\frac{1}{2i}(Z-\sigma(Z)),$$

where $Z+\sigma(Z)$, $(1/2i)\big(Z-\sigma(Z)\big) \in \mathfrak{g}_0$. If $Z = X+iY$ $(X, Y \in \mathfrak{g}_0)$, then $\sigma(Z) = X-iY$, so $X = \frac{1}{2}(Z+\sigma(Z))$, $Y = (1/2i)\,(Z-\sigma(Z))$. Hence every $Z \in \mathfrak{g}_1$ can be uniquely expressed as $Z = X+iY$ $(X, Y \in \mathfrak{g}_0)$, so $\mathfrak{g}_0$ is a real form of $\mathfrak{g}$. We have proved the following:

THEOREM 1. *Let $\mathfrak{g}_1$ be a complex Lie algebra and σ be a semi-involution of $\mathfrak{g}_1$, then the set of fixed points $\mathfrak{g}_0$ of σ is a real form of $\mathfrak{g}_1$. Conversely, every real form of $\mathfrak{g}_1$ is the set of fixed points of some semi-involution of $\mathfrak{g}_1$.*

We now study the relations between a real Lie algebra $\mathfrak{g}$ and its complex extension $[\mathfrak{g}]$. If $\mathfrak{h}$ is an ideal of $\mathfrak{g}$, then $[\mathfrak{h}]$ is an ideal of $[\mathfrak{g}]$ and

$$[\mathfrak{g}/\mathfrak{h}] \approx [\mathfrak{g}]/[\mathfrak{h}].$$

Thus $\mathfrak{g}$ is solvable iff $[\mathfrak{g}]$ is solvable. Moreover, we have the following:

THEOREM 2. *If $\mathfrak{r}$ is the radical of $\mathfrak{g}$, then $[\mathfrak{r}]$ is the radical of $[\mathfrak{g}]$.*

Proof. $[r]$ is naturally solvable. If $\mathfrak{n}$ is the radical of $[\mathfrak{g}]$, let

$$\sigma(\mathfrak{n}) = \{\sigma(X) | X \in \mathfrak{n}\},$$

then $\sigma(\mathfrak{n})$ is also a solvable ideal of $[\mathfrak{g}]$, thus $\sigma(\mathfrak{n}) \subset \mathfrak{n}$. Since $\dim \sigma(\mathfrak{n}) = \dim \mathfrak{n}$, $\sigma(\mathfrak{n}) = \mathfrak{n}$; thus σ is a semi-involution of $\mathfrak{n}$. By Theorem 1, $\mathfrak{n}$ is the complex extension of a real Lie algebra $\mathfrak{h}$, i.e. $\mathfrak{n} = [\mathfrak{h}]$. Hence $\mathfrak{h}$ is solvable and $\mathfrak{h} \subset \mathfrak{r}$, so $\mathfrak{n} = [\mathfrak{h}] \subset [\mathfrak{r}]$. This proves that $[\mathfrak{r}]$ is the radical of $[\mathfrak{g}]$.

COROLLARY. *$\mathfrak{g}$ is semisimple iff $[\mathfrak{g}]$ is semisimple.*

Let $\mathfrak{g}$ be a real Lie algebra and $A \in \mathfrak{g}$. Define

$$\operatorname{ad} AX = [A, X], \qquad X \in \mathfrak{g}$$

and call ad A an inner derivation of $\mathfrak{g}$. Let

$$(X, Y) = \operatorname{Tr} \operatorname{ad} X \operatorname{ad} Y, \qquad X, Y \in \mathfrak{g}$$

and call (X, Y) the Killing form of $\mathfrak{g}$.

THEOREM 3. *Let $\mathfrak{g}$ be a real Lie algebra, then $\mathfrak{g}$ is solvable iff $(X, X) = 0$ for all $X \in \mathfrak{g}' = [\mathfrak{g}, \mathfrak{g}]$.*

Proof. If $\mathfrak{g}$ is solvable, then $[\mathfrak{g}]$ is also solvable, so $(X, X) = 0$ for all $X \in [\mathfrak{g}]'$. Hence $(X, X) = 0$ for all $X \in \mathfrak{g}'$.

Conversely, if $(X, X) = 0$ for all $X \in \mathfrak{g}'$, then $(X, X) = 0$ for all $X, Y \in \mathfrak{g}'$. Since $[\mathfrak{g}'] = [\mathfrak{g}]'$, $(X, Y) = 0$ for all $X, Y \in [\mathfrak{g}]'$. In particular, $(X, X) = 0$ for all $X \in [\mathfrak{g}]'$, so $[\mathfrak{g}]$ is solvable.

THEOREM 4. *Let $\mathfrak{g}$ be a real Lie algebra, then $\mathfrak{g}$ is semisimple iff the Killing form of $\mathfrak{g}$ is non-degenerate.*

THEOREM 5. *Let $\mathfrak{g}$ be a real semisimple Lie algebra, then $\mathfrak{g}$ is the direct sum of all the non-abelian simple ideals. Conversely, a direct sum of simple Lie algebras is semisimple. If $\mathfrak{g}$ is semisimple, then $\mathfrak{g}' = \mathfrak{g}$.*

Theorem 4 is a consequence of Theorem 3 and Theorem 5 is a consequence of Theorem 4; the proofs are similar to the proofs for complex Lie algebras.

15.2. Compact Lie algebras

A real Lie algebra is said to be compact if a symmetric negative-definite bilinear form $B(X, Y)$ which is invariant, i.e.

$$B(\text{ad } AX, Y) + B(X, \text{ad } AY) = 0, \quad \text{for all } A \in \mathfrak{g}$$

is defined on it.

THEOREM 6. *A compact Lie algebra $\mathfrak{g}$ is the direct sum of its center $\mathfrak{c}$ and the unique maximal semisimple ideal $\mathfrak{g}_0$. Hence $\mathfrak{c}$ is the radical of $\mathfrak{g}$.*

Proof. Let

$$\mathfrak{g}_0 = \{X \in \mathfrak{g} \,|\, B(X, Z) = 0, \quad \text{for all } X \in \mathfrak{c}\}.$$

We show first that $\mathfrak{g}_0$ is an ideal of $\mathfrak{g}$ and $\mathfrak{g} = \mathfrak{c} \dot{+} \mathfrak{g}_0$. In fact, if $X \in \mathfrak{g}_0$ and $A \in \mathfrak{g}$, then for all $Z \in \mathfrak{c}$

$$B([A, X], Z) = -B(X, [A, Z]) = 0,$$

because $[A, Z] = 0$. This proves that $\mathfrak{g}_0$ is an ideal of $\mathfrak{g}$. Now let $X \in \mathfrak{g}_0 \cap \mathfrak{c}$, then $B(X, X) = 0$; since B is negative-definite, $X = 0$. Hence $\mathfrak{g}_0 \cap \mathfrak{c} = \{0\}$ and $\mathfrak{g} = \mathfrak{g}_0 \dot{+} \mathfrak{c}$.

We now show that $\mathfrak{g}_0$ is semisimple. Suppose $\mathfrak{g}_0$ is not semisimple, then $\mathfrak{g}_0$ contains a non-zero abelian ideal $\mathfrak{a}$. Now the restriction of $B(X, Y)$ to $\mathfrak{g}_0$ is an invariant symmetric negative-definite bilinear form. If

$$\mathfrak{a}_1 = \{X \in \mathfrak{g}_0 \,|\, B(X, Y) = 0, \quad \text{for all } Y \in \mathfrak{a}\}$$

then $\mathfrak{a}_1$ is an ideal of $\mathfrak{g}_0$ and $\mathfrak{g}_0 = \mathfrak{a} \dot{+} \mathfrak{a}_1$. Thus

$$\mathfrak{g} = \mathfrak{c} \dot{+} \mathfrak{a} \dot{+} \mathfrak{a}_1$$

and $\mathfrak{c} \dot{+} \mathfrak{a} \subseteq \mathfrak{c}$. Consequently, $\mathfrak{a} = \{0\}$ —— contradiction. Hence $\mathfrak{g}_0$ is semisimple.

Now let $\mathfrak{g}_1$ be a semisimple ideal of $\mathfrak{g}$. Since $\mathfrak{g}_1' = \mathfrak{g}_1$, any $X \in \mathfrak{g}_1$ can be written as

$$X = \sum_i [Y_i, Z_i], \quad Y_i, Z_i \in \mathfrak{g}_1.$$

For any $Z \in \mathfrak{c}$,

$$\begin{aligned} B(X, Z) &= B\Big(\sum_i (Y_i, Z_i], Z)\Big) = \sum_i B([Y_i, Z_i], Z) \\ &= -\sum_i B(Z_i, [Y_i, Z]) = 0, \quad \text{because} \quad [Y_i, Z] = 0. \end{aligned}$$

Hence $\mathfrak{g}_1 \subseteq \mathfrak{g}_0$ and $\mathfrak{g}_0$ is the unique maximal semisimple ideal of $\mathfrak{g}$.

COROLLARY. *Let $\mathfrak{c}$ be the center of a compact Lie algebra $\mathfrak{g}$, then*

$$\mathfrak{c} = \{X \in \mathfrak{g} \,|\, (X, X) = 0\} = \{X \in \mathfrak{g} \,|\, (X, Y) = 0, \quad \text{for all } Y \in \mathfrak{g}\}.$$

Proof. Let

$$\begin{aligned} \mathfrak{c}_1 &= \{X \in \mathfrak{g} \,|\, (X, X) = 0\}, \\ \mathfrak{c}_2 &= \{X \in \mathfrak{g} \,|\, (X, Y) = 0, \quad \text{for all } Y \in \mathfrak{g}\}. \end{aligned}$$

If $A \in \mathfrak{c}_2$ then $(A, Y) = 0$ for all $Y \in \mathfrak{g}$. If $X \in \mathfrak{g}_0$, then

$$([A, X], Y) = (A, [X, Y]) = 0, \quad \text{for all } Y \in \mathfrak{g}_0.$$

Now the Killing form of $\mathfrak{g}_0$ is the restriction of the Killing form of $\mathfrak{g}$ to $\mathfrak{g}_0$ and $\mathfrak{g}_0$ is semisimple, so the Killing form of $\mathfrak{g}_0$ is non-degenerate and $[A, X] = 0$. Thus $A \in \mathfrak{c}$. This proves that $\mathfrak{c}_2 \subseteq \mathfrak{c}$.

If $A \in \mathfrak{c}$, then

$$\operatorname{ad} A \operatorname{ad} X\, Y = 0, \quad \text{for all } X, Y \in \mathfrak{g},$$

so $(A, X) = 0$ for all $X \in \mathfrak{g}$. This proves that $\mathfrak{c} \subset \mathfrak{c}_2$. Hence $\mathfrak{c} = \mathfrak{c}_2$.

Obviously, $\mathfrak{c} = \mathfrak{c}_2 \subset \mathfrak{c}_1$. Let $X \in \mathfrak{c}_1$ and write

$$X = Z + Y, \quad Z \in \mathfrak{c}, \quad Y \in \mathfrak{g}_0,$$

then

$$0 = (X, X) = (Z, Z) + 2(Z, Y) + (Y, Y).$$

From $\mathfrak{c} \subset \mathfrak{c}_1$, it follows that $(Z, Z) = 0$; from $\mathfrak{c} = \mathfrak{c}_2$, it follows that $(Z, Y) = 0$. Hence $(Y, Y) = 0$, but $\mathfrak{g}_0$ is semisimple, so $Y = 0$. Consequently $X = Z \in \mathfrak{c}$. This proves that $\mathfrak{c}_1 \subset \mathfrak{c}$.

THEOREM 7. *A Lie algebra is compact and semisimple iff the Killing form is negative-definite.*

Proof. If the Killing form of $\mathfrak{g}$ is negative-definite, then it is non-degenerate, so $\mathfrak{g}$ is semisimple. Since the Killing form is invariant, $\mathfrak{g}$ is compact.

Conversely, let $\mathfrak{g}$ be a compact semisimple Lie algebra, then the Killing form of $\mathfrak{g}$ is non-degenerate. Since $\mathfrak{g}$ is compact, there is an invariant symmetric negative-definite bilinear form $B(X, Y)$ defined on it. Choose an orthonormal basis of $\mathfrak{g}$ relative to $B(X, Y)$, then all ad X ($X \in \mathfrak{g}$) are skew-symmetric with complex eigenvalues. Hence $(X, X) = \operatorname{Tr} \operatorname{ad} X \operatorname{ad} X \leqslant 0$ and (X, Y) is negative definite. The proof of Theorem 7 is now completed.

THEOREM 8. *Let* $\mathfrak{g}$ *be a compact semisimple Lie algebra, then* Aut $\mathfrak{g}$ *is a compact Lie group and* Ad $\mathfrak{g}$ *is the component of the identity of* Aut $\mathfrak{g}$, *so* Ad $\mathfrak{g}$ *is a compact connected Lie group. Moreover, the centers of* Aut $\mathfrak{g}$ *and* Ad $\mathfrak{g}$ *are* $\{e\}$.

Proof. Since an automorphism A of $\mathfrak{g}$ preserves the Killing form of $\mathfrak{g}$, i.e.

$$(AX, AY) = (X, Y), \qquad X, Y \in \mathfrak{g},$$

so Aut $\mathfrak{g}$ is a closed subgroup of the r-dimensional real orthogonal group, where dim $\mathfrak{g} = \mathfrak{r}$. Hence Aut $\mathfrak{g}$ is a compact Lie group.

The Lie algebra of Aut $\mathfrak{g}$ is the Lie algebra of derivations of $\mathfrak{g}$. Since $\mathfrak{g}$ is semisimple, all derivations of $\mathfrak{g}$ are inner, so the Lie algebra of Aut $\mathfrak{g}$ is Ad $\mathfrak{g}$. Hence the component of the identity of Aut $\mathfrak{g}$ is Ad $\mathfrak{g}$.

As in the case of a complex Lie algebra, it can be shown that the centers of Aut $\mathfrak{g}$ and Ad $\mathfrak{g}$ are $\{e\}$.

COROLLARY. Ad $\mathfrak{g}$ *is a compact connected Lie group with Lie algebra* $\mathfrak{g}$.

Proof. Since $\mathfrak{g}$ is semisimple, the Lie algebra ad $\mathfrak{g}$ of Ad $\mathfrak{g}$ is isomorphic to $\mathfrak{g}$.

THEOREM 9. *A real Lie algebra is compact iff it is the Lie algebra of a compact Lie group.*

Proof. Suppose $\mathfrak{g}$ is a compact Lie algebra, then by Lemma 6, $\mathfrak{g} = \mathfrak{c} \dotplus \mathfrak{g}_0$, where $\mathfrak{c}$ is the center of $\mathfrak{g}$ and $\mathfrak{g}_0$ is the maximal semisimple ideal of $\mathfrak{g}$. If dim $\mathfrak{c} = m$, then the toroidal group T^m is a compact connected Lie group with Lie algebra $\mathfrak{c}$. By the Corollary to Theorem 8, Ad $\mathfrak{g}_0$ is a compact connected Lie group with Lie algebra $\mathfrak{g}_0$. Hence $T^m \times \operatorname{Ad} \mathfrak{g}_0$ (direct sum as Lie groups) is a compact connected Lie group with Lie algebra $\mathfrak{g}$.

Conversely let $\mathfrak{g}$ be the Lie algebra of a compact Lie group $\mathfrak{G}$. Let $\varphi(X, Y)$ be a negative-definite bilinear form on $\mathfrak{g}$. For any $\sigma \in \mathfrak{G}$, the mapping

$$\alpha_\sigma : \tau \to \sigma\tau\sigma^{-1}$$

is an inner automorphism of $\mathfrak{G}$; the differential $d\alpha_\sigma$ of this mapping is an isomorphism of $\mathfrak{g}$. Let

$$\varphi_\sigma(X, Y) = \varphi(d\alpha_\sigma(X), d\alpha_\sigma(Y)),$$

then φ_σ is a negative-definite bilinear form on $\mathfrak{g}$ and for fixed X and Y, it is a continuous function of σ. Let

$$\psi(X, Y) = \int_{\sigma \in \mathfrak{G}} \varphi_\sigma(X, Y)\, d\sigma,$$

then $\psi(X, Y)$ is a negative-definite bilinear form on $\mathfrak{g}$ and

$$\psi(d\alpha_\sigma(X), d\alpha_\sigma(Y)) = \psi(X, Y), \quad \text{for all } \sigma \in \mathfrak{G}.$$

Therefore,

$$\psi(\operatorname{ad} AX, Y) + \psi(X, \operatorname{ad} AY) = 0, \quad \text{for all } A \in \mathfrak{g}.$$

This proves the invariance of ψ, so $\mathfrak{g}$ is compact.

15.3. Compact real forms of complex semisimple Lie algebras

THEOREM 10. *Let $\mathfrak{g}$ be a complex semisimple Lie algebra with real form $\mathfrak{g}_0$. If σ is the corresponding semi-involution, then $(X, \sigma(Y))$ $(X, Y \in \mathfrak{g})$ is a non-degenerate Hermitian bilinear form on $\mathfrak{g}$. Moreover, $\mathfrak{g}_0$ is compact iff $(X, \sigma(Y))$ is negative-definite.*

Proof. Since σ is a semi-involution, $(X, \sigma(Y))$ is a non-degenerate Hermitian bilinear form on $\mathfrak{g}$. If $(X, \sigma(Y))$ is negative-definite, (Z, Z) $(Z \in \mathfrak{g}_0)$ is also negative-definite, so $\mathfrak{g}_0$ is compact.

Conversely, if $\mathfrak{g}_0$ is compact, then (Z, Z) $(Z \in \mathfrak{g}_0)$ is negative-definite. For any $X \in \mathfrak{g}$, let $X = Y+iZ$, $Y, Z \in \mathfrak{g}_0$; if $X \neq 0$, then

$$(X, \sigma(X)) = (Y+iZ, Y-iZ) = (Y, Y)+(Z, Z) < 0.$$

Hence $(X, \sigma(Y))$ is negative-definite.

THEOREM 11.† *Let $\mathfrak{g}$ be a complex semisimple Lie algebra with real form $\mathfrak{g}_0$ and σ be the corresponding semi-involution, then*

(1) *There exists a Cartan subalgebra $\mathfrak{h}$ of $\mathfrak{g}$ such that $\sigma(\mathfrak{h}) = \mathfrak{h}$, $\sigma(\mathfrak{h}_0) = \mathfrak{h}_0$ and σ induces an orthogonal transformation on the roots of $\mathfrak{h}_0$.*

(2) *If $\mathfrak{g}_0$ is compact, then $\sigma(H) = -H$ $(\forall H \in \mathfrak{h}_0)$, and there exists a Weyl basis $\{E_\alpha\}$* of $\mathfrak{g}$ *such that $\sigma(E_\alpha) = -E_{-\alpha}$ for all $\alpha \in \Sigma$.*

(3) *Conversely, if $\{E_\alpha\}$ is a Weyl basis of $\mathfrak{g}$, then a semi-involution σ can be defined by:*

$$\sigma(H) = -H, \qquad \text{for all } H \in \mathfrak{h}_0,$$
$$\sigma(E_\alpha) = -E_{-\alpha}, \qquad \text{for all } \alpha \in \Sigma.$$

The real form corresponding to σ

$$\mathfrak{g}_u = \left\{\sum_1^n a_i H_i + \sum_{\alpha\in\Sigma} \sigma_\alpha E_\alpha \,|\, a_i\text{'s purely imaginariy}, \bar{\sigma}_\alpha = -\sigma_{-\alpha}\right\}$$

is compact.

Proof. (1) Let

$$|\lambda I - \operatorname{ad} X| = \lambda^r + \varphi_1(X)\lambda^{r-1} + \ldots + \varphi_n(X)\lambda^n$$

be the Killing polynomial of $X \in \mathfrak{g}$ and $\varphi_n(X) \not\equiv 0$. Since $\varphi_n(X)$ is not identically zero on $\mathfrak{g}$, it is not identically zero on the real form $\mathfrak{g}_0$ of $\mathfrak{g}$, so $\mathfrak{g}_0$ contains a regular element X_0. Since $\mathfrak{g}$ is semisimple,

$$\mathfrak{h} = \{H \in \mathfrak{g} \,|\, [H, X_0] = 0\}$$

is a Cartan subalgebra of $\mathfrak{g}$. If $H \in \mathfrak{h}$, then

$$[\sigma(H), X_0] = [\sigma(H), \sigma(X_0)] = \sigma([H, X_0]) = 0;$$

this proves that $\sigma(H) \in \mathfrak{h}$, i.e. $\sigma(\mathfrak{h}) = \mathfrak{h}$.

† See footnote for Theorem 6 in Section 5.3.

Let α be a root of $\mathfrak{h}$ and

$$\tilde{\alpha}(H) = \overline{\alpha(\sigma(H))}, \quad \text{for all } H \in \mathfrak{h},$$

then $\tilde{\alpha}$ is a linear function defined on $\mathfrak{h}$. We show that $\tilde{\alpha}$ is also a root and $\sigma(\mathfrak{g}^\alpha) = \mathfrak{g}^{\tilde{\alpha}}$. In fact, if $E_\alpha \in \mathfrak{g}^\alpha$, then

$$[H, E_\alpha] = \alpha(H)\, E_\alpha,$$

so

$$[\sigma(H), \sigma(E_\alpha)] = \overline{\alpha(H)}\, \sigma(E_\alpha) = \tilde{\alpha}(\sigma(H))\, \sigma(E_\alpha).$$

Since $\sigma(\mathfrak{h}) = \mathfrak{h}$,

$$[H, \sigma(E_\alpha)] = \tilde{\alpha}(H)\, \sigma(E_\alpha).$$

This proves that $\tilde{\alpha}$ is also a root and $\sigma(\mathfrak{g}^\alpha) = \mathfrak{g}^{\tilde{\alpha}}$.

Finally, from

$$(\sigma(H_\alpha), H) = \overline{(H_\alpha, \sigma(H))} = \overline{\alpha(\sigma(H))} = \tilde{\alpha}(H) = (H_{\tilde{\alpha}}, H),$$

it follows that $\sigma(H_\alpha) = H_{\tilde{\alpha}}$, for all $H \in \mathfrak{h}$. Hence σ induces a permutation on the roots, so $\sigma(\mathfrak{h}_0) = \mathfrak{h}_0$ and σ induces an orthogonal transformation on $\mathfrak{h}_0$.

Before we continue to prove Theorem 11, we prove the following:

LEMMA 1. *Let $\mathfrak{g}$ be a complex semisimple Lie algebra with a Cartan subalgebra $\mathfrak{h}$. Suppose φ is an orthogonal transformation on $\mathfrak{h}_0$ that induces a permutation $\varphi(H_\alpha) = H_{\tilde{\alpha}}$ $(\alpha \in \Sigma)$ on the roots and $\varphi^2 = 1$. If every $\alpha \in \Sigma$ is associated with a non-zero complex number ϱ_α and $\{E_\alpha\}$ $(\alpha \in \Sigma)$ is a Weyl basis of $\mathfrak{g}$, then φ can be extended to a semi-involution σ of $\mathfrak{g}$ satisfying $\sigma(E_\alpha) = \varrho_\alpha E_{\tilde{\alpha}}$* iff

$$\bar{\varrho}_\alpha \varrho_{\tilde{\alpha}} = 1, \tag{15.3}$$

$$\varrho_\alpha \varrho_{-\alpha} = 1, \tag{15.4}$$

$$\varrho_{\alpha+\beta} N_{\alpha\beta} = \varrho_\alpha \varrho_\beta N_{\tilde{\alpha},\tilde{\beta}}. \tag{15.5}$$

Proof. Since $\dim_R \mathfrak{h}_0 = \dim_c \mathfrak{h}$, $\mathfrak{h}_0$ and all E_α $(\alpha \in \Sigma)$ generate $\mathfrak{g}$, so φ can be extended to a semi-linear mapping σ such that

$$\sigma(H) = \varphi(H), \quad \text{for all } H \in \mathfrak{h}_0,$$

$$\sigma(E_\alpha) = \varrho_\alpha E_{\tilde{\alpha}}, \quad \text{for all } \alpha \in \Sigma.$$

Now consider the condition $\sigma^2 = 1$. For any $H \in \mathfrak{h}$, write $H = H_1 + iH_2$. Since $\varphi^2 = 1$,

$$\begin{aligned}\sigma^2(H) &= \sigma(\sigma(H_1) - i\sigma(H_2)) = \sigma(\varphi(H_1) - i\varphi(H_2)) \\ &= \varphi^2(H_1) + i\varphi^2(H_2) = H_1 + iH_2 = H,\end{aligned}$$

i.e. $\sigma^2 = 1$ on $\mathfrak{h}$. Now

$$\sigma^2(E_\alpha) = \sigma(\varrho_\alpha E_{\tilde{\alpha}}) = \bar{\varrho}_\alpha \varrho_{\tilde{\alpha}} E_{\tilde{\tilde{\alpha}}};$$

since $\varphi^2 = 1$ on $\mathfrak{h}_0$, $\varphi^2(H_\alpha) = H_{\tilde{\tilde{\alpha}}} = H_\alpha$, i.e. $\tilde{\tilde{\alpha}} = \alpha$. Therefore, $\sigma^2 = 1$ iff

$$\bar{\varrho}_\alpha \varrho_{\tilde{\alpha}} = 1, \quad \text{for all } \alpha \in \Sigma.$$

Now suppose $\sigma([X, Y]) = [\sigma(X), \sigma(Y)]$ for all $Y, X \in \mathfrak{g}$. For any $H, H' \in \mathfrak{h}$, since $\sigma(H)$, $\sigma(H)' \in \mathfrak{h}$,

$$[\sigma(H), \sigma(H')] = 0 = \sigma(0) = \sigma([H, H']).$$

Moreover,

$$\sigma([E_\alpha, E_{-\alpha}]) = \sigma(H_\alpha) = H_{\tilde{\alpha}},$$

and

$$[\sigma(E_\alpha), \sigma(E_{-\alpha})] = [\varrho_\alpha E_{\tilde{\alpha}}, \varrho_{-\alpha} E_{-\tilde{\alpha}}] = \varrho_\alpha \varrho_{-\alpha} H_{\tilde{\alpha}},$$

so

$$\sigma([E_\alpha, E_{-\alpha}]) = [\sigma(E_\alpha), \sigma(E_{-\tilde{\alpha}})] \text{ iff}$$

$$\varrho_\alpha \varrho_{-\alpha} = 1.$$

Now

$$\sigma([H, E_\alpha]) = \sigma(\alpha(H) E_\alpha) = \overline{\alpha(H)}\, \sigma(E_\alpha) = \overline{\alpha(H)}\, \varrho_\alpha E_{\tilde{\alpha}}$$

and

$$[\sigma(H), \sigma(E_\alpha)] = [\sigma(H), \varrho_\alpha E_{\tilde{\alpha}}] = \varrho_\alpha [\sigma(H), E_\alpha]$$
$$= \tilde{\alpha}(\sigma(H))\, \varrho_\alpha E_{\tilde{\alpha}},$$

thus

$$\sigma([H, E_\alpha]) = [\sigma(H), (E_\alpha)] \text{ iff}$$

$$\overline{\alpha(H)} = \tilde{\alpha}(\sigma(H))$$

i.e.

$$\tilde{\alpha}(H) = \overline{\alpha(\sigma(H))}.$$

Since

$$\sigma(H_\alpha) = H_{\tilde{\alpha}},$$

$$\tilde{\alpha}(H) = (H, H_{\tilde{\alpha}}) = (H, \sigma(H_\alpha)) = \overline{(\sigma(H), H_\alpha)} = \overline{\alpha(\sigma(H))}.$$

Finally,

$$\sigma([E_\alpha, E_\beta]) = \sigma(N_{\alpha\beta} E_{\alpha+\beta}) = N_{\alpha\beta} \varrho_{\alpha+\beta} E_{\tilde{\alpha}+\tilde{\beta}}$$

and

$$[\sigma(E_\alpha), \sigma(E_\beta)] = [\varrho_\alpha E_{\tilde{\alpha}}, \varrho_\beta E_{\tilde{\beta}}] = \varrho_\alpha \varrho_\beta N_{\tilde{\alpha}\beta}\, E_{\tilde{\alpha}+\tilde{\beta}},$$

so

$$\sigma([E_\alpha, E_\beta]) = [\sigma(E_\alpha), \sigma(E_\beta)] \text{ iff}$$

$$\varrho_{\alpha+\beta} N_{\alpha\beta} = \varrho_\alpha \varrho_\beta N_{\tilde{\alpha}\tilde{\beta}}.$$

The proof of Lemma 1 is now completed.

(2) Suppose $\mathfrak{g}_0$ is compact, i.e. $(X, \sigma(X)) < 0$ for any $X \neq 0$. Since σ induces an orthogonal transformation on $\mathfrak{h}_0$ and $\sigma^2 = 1$, we have $\mathfrak{h}_0 = \mathfrak{h}_0^+ \dotplus \mathfrak{h}_0^-$, where

$$\mathfrak{h}_0^\pm = \{H_0 \in \mathfrak{h}_0 \,|\, \sigma(H_0) = \pm H_0\}.$$

If $H_0 \in \mathfrak{h}_0^+$, then

$$(H_0, \sigma(H_0)) = (H_0, H_0) \geqslant 0;$$

since $(X, \sigma(X))$ is negative-definite on $\mathfrak{g}$, $H_0 = 0$. Hence $\mathfrak{h}_0 = \mathfrak{h}_0^-$, i.e. $\sigma(H_0) = -H_0$ for all $H_0 \in \mathfrak{h}_0$. From $\sigma(H_\alpha) = -H_\alpha = H_{-\alpha}$, it follows that $\tilde{\alpha} = -\alpha$.

Choose a Weyl basis $\{E_\alpha\}$ and set $\sigma(E_\alpha) = \varrho_\alpha E_{-\alpha}$. Since $N_{\alpha,\beta} = -N_{-\alpha,-\beta} = -N_{\bar{\alpha},\bar{\beta}}$, by Lemma 1, we have

$$\bar{\varrho}_\alpha \varrho_{-\alpha} = \varrho_\alpha \varrho_{-\alpha} = 1,$$

and

$$\varrho_{\alpha+\beta} = -\varrho_\alpha \varrho_\beta,$$

From the first of these equations, it follows that any ϱ_α is real. Now

$$\big(E_\alpha, \sigma(E_\alpha)\big) = \varrho_\alpha(E_\alpha, E_{-\alpha}) = \varrho_\alpha < 0,$$

so $\varrho_\alpha < 0$. Let $\varrho_{\alpha'} = (-\varrho_\alpha)^{1/2}$ (take the positive square root) and $E'_\alpha = \varrho'_\alpha E_\alpha$, then

$$(E'_\alpha, E'_{-\alpha}) = \varrho'_\alpha \varrho'_{-\alpha} = 1$$

and

$$[E'_\alpha, E'_\beta] = \varrho'_\alpha \varrho'_\beta N_{\alpha\beta} E_{\alpha+\beta} = N'_{\alpha\beta} E'_{\alpha+\beta}.$$

Hence

$$\begin{aligned} N'_{\alpha\beta} &= \frac{\varrho'_\alpha \varrho'_\beta}{\varrho'_{\alpha+\beta}} N_{\alpha\beta} = \frac{(-\varrho_\alpha)^{-1/2}(-\varrho_\beta)^{-1/2}}{(-\varrho_{\alpha+\beta})^{-1/2}} N_{\alpha\beta} \\ &= N_{\alpha\beta} = -N_{-\alpha-\beta} = -N'_{-\alpha-\beta}, \end{aligned}$$

i.e. $\{E'_\alpha\}$ $(\alpha \in \Sigma)$ also forms a Weyl basis. We have

$$\begin{aligned} \sigma(E'_\alpha) &= \varrho'_\alpha \sigma(E_\alpha) = \varrho'_\alpha \varrho_\alpha E_{-\alpha} = \varrho'_\alpha \varrho_\alpha (\varrho'_{-\alpha})^{-1} E'_{-\alpha} \\ &= (-\varrho_\alpha)^{1/2} \varrho_\alpha (-\varrho_{-\alpha})^{1/2} E'_{-\alpha} = -E'_{-\alpha}. \end{aligned}$$

This proves (2).

(3) By Lemma 1, σ is a semi-involution of $\mathfrak{g}$. Let

$$\mathfrak{g}_u = \{X \in \mathfrak{g} \mid \sigma(X) = X\},$$

then $\mathfrak{g}_u$ is a real form of $\mathfrak{g}$. It is easy to show that $\sqrt{-1}\, H_{\alpha_1}, \ldots, \sqrt{-1}\, H_{\alpha_n}$, $E_\alpha - E_{-\alpha}$ and $\sqrt{-1}\,(E_\alpha + E_{-\alpha})$ $(\alpha \in \Sigma_+)$ form a basis of $\mathfrak{g}_u$ (over the real numbers). We need to show that $\mathfrak{g}_u$ is a compact Lie algebra. Let

$$X = \sum_1^n x_i \sqrt{-1}\, H_{\alpha_i} + \sum_{\alpha\in\Sigma} \sigma_\alpha E_\alpha, \quad x_i'\text{'s are real}, \ \bar{\sigma}_\alpha = -\sigma_{-\alpha},$$

then

$$\begin{aligned} (X, X) &= -\left(\sum_1^n x_i H_{\alpha_i}, \sum_1^n x_i H_{\alpha_i}\right) + \sum_{\alpha\in\Sigma} \sigma_\alpha \sigma_{-\alpha} \\ &= -\left(\sum_1^n x_i H_{\alpha_i} + \sum_1^n x_i H_{\alpha_i}\right) - \sum_{\alpha\in\Sigma} \sigma_\alpha \bar{\sigma}_\alpha. \end{aligned}$$

Hence (X, X) is negative definite. This proves that $\mathfrak{g}_u$ is a compact semisimple Lie algebra. The proof of Theorem 10 is now completed.

$\mathfrak{g}_u$ is said to be the unitary restriction of $\mathfrak{g}$.

To determine all the compact semisimple Lie algebras, it suffices to determine all the compact real forms of all the complex semisimple Lie algebras. By Theorem 10, these compact real forms can be obtained by the method in (3). We also have the following:

THEOREM 12. *Any two compact real forms of a complex semisimple Lie algebra are isomorphic. Moreover, there exists an inner automorphism of* $\mathfrak{g}$ *that transforms one of the real forms to the other.*

Proof. By conjugacy of Cartan subalgebras, we may assume that the two compact real forms are determined by the same Cartan subalgebra $\mathfrak{h}$. Let $\{E_\alpha\}$ and $\{E'_\alpha\}$ $(\alpha \in \Sigma)$ be parts of the two Weyl basis that give rise to these two compact real forms; suppose

$$[E_\alpha, E_\beta] = N_{\alpha\beta}E_{\alpha+\beta}, \ [E'_\alpha, E'_\beta] = N'_{\alpha\beta}E'_{\alpha+\beta}.$$

Since $N^2_{\alpha\beta} = N'^2_{\alpha\beta} = R_{\alpha\beta} > 0$, $N_{\alpha\beta} = \pm N'_{\alpha\beta}$. By Theorem 6 of Chapter 5, there exists an isomorphism σ of $\mathfrak{g}$ such that

$$\sigma(H) = H, \qquad H \in \mathfrak{h}$$
$$\sigma(E_\alpha) = \pm E'_\alpha, \qquad \alpha \in \Sigma.$$

By Lemma 2 of Chapter 8, σ is an inner automorphism of the form exp ad H, where $H \in \mathfrak{h}$. Theorem 12 follows from this.

Now from Theorem 11, it follows that the compact Lie algebra $\mathfrak{g}_u$ is simple iff $[\mathfrak{g}_u]$ is simple. From Theorem 11 and Theorem 12, it follows that to determine all the compact simple Lie algebras, it suffices to determine a compact real form for each of the complex simple Lie algebras; such a compact real form can be obtained by the method in (3) of Theorem 11 for a fixed Cartan subalgebra.

We now determine compact real forms for the classical Lie algebras.

(A_n) A_n consists of all $(n+1)\times(n+1)$ matrices with zero trace and

$$\mathfrak{h}_0 = H(\lambda_1, \ldots, \lambda_{n+1}) = \left\{\begin{bmatrix} \lambda_1 & & \\ & \ddots & \\ & & \lambda_{n+1}\end{bmatrix} \middle|\, \Sigma\lambda_i = 0\right\}$$

is a Cartan subalgebra of A_n. The roots of A_n are

$$\lambda_i - \lambda_k, \qquad 1 \leq i, k \leq n+1, \quad i \neq k.$$

The root $\lambda_i - \lambda_k$ can be embedded into $\mathfrak{h}$ and

$$H_{\lambda_i-\lambda_k} = \frac{1}{2(n+1)}(E_{ii}-E_{kk});$$

E_{ik} is a root vector corresponding to the root $\lambda_i - \lambda_k$ and

$$[E_{ik}, E_{ki}] = E_{ii}-E_{kk} = 2(n+1)H_{\lambda_i-\lambda_k}.$$

If

$$E_{\lambda_i-\lambda_k} = \frac{1}{\sqrt{[2(n+1)]}}E_{ik},$$

then

$$[E_{\lambda_i-\lambda_k}, E_{-(\lambda_i-\lambda_k)}] = H_{\lambda_i-\lambda_k}.$$

It is easy to show that $\{E_{\lambda_i-\lambda_k} | 1 \leq i,\ k \leq n+1,\ i \neq k\}$ form part of a Weyl basis of A_n. Hence the compact real form $(A_n)_u$ of A_n is the set of all real linear combinations of

$$\sqrt{-1}\, H_{\lambda_i-\lambda_k}, \quad E_{ik}-E_{ki}, \quad \sqrt{-1}\,(E_{ik}+E_{ki}), \qquad 1 \leq i,\ k \leq n+1,\ i \neq k.$$

Thus $(A_n)_u$ is the set of zero trace skew-symmetric Hermitian matrices.

(B_n) Let

$$\mathfrak{h} = \left\{ \begin{bmatrix} 0 & & & & & & \\ & \lambda_1 & & & & & \\ & & \ddots & & & & \\ & & & \lambda_n & & & \\ & & & & -\lambda_1 & & \\ & & & & & \ddots & \\ & & & & & & -\lambda_n \end{bmatrix} \middle| \lambda_1, \ldots, \lambda_n \in C \right\}.$$

The roots of $\mathfrak{h}$ are

$$\pm\lambda_i \pm \lambda_k \quad (i < k), \qquad \pm\lambda_i,\ i, k = 1, \ldots, n.$$

Embedding these roots into $\mathfrak{h}$, we have

$$H_{\pm\lambda_i\pm\lambda_k} = \frac{1}{4n-2}(\pm H_i \pm H_k), \qquad H_{\pm\lambda_i} = \frac{1}{4n-2}(\pm H_i).$$

The vectors

$$E_{\lambda_i-\lambda_k} = \begin{bmatrix} 0 & & \\ & E_{ik} & \\ & & -E_{ki} \end{bmatrix}, \quad E_{-\lambda_i+\lambda_k} = \begin{bmatrix} 0 & & \\ & E_{ki} & \\ & & -E_{ik} \end{bmatrix}, \qquad i < k,$$

$$E_{\lambda_i+\lambda_k} = \begin{bmatrix} 0 & & \\ & 0 & E_{ik}-E_{ki} \\ & & 0 \end{bmatrix}, \quad E_{-\lambda_i-\lambda_k} = \begin{bmatrix} 0 & & \\ & 0 & \\ & -E_{ik}+E_{ki} & 0 \end{bmatrix}, \qquad i < k$$

$$E_{\lambda_i} = \begin{bmatrix} 0 & 0 & e_i \\ -e_i' & 0 & 0 \\ 0 & 0 & 0 \end{bmatrix}, \qquad E_{-\lambda_i} = \begin{bmatrix} 0 & -e_i & 0 \\ 0 & 0 & 0 \\ e_i' & 0 & 0 \end{bmatrix}$$

are weight vectors. We have

$$[E_{\lambda_i-\lambda_k}, E_{-(\lambda_i-\lambda_k)}] = H_i - H_k,$$
$$[E_{\lambda_i+\lambda_k}, E_{-(\lambda_i+\lambda_k)}] = H_i + H_k,$$
$$[E_{\lambda_i}, E_{-\lambda_i}] = H_i.$$

If

$$\tilde{E}_{\lambda_i-\lambda_k} = \frac{1}{\sqrt{(4n-2)}} E_{\lambda_i-\lambda_k}, \quad \tilde{E}_{-(\lambda_i-\lambda_k)} = \frac{1}{\sqrt{(4n-2)}} E_{-(\lambda_i-\lambda_k)},$$

$$\tilde{E}_{\lambda_i+\lambda_k} = \frac{1}{\sqrt{(4n-2)}} E_{\lambda_i+\lambda_k}, \quad \tilde{E}_{-(\lambda_i+\lambda_k)} = \frac{1}{\sqrt{(4n-2)}} E_{-(\lambda_i-\lambda_k)},$$

$$\tilde{E}_{\lambda_i} = \frac{1}{\sqrt{(4n-2)}} E_{\lambda_i}, \quad \tilde{E}_{-\lambda_i} = \frac{1}{\sqrt{(4n-2)}} E_{-\lambda_i},$$

then $\{\tilde{E}_{\pm\lambda_i\pm\lambda_k}, \tilde{E}_{\pm\lambda_i} | \pm\lambda_i\pm\lambda_k, \pm\lambda_i, i<k, \quad i, k = 1, \ldots, n\}$ is part of a Weyl basis. The compact real form of B_n determined by this Weyl basis is the set of all skew-symmetric Hermitian matrices in B_n.

Similarly, the set of skew-symmetric Hermitian matrices in C_n is a compact real form of C_n and the set of skew-symmetric Hermitian matrices in D_n is a compact real form of D_n.

15.4. Roots and weights of compact semisimple Lie algebras

Since any compact semisimple Lie algebra is the unitary restriction of its complex extension, it suffices to consider the unitary restriction $\mathfrak{g}_u$ of a semisimple Lie algebra $\mathfrak{g}$.

Let $\{X_1, \ldots, X_r\}$ be a basis of $\mathfrak{g}_u$ and ϱ be a representation of $\mathfrak{g}_u$, then

$$\sum_1^r c_i X_i \to \sum_1^r c_i \varrho(X_i) \quad (c_i\text{'s are complex})$$

is a representation of $\mathfrak{g}$. Conversely, the restriction of any representation ϱ of $\mathfrak{g}$ to $\mathfrak{g}_u$ is a representation of $\mathfrak{g}_u$. Since $\mathfrak{g}$ and $\mathfrak{g}_u$ have the same basis, ϱ is irreducible iff $\varrho|_{\mathfrak{g}_u}$ is irreducible and ϱ is completely reducible iff $\varrho|_{\mathfrak{g}_u}$ is completely reducible. Therefore, we have the following:

THEOREM 13. *Any representation of a compact semisimple Lie algebra is completely reducible.*

We now give another proof (integral method) of Theorem 13. Let $\mathfrak{g}_u$ be a compact semisimple Lie algebra, then the simply connected Lie group $\mathfrak{G}_u$ with Lie algebra $\mathfrak{g}_u$ is compact. An arbitrary representation ϱ is the differential $d\Phi$ of some representation Φ of $\mathfrak{G}_u$. Conversely, the differential $d\Phi$ of any representation Φ of $\mathfrak{G}_u$ is a representation of $\mathfrak{g}_u$. Obviously, ϱ is irreducible iff Φ is irreducible and ϱ is completely reducible iff Φ is completely reducible. Since any representation of a compact group is completely reducible, an arbitrary representation of $\mathfrak{g}_u$ is completely reducible.

Remark. With this proof, it can be shown that an arbitrary representation of a complex semisimple Lie algebra is completely reducible. This is the original proof of H. Weyl.

Let $\mathfrak{h}$ be a Cartan subalgebra of $\mathfrak{g}$ and $\{\alpha_1, \ldots, \alpha_n\}$ be a fundamental system of roots, then $\{\sqrt{-1}H_{\alpha_1}, \ldots, \sqrt{-1}H_{\alpha_n}\}$ is a basis of $\mathfrak{h}_u = \mathfrak{h} \cap \mathfrak{g}_u$. Clarly, $\mathfrak{h}_u$ is a maximal abelian subalgebra of $\mathfrak{g}_u$; it is said to be a Cartan subalgebra of $\mathfrak{g}_u$.

Let ω be a weight of a representation ϱ of $\mathfrak{g}$ with representation space V and $x\,(\neq 0) \in V$ be of weight ω, i.e.

$$\varrho(H)x = \omega(H)x, \quad \forall H \in \mathfrak{h},$$

then

$$\varrho(H')x = \omega(H')x, \quad \forall H' \in \mathfrak{h}_u.$$

Hence ω (as a linear function on $\mathfrak{h}_u$) is a weight of $\varrho\,|_{\mathfrak{g}_u}$.

LEMMA 2. *The weights of* $\mathfrak{g}_u$ *are purely imaginary linear functions on* $\mathfrak{h}_u$.

Proof. Since any weight is a linear combination of the roots of $\mathfrak{g}_u$ with rational coefficients, it suffices to consider the roots of $\mathfrak{g}_u$. It is known that the Killing form of $\mathfrak{g}_u$ is negative-definite and any ad A ($A \in \mathfrak{g}_u$) is a skew-symmetric Hermitian transformation on $\mathfrak{g}_u$, so the eigenvalues of ad A are purely imaginary. In particular, the eigenvalues of any ad $H'(H' \in \mathfrak{h})$ are purely imaginary. This completes the proof.

Let ω be a weight, so there exists $H_\omega \in \mathfrak{h}$ such that

$$(H, H_\omega) = \omega(H), \quad \text{for all } H \in \mathfrak{h}.$$

In particular,

$$\omega(H') = (H', H_\omega), \quad \text{for all } H' \in \mathfrak{h}_u.$$

Let $H_\omega = 2\pi i H_{\tilde{\omega}}$, then $H_{\tilde{\omega}} \in \mathfrak{h}_u$ and

$$\omega(H') = 2\pi i(H', H_{\tilde{\omega}}), \quad \text{for all } H' \in \mathfrak{h}_u.$$

Let $\tilde{\omega}(H') = (H', H_{\tilde{\omega}})$ and call $\tilde{\omega}$ a weight of $\mathfrak{h}_u$. In particular, for any root α, there exists $H_{\tilde{\alpha}} \in \mathfrak{h}_u$ such that

$$\alpha(H') = 2\pi i(H', H_{\tilde{\alpha}}), \quad \text{for any } H' \in \mathfrak{h}_u.$$

Let

$$\tilde{\alpha}(H') = (H', H_{\tilde{\alpha}})$$

and call $\tilde{\alpha}$ a root of $\mathfrak{g}_u$.

Suppose $\{E_{\tilde{\alpha}} \mid \alpha \in \Sigma\}$ form part of a Weyl basis of $\mathfrak{g}$, then the structure formulas of $[\mathfrak{g}_u]$ can be written as

$$\begin{aligned}
[H_1', H_2'] &= 0, \quad H_1', \; H_2' \in \mathfrak{h}_u,\\
[H', E_{\tilde{\alpha}}] &= 2\pi i \tilde{\alpha}(H') E_{\tilde{\alpha}},\\
[E_{\tilde{\alpha}}, E_{\tilde{\alpha}}] &= 2\pi i H_{\tilde{\alpha}},\\
[E_{\tilde{\alpha}}, E_{\tilde{\beta}}] &= N_{\tilde{\alpha},\tilde{\beta}} E_{\tilde{\alpha}+\tilde{\beta}}, \quad N_{\tilde{\alpha},\tilde{\beta}} = 0 \quad \text{iff } \alpha+\beta \notin \Sigma.
\end{aligned}$$

If σ is the semi-involution corresponding to $\mathfrak{g}_u$, then

$$\sigma(E_{\tilde{\alpha}}) = -E_{-\tilde{\alpha}}.$$

It is easy to see that $\mathfrak{h}_u$ is the set of linear combinations of $H_{\tilde{\alpha}}$ ($\alpha \in \Sigma$) with real coefficients. The restriction of the Killing form of $\mathfrak{g}_u$ to $\mathfrak{h}_u$ gives rise to a negative-definite bilinear form on $\mathfrak{h}_u$. Define

$$(\tilde{\alpha}, \tilde{\beta}) = -(H_{\tilde{\alpha}}, H_{\tilde{\beta}}),$$

then the set of roots of $\mathfrak{g}_u$ form a σ-system. With respect to an ordering on $\mathfrak{h}_u$, the prime roots of $\mathfrak{g}_u$ form a π system. Obviously, $\mathfrak{g}$ and $\mathfrak{g}_u$ have similar systems of roots and fundamental systems of roots and

$$(\alpha, \beta) = (2\pi)^2 (\tilde{\alpha}, \tilde{\beta}).$$

Hence the Dynkin diagram of $\mathfrak{g}$ can be considered as the Dynkin diagram of $\mathfrak{g}_u$.

From results in Chapter 10, we have the following:

THEOREM 14. (1) *Let ϱ be an irreducible representation of $\mathfrak{g}_u$ with representation space V. Then V is generated by weight vectors and there exists a unique weight $\tilde{\omega}_0$ (relative to some fundamental system of roots), called the highest weight, such that any weight of ϱ is of the form*

$$\tilde{\omega}_0 - \tilde{\alpha}_{i_1} - \ldots - \tilde{\alpha}_{i_k}, \quad \tilde{\alpha}_{i_1}, \ldots, \tilde{\alpha}_{i_k} \in \Pi,$$

and dim $V_{\tilde{\omega}_0} = 1$. *If $\tilde{\omega}$ is a weight and $\tilde{\alpha}$ is a root, then*

$$\frac{2(\tilde{\omega}, \tilde{\alpha})}{(\tilde{\alpha}, \tilde{\alpha})} \text{ is an integer, } \tilde{\omega} - \frac{2(\tilde{\omega}, \tilde{\alpha})}{(\tilde{\alpha}, \tilde{\alpha})} \tilde{\alpha} \text{ is also a weight}$$

vnd these two weights have the same multiplicity.

(2) *Any irreducible representation of $\mathfrak{g}_u$ is determined uniquely by the highest weight.*

(3) *A real linear function $\tilde{\omega}$ defined on $\mathfrak{h}_u$ is the highest weight of some irreducible representation of $\mathfrak{g}_u$ iff all*

$$\frac{2(\tilde{\omega}, \tilde{\alpha}_i)}{(\tilde{\alpha}_i, \tilde{\alpha}_i)}$$

are non-negative integers.

(4) *$\mathfrak{g}_u$ has n fundamental representations $\varrho_1, \ldots, \varrho_n$ with highest weights $\tilde{\omega}_1, \ldots, \tilde{\omega}_n$ respectively and*

$$\frac{2(\tilde{\omega}_i, \tilde{\alpha}_j)}{(\tilde{\alpha}_j, \tilde{\alpha}_j)} = \delta_{ij}, \qquad i, j = 1, \ldots, n.$$

15.5. Real forms of complex semisimple Lie algebras

LEMMA 3. (1) *Let $\mathfrak{g}$ be a complex Lie algebra, $\mathfrak{g}_0$ be a real form of $\mathfrak{g}$ and σ be the corresponding semi-involution. If $\mathfrak{h}$ is a subspace of $\mathfrak{g}$ and $\mathfrak{h}_0 = \mathfrak{h} \cap \mathfrak{g}_0$, then $\mathfrak{h}$ is generated by $\mathfrak{h}_0$ (over C) iff $\sigma(\mathfrak{h}) = \mathfrak{h}$. If $\sigma(\mathfrak{h}) = \mathfrak{h}$, then $\mathfrak{h}$ is the complex extension of $\mathfrak{h}_0$.*

(2) *Suppose τ is another semi-involution of $\mathfrak{g}$ satisfying $\sigma\tau = \tau\sigma$. Let $\mathfrak{g}_\tau$ be the real form corresponding to τ and*

$$\mathfrak{g}_\tau^\pm = \{X \in \mathfrak{g}_\tau \,|\, \sigma(X) = \pm X\},$$

then

$$\mathfrak{g}_\tau = \mathfrak{g}_\tau^+ + \mathfrak{g}_\tau^-, \quad \mathfrak{g}_0 = \mathfrak{g}_\tau^+ + i\mathfrak{g}_\tau^-.$$

Proof. (1) Suppose $\mathfrak{h}$ is generated by $\mathfrak{h}_0$(over C), then any $Z \in \mathfrak{h}$ can be written as $Z = X + iY$ $(X, Y \in \mathfrak{h}_0)$, so $\sigma(Z) = X - iY$ and $\sigma(\mathfrak{h}) \subseteq \mathfrak{h}$. Since $\sigma^2 = 1$, $\sigma(\mathfrak{h}) = \mathfrak{h}$. Conversely, suppose $\sigma(\mathfrak{h}) = \mathfrak{h}$ and $Z \in \mathfrak{h}$, then $X = \frac{1}{2}(Z+\sigma(Z))$ and $Y = (1/2i)(Z-\sigma(Z))$ are in $\mathfrak{h}_0 = \mathfrak{h} \cap \mathfrak{g}_0$ and $Z = X+iY$. Hence $\mathfrak{h}$ is generated by $\mathfrak{h}_0$.

If $\sigma(\mathfrak{h}) = \mathfrak{h}'$, then $\mathfrak{h}$ is clearly the complex extension of $\mathfrak{h}_0$.

(2) If $X \in \mathfrak{g}_\tau$, then $\tau(\sigma(X)) = \sigma(\tau(X)) = \sigma(X)$, so $\sigma(\mathfrak{g}_\tau) = \mathfrak{g}_\tau$. Therefore, if $\mathfrak{g}_\tau$ is considered as a real vector space, then σ is a linear transformation on $\mathfrak{g}_\tau$ and $\sigma^2 = 1$. Hence

$$\mathfrak{g}_\tau = \mathfrak{g}_\tau^+ \dot{+} \mathfrak{g}_\tau^-.$$

Now $\mathfrak{g} = \mathfrak{g}_\tau \dot{+} i\mathfrak{g}_\tau = \mathfrak{g}_\tau^+ \dot{+} \mathfrak{g}_\tau^- \dot{+} i\mathfrak{g}_\tau^+ \dot{+} i\mathfrak{g}_\tau^-$, so

$$\mathfrak{g}_0 = \mathfrak{g}_\tau^+ + i\mathfrak{g}_\tau^-.$$

THEOREM 15. *Let $\mathfrak{g}$ be a complex semisimple Lie algebra, $\mathfrak{g}_0$ be a real form of $\mathfrak{g}$ and σ be the corresponding semi-involution, then*

(1) *There is a compact real form $\mathfrak{g}_u$ of $\mathfrak{g}$ satisfying $\sigma(\mathfrak{g}_u) = \mathfrak{g}_u$. Hence σ is an automorphism of $\mathfrak{g}_u$ and $\sigma^2 = 1$ (σ is said to be an involutive automorphism).*

(2) *Let $\mathfrak{g}_u^+ = \mathfrak{g}_0 \cap \mathfrak{g}_u$, then $\mathfrak{g}_u^+$ is the set of fixed points of $\mathfrak{g}_u$. Let $\mathfrak{g}_u^- = \{X \in \mathfrak{g}_u \mid \sigma(X) = -X\}$, then $\mathfrak{g}_u^+$ is a compact subalgebra of $\mathfrak{g}_0$ and*

$$\mathfrak{g}_0 = \mathfrak{g}_u^+ + i\mathfrak{g}_u^-, \quad \mathfrak{g}_u = \mathfrak{g}_u^+ + \mathfrak{g}_u^-.$$

Proof. (1) By Theorem 12, there exists a Cartan subalgebra $\mathfrak{h}$ satisfying $\sigma(\mathfrak{h}) = \mathfrak{h}$ and σ is a permutation on the roots of $\mathfrak{g}$, i.e.

$$\sigma(H_\alpha) = H_{\tilde{\alpha}}, \quad \tilde{\alpha}(H) = \overline{\alpha(\sigma(H))}.$$

Let $\{E_\alpha \mid \alpha \in \Sigma\}$ be part of a Weyl basis of $\mathfrak{g}$. From

$$[H, E_\alpha] = \alpha(H) E_\alpha,$$

it follows that

$$[\sigma(H), \sigma(E_\alpha)] = \overline{\alpha(H)}\, \sigma(E_\alpha),$$

i.e.

$$[H, \sigma(E_\alpha)] = \tilde{\alpha}(H)\, \sigma(E_\alpha).$$

Hence $\sigma(E_\alpha) = \varrho_\alpha E_{\tilde{\alpha}}$. By Lemma 1, the ϱ_α ($\alpha \in \Sigma$) satisfying

$$\begin{aligned} \bar{\varrho}_\alpha \varrho_{\tilde{\alpha}} &= 1, \\ \varrho_\alpha \varrho_{-\alpha} &= 1, \\ \varrho_{\alpha+\beta} N_{\alpha\beta} &= \varrho_\alpha \varrho_\beta N_{\tilde{\alpha}\tilde{\beta}}. \end{aligned}$$

Define a semi-linear mapping on $\mathfrak{g}$ by:

$$\begin{aligned} \tau(H_0) &= -H_0, & H_0 &\in \mathfrak{h}_0, \\ \tau(E_\alpha) &= -|\varrho_\alpha| E_\alpha, & \alpha &\in \Sigma. \end{aligned}$$

We show that conditions (1), (2) and (3) of Lemma 1 are satisfied. We have

$$\begin{aligned} \overline{(-|\varrho_\alpha|)}(-|\varrho_{-\alpha}|) &= |\varrho_\alpha \varrho_{-\alpha}| = 1, \\ (-|\varrho_\alpha|)(-|\varrho_{-\alpha}|) &= |\varrho_\alpha \varrho_{-\alpha}| = 1. \end{aligned}$$

These conditions are (1) and (2). Since $N_{\alpha\beta} = -N_{-\alpha,-\beta}$, (3) is equivalent to

$$|\varrho_{\alpha+\beta}| = |\varrho_\alpha||\varrho_\beta|.$$

We know that

$$(H_\alpha, H_\beta) = \overline{(\sigma(H_\alpha), \sigma(H_\beta))} = \overline{(H_{\tilde{\alpha}}, H_{\tilde{\beta}})} = (H_{\tilde{\alpha}}, H_{\tilde{\beta}}),$$

i.e.

$$(\alpha, \beta) = (\tilde{\alpha}, \tilde{\beta}),$$

so $\alpha \to \tilde{\alpha}$ is an equivalent mapping on Σ. Hence

$$N_{\alpha\beta}^2 = \tfrac{1}{2}(\alpha, \alpha)\, q(1+p) = \tfrac{1}{2}(\tilde{\alpha}, \tilde{\alpha})\, q(1+p) = N_{\tilde{\alpha}\tilde{\beta}}^2.$$

Taking the absolute values of both sides of $\varrho_{\alpha+\beta}N_{\alpha,\beta} = \varrho_\alpha\varrho_\beta N_{\tilde{\alpha},\tilde{\beta}}$, we get

$$|\varrho_{\alpha+\beta}| = |\varrho_\alpha||\varrho_\beta|.$$

By Lemma 1, τ is a semi-involution of $\mathfrak{g}$. Let $\mathfrak{g}_u$ be the set of fixed points of τ; we now show that $\mathfrak{g}_u$ is compact. Let $X \in \mathfrak{g}$ and write

$$X = \sum_1^n c_i H_{\alpha_i} + \sum_{\alpha\in\Sigma} \tau_\alpha E_\alpha, \quad c_i \ (i = 1, \ldots, n) \text{ and } \tau_\alpha \ (\alpha \in \Sigma) \text{ are complex,}$$

then

$$\tau(X) = -\sum_1^n \bar{c}_i H_{\alpha_i} - \sum_{\alpha\in\Sigma} \bar{\tau}_\alpha |\varrho_\alpha| E_{-\alpha}.$$

Therefore,

$$\begin{aligned}(X, \tau(X)) &= -\left(\sum_1^n c_i H_{\alpha_i}, \sum_1^n \bar{c}_i H_{\alpha_i}\right) - \sum_{\alpha\in\Sigma} \tau_\alpha\bar{\tau}_\alpha|\varrho_\alpha| \\ &= -\sum_{i,j=1}^n c_i\bar{c}_j(H_{\alpha_i}, H_{\alpha_j}) - \sum_{\alpha\in\Sigma}\tau_\alpha\bar{\tau}_\alpha|\varrho_\alpha| \leqslant 0,\end{aligned}$$

i.e. $(X, \tau(X))$ is negative-definite. By Lemma 11, $\mathfrak{g}_u$ is compact.

We now show that $\sigma\tau = \tau\sigma$. Let $H_0 \in \mathfrak{h}_0$, then

$$\begin{aligned}\sigma\tau(H_0) &= \sigma(-H_0) = -\sigma(H_0),\\ \tau\sigma(H_0) &= -\sigma(H_0), \quad \text{because } \sigma(\mathfrak{h}_0) = \mathfrak{h}_0.\end{aligned}$$

$$\sigma\tau(E_\alpha) = \sigma(-|\varrho_\alpha|E_{-\alpha}) = -|\varrho_\alpha|\varrho_{-\alpha}E_{-\tilde{\alpha}} = -\frac{|\varrho_\alpha|}{\varrho_\alpha}E_{-\tilde{\alpha}},$$

$$\begin{aligned}\tau\sigma(E_\alpha) &= \tau(\varrho_\alpha E_{\tilde{\alpha}}) = \bar{\varrho}_\alpha(-|\varrho_{\tilde{\alpha}}|)E_{-\tilde{\alpha}}\\ &= -\frac{\bar{\varrho}_\alpha}{|\bar{\varrho}_\alpha|}E_{-\tilde{\alpha}} = -\frac{|\varrho_\alpha|}{\varrho_\alpha}E_{-\tilde{\alpha}}\end{aligned}$$

(since $\varrho_\alpha\bar{\varrho}_\alpha$ is a positive real number). Hence $\sigma\tau = \tau\sigma$.

Now let $X \in \mathfrak{g}_u$, then

$$\sigma(X) = \sigma\tau(X) = \tau\sigma(X),$$

so $\sigma(X) \in \mathfrak{g}_u$. This proves that $\sigma(\mathfrak{g}_u) = \mathfrak{g}_u$, i.e. (1) is true.

By noticing that $\sigma\tau = \tau\sigma$, (2) can be obtained as a consequence of Lemma 1. The proof of Theorem 15 is now completed.

Let $\mathfrak{g}_0$ be a real semisimple Lie algebra, a decomposition

$$\mathfrak{g}_0 = \mathfrak{g}_1 \dot{+} \mathfrak{g}_2$$

of $\mathfrak{g}_0$ is said to be a Cartan decomposition if $\mathfrak{g}_1$ is a subalgebra of $\mathfrak{g}_0$ and

$$\mathfrak{g}_u = \mathfrak{g}_1 \dot{+} i\mathfrak{g}_2$$

is a compact real form of $[\mathfrak{g}_0]$. $\mathfrak{g}_1$ is said to be a characteristic subalgebra of $\mathfrak{g}_0$; it is obviously compact; $\delta = \dim \mathfrak{g}_1 - \dim \mathfrak{g}_2$ is said to be the character of $\mathfrak{g}_0$. Theorem 15 ensures the existence of a Cartan decomposition $\mathfrak{g}_0 = \mathfrak{g}_1 + \mathfrak{g}_2$ of $\mathfrak{g}_0$; furthermore, the mapping

$$\sigma(X) = X, \qquad X \in \mathfrak{g}_1$$
$$\sigma(X) = -X, \qquad X \in i\mathfrak{g}_2$$

is an involutive automorphism.

Conversely, suppose σ is an involutive automorphism of $\mathfrak{g}_u$. In $[\mathfrak{g}_u]$, let

$$\mathfrak{g}_0 = \mathfrak{g}_u^+ \dot{+} i\mathfrak{g}_u^-, \quad \mathfrak{g}_u^\pm = \{X \in \mathfrak{g}_u \mid \sigma(X) = \pm X\}.$$

It can be shown that $\mathfrak{g}_0$ is a real form of $[\mathfrak{g}_u]$ and $\mathfrak{g}_0 = \mathfrak{g}_u^+ \dot{+} i\mathfrak{g}_u^-$ is a Cartan decomposition of $\mathfrak{g}_0$. We have the following:

THEOREM 16. *Let $\mathfrak{g}_u$ be a compact semisimple Lie algebra and τ be an involutive automorphism of $\mathfrak{g}_u$. Let $\mathfrak{g}_u^\pm = \{X \in \mathfrak{g} \mid \tau(X) = \pm X\}$, then $\mathfrak{g}_0 = \mathfrak{g}_u \dot{+} i\mathfrak{g}_u^-$ is a real semisimple Lie algebra and $[\mathfrak{g}_0] = [\mathfrak{g}_u]$. Conversely, if $\mathfrak{g}_0$ is any real semisimple Lie algebra, then there exists a compact semisimple Lie algebra $\mathfrak{g}_u$ and an involutive automorphism τ of $\mathfrak{g}_u$, such that $\mathfrak{g}_0 = \mathfrak{g}_u^+ \dot{+} i\mathfrak{g}_u^-$.*

Therefore, the problem of determining all the real semisimple Lie algebras is reduced to the problem of determining all the involutive automorphisms of all the compact semisimple Lie algebras.†

† For the solution of this problem and applications to the classification of real simple Lie algebras, see F. Gantmacher, Canonical representation of automorphisms of a complex semisimple Lie group, *Mat. Sbornik* **5** (1939), 101–46; F. Gantmacher, On the classification of real simple Lie groups, *Mat. Sbornik* **5** (1939), 217–49; and C. T. Yen, *Lie Groups and Differential Geometry*, Chapter 5, Peking, 1961.

INDEX

OTHER TITLES IN THE SERIES IN PURE AND APPLIED MATHEMATICS

Vol. 1. WALLACE—*An Introduction to Algebraic Topology*
Vol. 2. PEDOE—*Circles*
Vol. 3. SPAIN—*Analytical Conics*
Vol. 4. MIKHLIN—*Integral Equations*
Vol. 5. EGGLESTON—*Problems in Euclidean Space: Application of Convexity*
Vol. 6. WALLACE—*Homology Theory on Algebraic Varieties*
Vol. 7. NOBLE—*Methods Based on the Wiener-Hopf Technique for the Solution of Partial Differential Equations*
Vol. 8. MIKUSIŃSKI—*Operational Calculus*
Vol. 9. HEINE—*Group Theory in Quantum Mechanics*
Vol. 10. BLAND—*The Theory of Linear Viscoelasticity*
Vol. 11. KURTH—*Axiomatics of Classical Statistical Mechanics*
Vol. 12. FUCHS—*Abelian Groups*
Vol. 13. KURATOWSKI—*Introduction to Set Theory and Topology*
Vol. 14. SPAIN—*Analytical Quadrics*
Vol. 15. HARTMAN and MIKUSIŃSKI—*Theory of Lebesgue Measure and Integration*
Vol. 16. KULCZYCKI—*Non-Euclidean Geometry*
Vol. 17. KURATOWSKI—*Introduction to Calculus (2nd edition)*
Vol. 18. GERONIMUS—*Polynomials Orthogonal on a Circle and Interval*
Vol. 19. ELSGOLC—*Calculus of Variations*
Vol. 20. ALEXITS—*Convergence Problems of Orthogonal Series*
Vol. 21. FUCHS and LEVIN—*Functions of a Complex Variable, Volume II*
Vol. 22. GOODSTEIN—*Fundamental Concepts of Mathematics*
Vol. 23. KEENE—*Abstract Sets and Finite Ordinals*
Vol. 24. DITKIN and PRUDNIKOV—*Operational Calculus in Two Variables and its Applications*
Vol. 25. VEKUA—*Generalized Analytic Functions*
Vol. 26. AMIR-MOÉZ and FASS—*Elements of Linear Spaces*
Vol. 27. GRADSHTEIN—*Direct and Converse Theorems*
Vol. 28. FUCHS—*Partially Ordered Algebraic Systems*
Vol. 29. POSTNIKOV—*Foundations of Galois Theory*
Vol. 30. BERMANT—*A Course of Mathematical Analysis, Part II*
Vol. 31. ŁUKASIEWICZ—*Elements of Mathematical Logic*
Vol. 32. VULIKH—*Introduction to Functional Analysis for Scientists and Technologists*
Vol. 33. PEDOE—*An Introduction to Projective Geometry*
Vol. 34. TIMAN—*Theory of Approximation of Functions of a Real Variable*
Vol. 35. CSÁSZÁR—*Foundations of General Topology*
Vol. 36. BRONSHTEIN and SEMENDYAYEV—*A Guide-Book to Mathematics for Technologists and Engineers*
Vol. 37. MOSTOWSKI and STARK—*Introduction to Higher Algebra*
Vol. 38. GODDARD—*Mathematical Techniques of Operational Research*
Vol. 39. TIKHONOV and SAMARSKII—*The Equations of Mathematical Physics*
Vol. 40. MCLEOD—*Introduction to Fluid Dynamics*
Vol. 41. MOISIL—*The Algebraic Theory of Switching Circuits*
Vol. 42. OTTO—*Nomography*
Vol. 43. RANKIN—*An Introduction to Mathematical Analysis*
Vol. 44. BERMANT—*A Course of Mathematical Analysis, Part I*

Vol. 92. GEL'FOND and LINNIK—*Elementary Methods in the Analytic Theory of Numbers*
Vol. 93. GUREVICH—*The Theory of Jets in an Ideal Fluid*
Vol. 94. LANCASTER—*Lambda-matrices and Vibrating Systems*
Vol. 95. DINCULEANU—*Vector Measures*
Vol. 96. SŁUPECKI and BORKOWSKI—*Elements of Mathematical Logic and Set Theory*
Vol. 97. REDEI—*Foundations of Euclidean and Non-euclidean Geometries*
Vol. 98. MACROBERT—*Spherical Harmonics*
Vol. 99. KUIPERS and TIMMAN—*Handbook of Mathematics*
Vol. 100. SALOMAA—*Theory of Automata*
Vol. 101. KURATOWSKI—*Introduction to Set Theory and Topology*
Vol. 102. BLYTH and JANOWITZ—*Residuation Theory*
Vol. 103. KOSTEN—*Stochastic Theory of Service Systems*